Lecture Notes in Electrical Engineering

Volume 320

AF183925

Board of Series editors

Leopoldo Angrisani, Napoli, Italy
Marco Arteaga, Coyoacán, México
Samarjit Chakraborty, München, Germany
Jiming Chen, Hangzhou, P.R. China
Tan Kay Chen, Singapore, Singapore
Rüdiger Dillmann, Karlsruhe, Germany
Gianluigi Ferrari, Parma, Italy
Manuel Ferre, Madrid, Spain
Sandra Hirche, München, Germany
Faryar Jabbari, Irvine, USA
Janusz Kacprzyk, Warsaw, Poland
Alaa Khamis, New Cairo City, Egypt
Torsten Kroeger, Stanford, USA
Tan Cher Ming, Singapore, Singapore
Wolfgang Minker, Ulm, Germany
Pradeep Misra, Dayton, USA
Sebastian Möller, Berlin, Germany
Subhas Mukhopadyay, Palmerston, New Zealand
Cun-Zheng Ning, Tempe, USA
Toyoaki Nishida, Sakyo-ku, Japan
Federica Pascucci, Roma, Italy
Tariq Samad, Minneapolis, USA
Gan Woon Seng, Nanyang Avenue, Singapore
Germano Veiga, Porto, Portugal
Haitao Wu, Beijing, China
Junjie James Zhang, Charlotte, USA

About this Series

"Lecture Notes in Electrical Engineering (LNEE)" is a book series which reports the latest research and developments in Electrical Engineering, namely:

- Communication, Networks, and Information Theory
- Computer Engineering
- Signal, Image, Speech and Information Processing
- Circuits and Systems
- Bioengineering

LNEE publishes authored monographs and contributed volumes which present cutting edge research information as well as new perspectives on classical fields, while maintaining Springer's high standards of academic excellence. Also considered for publication are lecture materials, proceedings, and other related materials of exceptionally high quality and interest. The subject matter should be original and timely, reporting the latest research and developments in all areas of electrical engineering.

The audience for the books in LNEE consists of advanced level students, researchers, and industry professionals working at the forefront of their fields. Much like Springer's other Lecture Notes series, LNEE will be distributed through Springer's print and electronic publishing channels.

More information about this series at http://www.springer.com/series/7818

Krzysztof J. Latawiec · Marian Łukaniszyn
Rafał Stanisławski
Editors

Advances in Modelling and Control of Non-integer Order Systems

6th Conference on Non-integer Order Calculus and Its Applications, 2014 Opole, Poland

Springer

Editors
Krzysztof J. Latawiec
Department of Electrical, Control and Computer Engineering
Opole University of Technology
Opole
Poland

Rafał Stanisławski
Department of Electrical, Control and Computer Engineering
Opole University of Technology
Opole
Poland

Marian Łukaniszyn
Department of Electrical, Control and Computer Engineering
Opole University of Technology
Opole
Poland

ISSN 1876-1100 ISSN 1876-1119 (electronic)
ISBN 978-3-319-36047-8 ISBN 978-3-319-09900-2 (eBook)
DOI 10.1007/978-3-319-09900-2

Springer Cham Heidelberg New York Dordrecht London

© Springer International Publishing Switzerland 2015
Softcover reprint of the hardcover 1st edition 2015
This work is subject to copyright. All rights are reserved by the Publisher, whether the whole or part of the material is concerned, specifically the rights of translation, reprinting, reuse of illustrations, recitation, broadcasting, reproduction on microfilms or in any other physical way, and transmission or information storage and retrieval, electronic adaptation, computer software, or by similar or dissimilar methodology now known or hereafter developed. Exempted from this legal reservation are brief excerpts in connection with reviews or scholarly analysis or material supplied specifically for the purpose of being entered and executed on a computer system, for exclusive use by the purchaser of the work. Duplication of this publication or parts thereof is permitted only under the provisions of the Copyright Law of the Publisher's location, in its current version, and permission for use must always be obtained from Springer. Permissions for use may be obtained through RightsLink at the Copyright Clearance Center. Violations are liable to prosecution under the respective Copyright Law.
The use of general descriptive names, registered names, trademarks, service marks, etc. in this publication does not imply, even in the absence of a specific statement, that such names are exempt from the relevant protective laws and regulations and therefore free for general use.
While the advice and information in this book are believed to be true and accurate at the date of publication, neither the authors nor the editors nor the publisher can accept any legal responsibility for any errors or omissions that may be made. The publisher makes no warranty, express or implied, with respect to the material contained herein.

Printed on acid-free paper

Springer is part of Springer Science+Business Media (www.springer.com)

Preface

Although the birth of noninteger- or fractional-order systems can be traced back to the celebrated communication between Leibnitz and l'Hôpital in 1695, it is not until the advent of the modern computing machinery that the sophisticated theories that followed could be applied and implemented. Still, the complicated matter of the noninteger-order systems results in the fact that it is mainly theoretically-oriented papers that have dominated the publishing market of control and systems sciences. This is also reflected in the contents of both the previous[1] and this conference volumes whose prevailing contributions concern the theory, modeling and simulations, rather than implementations of noninteger-order systems. Such an unsurprising situation of an excess of theoretical over practically-oriented papers illustrates the vast mathematical potential of otherwise practically attractive field of noninteger-order systems. And this holds true even though the practical attractiveness of the field can be visualized by a tremendous increase in a number of subject areas (applications) for 'fractional calculus' as reported in the WoK platform[2] to have reached the figure of 136.

This way or another, the fractional calculus has recently attracted an unprecedented research interest both from the academia and various application-related environments. The field has experienced a publication explosion and apparently matured but it is still far from conceptual completeness both in the theory and, in particular, applications and implementations. This volume presents one small step ahead in the development of the theory and applications of the fractional calculus. Divided into six parts, it provides a bunch of new results in, consecutively, mathematical fundamentals, approximation, modeling and simulations, controllability and control, stability analysis and applications of various noninteger-order methodologies.

---

[1] Mitkowski W., Kacprzyk J., Baranowski J.: Advances in the Theory and Applications of Non-integer Order Systems, the 5th Conference on Non-integer Order Calculus and Its Applications, Cracow, Poland; Springer, Lecture Notes in Electrical Engineering, vol. 257.

[2] Chen, YangQuan: Fractional Order Thinking - from control, signal processing to energy informatics and beyond. Conference TOK'2013, Malatya, Turkey; http://mechatronics.ucmerced.edu/sites/mechatronics.ucmerced.edu/files/page/documents/tok 2013-plenary-lecture-09-27-2013-new-v2.pdf

Part 1 constitutes a single, invited paper by a honored member of the Polish, so to say, 'fractional society'. *Tadeusz Kaczorek* (Perfect observers of fractional descriptor continuous-time linear system) introduces fractional-order descriptor observers for fractional-order descriptor continous LTI systems. He presents new necessary and sufficient conditions for the existence of the observers and offers an original design procedure for the observers. The theoretical contributions are illustrated with a numerical example.

Part 2 provides a series of valuable formalisms that contribute to the further development of the mathematical theory of fractional-order systems.

Ewa Girejko, Dorota Mozyrska and *Małgorzata Wyrwas* (Viable solutions to fractional difference and differential equations) seek for the existence conditions for viable solutions to a discrete-time fractional-order equation making use of the viability properties of fractional-order differential equations. It is shown that it is sufficient for the existence of viable solutions to the fractional-order differential equation that viable solutions to the fractional-order difference equation are provided.

Małgorzata Klimek and *Marek Blasik* (Regular Sturm-Liouville problem with Riemann-Liouville derivatives of order in (1,2): discrete spectrum, solutions and applications) analyze the regular fractional Sturm-Liouville problem formulated using the left and right Riemann-Liouville derivatives of order in the range of (1,2). They prove a theorem describing the eigenvalues and eigenfunctions in the problem considered on the space of functions continuously differentiable in a finite interval and obeying the vanishing Dirichlet and fractional Neumann boundary conditions. It is found out that the spectrum of eigenvalues is discrete and that the eigenfunctions form a basis in the space of square-integrable functions. They also show applications of the derived eigenfunctions in the theory of partial fractional differential equations.

Agnieszka B. Malinowska and *Tatiana Odzijewicz* (Noether's second theorem for variable order fractional variational problems) succeed to prove an analog of the second Noether theorem for variable order fractional variational problems. From this theorem, they obtain useful identities between the Euler-Lagrange expressions and their variable order fractional derivatives.

Dorota Mozyrska and *Małgorzata Wyrwas* (Fractional linear equations with discrete operators of positive order) consider the Caputo- and Riemann-Liouville-type fractional order difference initial value problems for linear and semilinear fractional equations. They analyze possible solutions using the classical Z-transform method for any positive order. The formulae using the concept of the discrete Mittag-Leffler fractional function are highlighted.

Piotr Ostalczyk (The fractional-order backward-difference of a product of two discrete-variable functions (discrete fractional Leibnitz Rule)) studies a discrete fractional version of the so called Leibnitz Rule and he derives the fractional-order backward difference of a product of two discrete-variable functions, finally obtaining a generalisation to the first-order backward difference of a product. The new formula can be useful in evaluation of the fractional-order backward differences for selected functions.

In Part 3, a bunch of new results in approximation, modeling and simulations of fractional-order systems is provided.

Wiktor Malesza, *Michał Macias* and *Dominik Sierociuk* (Matrix approach and analog modeling for solving fractional variable order differential equations) introduce a matrix approach for solving fractional variable order linear differential equations of an additive-switching type. The method is based on a duality property between additive and recursive definitions of variable order differential equations. The obtained solutions will be validated by comparing them with analog model results.

Paweł Piątek, *Jerzy Baranowski*, *Marta Zagórowska*, *Waldemar Bauer* and *Tomasz Dziwiński* (Bi-fractional filters, Part 1: Left half-plane case) introduce a new non-integer α-order filter, called bi-fractional filter, given by a transfer function with parameters b and c selected in such a way that eigenvalues of the system are located in the open left half complex plane. Frequency characteristics of the system related to the parameters α, b and c are analyzed. Also, a method for realisation of the filter in the form of non-integer order differential equations is discussed.

Marek Rydel, *Rafał Stanisławski*, *Włodzimierz Stanisławski* and *Krzysztof J. Latawiec* (A comparative analysis of selected integer-order and noninteger-order linear models of complex dynamical systems) compare four model order reduction algorithms applied to modeling of evaporating tubes system in the BP-1150 steam boiler. The following model reduction techniques are analyzed: Frequency Weighted, Rational Krylov, Frequency Weighted with time delay and non-integer order transfer function with and without time delay. Optimal reduction parameters and values of f-zeros and f-poles of non-integer order transfer functions are obtained using evolutionary algorithm. The non-integer order model with time delay is found most accurate.

Rafał Stanisławski, *Krzysztof J. Latawiec*, *Marcin Gałek* and *Marian Łukaniszyn* (Modeling and identification of fractional-order discrete-time Laguerre-based feedback-nonlinear systems) present a new implementable strategy for modeling and identification of a fractional-order discrete-time block-oriented feedback-nonlinear system. It is shown that the inverse orthonormal basis functions (IOBF) concept enables to separate linear and nonlinear submodels, which leads to a linear regression formulation of the parameter estimation problem, with the detrimental bilinearity effect totally eliminated. Finally, the Laguerre filters are uniquely embedded in modeling of the fractional-order dynamics. Simulation experiments show a very good identification performance for an IOBF-structured, fractional-order Laguerre-based feedback-nonlinear model.

Marta Zagórowska, *Jerzy Baranowski*, *Piotr Bania*, *Paweł Piątek*, *Waldemar Bauer* and *Tomasz Dziwiński* (Impulse response approximation method for 'fractional order lag') present a new method for approximation of the 'fractional order lag' system based on its impulse response. Certain assumptions of the approximation method are verified and a new algorithm is presented. Also, some problems in analysis of this system are discussed, in particular its realisation in the form of fractional order differential equations.

Part 4 deals with the problems in controllability and control of noninteger order systems, in particular fractional PID-like control.

Waldemar Bauer, *Tomasz Dziwiński*, *Jerzy Baranowski*, *Paweł Piątek* and *Marta Zagórowska* (Comparison of performance indices for tuning of $PI^{\lambda}D^{\mu}$ controller for magnetic levitation system) discuss various performance indices for tuning of a fractional $PI^{\lambda}D^{\mu}$ controller for a highly nonlinear, integer-order system, that is a laboratory magnetic levitation system. Some tuning rules for the fractional controller are proposed.

Stefan Domek (Fractional-order model predictive control with small set of coincidence points) analyzes the abilities for synthesizing fractional-order model predictive control in case of a small set of coincidence points. In a design of a predictive control algorithm, the fractional-order option offers an additional degree of freedom in tuning the controller, which is illustrated in simulation examples.

Wojciech Mitkowski and *Krzysztof Oprzedkiewicz* (Tuning of the half-order robust PID controller dedicated to oriented PV system) design tuning rules for robust half-order PID controller dedicated to control an oriented PV system described by an interval transfer function. Simulation examples illustrate the usefulness of the tuning procedure.

Piotr W. Ostalczyk, Piotr Duch, Dariusz W. Brzeziński and *Dominik Sankowski* (Order functions selection in the variable-, fractional-order PID controller) discuss important problems related with microprocessor implementation of a variable fractional order PID (VFOPID) controller. In such controllers, the variable fractional order backward differences and sums (VFOBD/S) are used to perform discrete-time differentiation and integration of a closed-loop system error. In practice, all digitally differentiated and integrated signals are corrupted with noise, so there is a necessity of pre-filtering of a digital signal. This additionally loads the DSP system. A solution to this problem is proposed. Also, the abilities for DSP realizations of the VFOPID controller are presented and compared with computer simulation results.

Łukasz Sajewski (Minimum energy control of fractional positive continuous-time linear systems with two different fractional orders and bounded inputs) states and solves the problem of minimum energy control for fractional positive continuous-time linear systems with two different fractional orders and bounded inputs. A control algorithm is offered and illustrated with a simulation example.

Zbigniew Zaczkiewicz (Relative controllability of differential-algebraic systems with delay within Riemann-Liouville fractional derivatives) discusses the problem of relative controllability for linear fractional differential-algebraic systems with delay (FDAD). Firstly, the solutions to FDAD are represented as a series of determining equations solutions. Then, effective parametric rank criteria for relative controllability are introduced.

Part 5 analyzes the stability of noninteger order systems and some new results are offered in this important respect, in particular for discrete-time systems.

Mikołaj Busłowicz and *Andrzej Ruszewski* (Robust stability check of fractional discrete-time linear systems with interval uncertainties) discuss the problems in robust practical stability and robust asymptotic stability of fractional-order discrete-time linear systems with uncertainty. It is assumed that the system matrix is the interval matrix which elements are convex combinations of elements of specified bounded matrices and the fractional order α satisfies $0 < \alpha < 1$. Robust stability conditions are given using the matrix measure. The considerations are illustrated by numerical examples.

Małgorzata Wyrwas and *Dorota Mozyrska* (On Mittag-Leffler stability of fractional order difference systems) analyze the Mittag-Leffler stability of fractional order difference systems and study conditions for such a stability both in case of the Riemann-Liouville and Caputo-type operators used.

Marta Zagórowska, Jerzy Baranowski, Waldemar Bauer, Tomasz Dziwiński, Paweł Piątek and *Wojciech Mitkowski* (Lyapunov direct method for non-integer order systems) give an extension of the Lyapunov direct method to non-integer order systems. The extension enables to analyze a special case of the classical stability theory, that is the Mittag-Leffler stability. Some differences between the two are analyzed and illustrated in the presented examples.

Final Part 6 of this volume presents a spectrum of applications of the noninteger order calculus, ranging from bi-fractional filtering, in particular of electromyographic signals, through the thermal diffusion and advection diffusion processes to the SIEMENS platform implementation.

Jerzy Baranowski, Paweł Piątek, Aleksandra Kawala-Janik, Marta Zagórowska, Waldemar Bauer and *Tomasz Dziwiński* (Non-integer order filtration of electromyographic signals) present a new approach to filtering of electromyographic (EMG) signals, whose processing has recently attracted a growing interest from the medical environment seeking for a more reliable tool for muscle performance verification. To this end, non-integer order filtering is employed, in particular a bi-fractional filter is designed and effectively applied.

Rafał Brociek, Damian Słota and *Roman Wituła* (Reconstruction of the thermal conductivity coefficient in the time fractional diffusion equation) succeed in the reconstruction of the thermal conductivity coefficient in the time fractional diffusion equation. An additional information for the considered inverse problem is acquired from temperature measurements at selected points of the domain. The direct problem is solved by using the finite difference method. The Fibonacci search algorithm is used to minimize a functional defining the error of the approximate solution.

Tomasz Dziwiński, Waldemar Bauer, Jerzy Baranowski, Paweł Piątek and Marta Zagórowska (Robust non-integer order controller for air heating process trainer) design a robust non-integer order controller providing the prespecified flat-phase and stability margins. The controller is applied in an air heating process trainer system installed at the Department of Automatics and Biomedical Engineering, AGH University of Science and Technology in Cracow.

Ryszard Kopka (Model-Based Fault Diagnosis with Fractional Models) presents new results of experimental research on fault detection for the plant simulated by electronic circuit with supercapacitor 'damaged' by discharging it by the current source. The model of the plant is described by fractional order differential equations. Model parameters are estimated using identification procedures based on a step response signal. The fault detection problem of this paper can be used as a simple benchmark example to test new fault detection rules before applying them to real systems.

Krzysztof Oprzedkiewicz, Marek Chochol, Waldemar Bauer and *Tomasz Meresinski* (Modeling of elementary fractional order plants at PLC SIEMENS platform) effectively solve the problem of modeling of elementary fractional order plants at a PLC platform. The models are implemented at the SIEMENS SIMATIC S7 300 platform with the use of the STEP7 SCL language. Comparative simulations illustrate the usefullness of the proposed method.

Yuriy Povstenko (Space-Time-Fractional Advection Diffusion Equation in a Plane) considers the fundamental solution to the Cauchy problem for the space-time-fractional advection diffusion equation with the Caputo time-fractional derivative and the Riesz fractional Laplace operator in a case of two spatial variables. The solution is obtained using the Laplace integral transform with respect to time t and the double Fourier transform with respect to space variables x and y. Several particular cases of the solution are analyzed in details. Numerical results are illustrated in a graphical way.

Dominik Sierociuk and *Pawel Ziubinski* (Variable Order Fractional Kalman Filters for estimation over lossy network) present a generalization of Fractional Variable Order Kalman Filter (FvoKF) and Improved Fractional Variable Order Kalman Filter (ExFvoKF) algorithms for estimation of fractional variable order state-space systems over lossy network. This generalization is obtained for a state-space system based on one type of fractional variable order difference (A-type) and assuming the knowledge about packets being lost. The generalization of the ExFKF algorithm is based on the infinite dimensional form of a linear discrete fractional variable order state-space system, including state vector augmentation. The generalized algorithm requires less restrictive assumptions than derivation of FKF. Finally, numerical simulations of the proposed algorithms are presented and compared.

This volume is a result of productive and stimulating discussions during the RRNR'2014, the 6th Conference on Non-integer Order Calculus and Its Applications that was organized by the Department of Electrical, Control and Computer Engineering, Opole University of Technology, Opole, Poland. The Conference attracted a number of recognized researchers from various fields of science and engineering, both those interested in sophisticated mathematical machinery of the fractional order calculus and those involved in its application and implementation issues. Such a wide spectrum of the outstanding participants' interests exploded in stimulation of eventful discussions across the field and contributed to the success of the Conference. We would like to acknowledge the attendees to the Conference for their active participation and inspiring contributions. We are grateful to the anonymous reviewers whose helpful comments contributed to the final form of the papers. Finally, the assistance and support from Dr. Thomas Ditzinger and Holger Schäpe from Applied Sciences and Engineering at Springer is gratefully acknowledged.

Opole, Autumn 2014

Krzysztof J. Latawiec
Marian Łukaniszyn
Rafał Stanisławski

Contents

Part I: Invited Paper

Part II: Mathematical Fundamentals

Part III: Approximation, Modeling and Simulations

Part IV: Controllability and Control

Part V: Stability Analysis

Part VI: Applications

Part I
Invited Paper

Perfect Observers of Fractional Descriptor Continuous-Time Linear System

Tadeusz Kaczorek

Bialystok University of Technology
Faculty of Electrical Engineering
Wiejska 45D, 15-351 Bialystok
kaczorek@isep.pw.edu.pl

Abstract. Fractional descriptor observers for fractional descriptor continuous-time linear systems are proposed. Necessary and sufficient conditions for the existence of the observers are established. The design procedure of the observers is given and is demonstrated on a numerical example.

Keywords: fractional descriptor linear systems, design, perfect, observer.

1 Introduction

The fractional linear systems have been considered in many papers and books [8, 9, 11, 15, 23]. Positive linear systems consisting of n subsystems with different fractional orders have been proposed in [14, 15]. Descriptor (singular) linear systems have been investigated in [1-6, 12, 13, 18-21, 24, 25]. The eigenvalues and invariants assignment by state and input feedbacks have been addressed in [4, 12, 18]. The computation of Kronecker's canonical form of a singular pencil has been analyzed in [24].

A new concept of perfect observers for linear continuous-time systems has been proposed in [10, 22]. Observers for fractional linear systems have been addressed in [17, 22]. Fractional descriptor full-order observers for fractional descriptor continuous-time linear systems have been proposed in [16].

In this paper perfect fractional descriptor observers for fractional descriptor continuous-time linear systems will be proposed and necessary and sufficient conditions for the existence of the observer will be established.

The paper is organized as follows. In section 2 the basic definitions and theorems of fractional descriptor linear continuous-time systems are recalled and their perfect fractional descriptor observers are defined. In section 3 necessary and sufficient conditions for the existence of the perfect observers are established and design procedure of the perfect observer is proposed. An illustrating example is given in section 4. Concluding remarks are given in section 5.

© Springer International Publishing Switzerland 2015

K.J. Latawiec et al. (eds.), *Advances in Modeling and Control of Non-integer Order Systems*,
Lecture Notes in Electrical Engineering 320, DOI: 10.1007/978-3-319-09900-2_1

2 Fractional Descriptor Systems and Their Perfect Observers

Consider the fractional descriptor continuous-time linear system

$$E\frac{d^{\alpha}x(t)}{dt^{\alpha}} = Ax(t) + Bu(t), \quad x_0 = x(0), \tag{2.1a}$$

$$y = Cx, \tag{2.1b}$$

where $\frac{d^{\alpha}x(t)}{dt^{\alpha}}$ is the fractional α order derivative defined by Caputo [15, 23]

$${}_0D_t^{\alpha}x(t) = \frac{d^{\alpha}x(t)}{dt^{\alpha}} = \frac{1}{\Gamma(n-\alpha)}\int_0^t \frac{\frac{d^n x(\tau)}{d\tau^n}}{(t-\tau)^{\alpha-n+1}}d\tau, \quad n-1<\alpha<n\in N=\{1,2,\dots\}, \tag{2.2}$$

$\Gamma(x) = \int_0^{\infty} e^{-t}t^{x-1}dt$ is the gamma function, $x(t)\in\Re^n$, $u(t)\in\Re^m$, $y(t)\in\Re^p$ are the state, input and output vectors, $E, A\in\Re^{n\times n}$, $B\in\Re^{n\times m}$, $C\in\Re^{p\times n}$. It is assumed that $\det E = 0$ and

$$\det[E\lambda - A] \neq 0 \text{ for some } \lambda\in C \text{ (the field of complex number).} \tag{2.3}$$

Let U be the set of admissible inputs $u(t)\in U\subset\Re^m$ and $X_0\subset\Re^n$ be the set of consistent initial conditions $x_0\in X_0$ for which the equation (2.1) has a solution $x(t)$ for $u(t)\in U$.

The solution of the equation (2.1a) for $x_0\in X_0$ has been derived in [16].

Definition 2.1. The fractional descriptor continuous-time linear system

$$E\frac{d^{\alpha}\hat{x}(t)}{dt^{\alpha}} = F\hat{x}(t) + Gu(t) + Hy(t) \tag{2.4}$$

where $\hat{x}(t)\in\Re^n$ is the estimate of $x(t)$, $u(t)\in\Re^m$ and $y(t)\in\Re^p$ are the same input and output vectors as in (2.1), $E,F\in\Re^{n\times n}$, $G\in\Re^{n\times m}$, $H\in\Re^{n\times p}$, $\det E = 0$ is called a (full-order) perfect state observer for the system (2.1) if

$$x(t) = \hat{x}(t) \text{ for } t>0. \tag{2.5}$$

3 Design of Perfect Fractional Descriptor Observers

The following elementary row (column) operations will be used [13, 15]:

1. Multiplication of the *i*th row (column) by a real number *c*. This operation will be denoted by $L[i\times c]$ ($R[i\times c]$).
2. Addition to the *i*th row (column) of the *j*th row (column) multiplied by a real number *c*. This operation will be denoted by $L[i+j\times c]$ ($R[i+j\times c]$).
3. Interchange of the *i*th and *j*th rows (columns). This operation will be denoted by $L[i,j]$ ($R[i,j]$).

Lemma 3.1. If

$$\text{rank } E = r < n \tag{3.1}$$

then by the use of the elementary row and column operations the matrix E can be reduced to the following upper triangular form

$$P_1EQ_1 = \begin{bmatrix} 0 & E_{12} \\ 0 & 0 \end{bmatrix}, \quad E_{12} = \begin{bmatrix} e_{11} & e_{12} & \dots & e_{1r} \\ 0 & e_{22} & \dots & e_{2r} \\ \vdots & \vdots & \ddots & \vdots \\ 0 & 0 & \dots & e_{rr} \end{bmatrix} \tag{3.2a}$$

or lower triangular form

$$P_2EQ_2 = \begin{bmatrix} 0 & 0 \\ E_{21} & 0 \end{bmatrix}, \quad E_{21} = \begin{bmatrix} e_{11} & 0 & \dots & 0 \\ e_{21} & e_{22} & \dots & 0 \\ \vdots & \vdots & \ddots & \vdots \\ e_{r1} & e_{r2} & \dots & e_{rr} \end{bmatrix} \tag{3.2b}$$

where P_k and Q_k, $k=1,2$ are the matrices of elementary row and column operations.

Proof. If the condition (3.1) is satisfied then by elementary row and column operations the matrix E can be reduced to the form

$$\begin{bmatrix} 0 & E'_{12} \\ 0 & 0 \end{bmatrix}, \quad E'_{12} \in \Re^{r\times r}. \tag{3.3}$$

Next applying the elementary column operations we can reduced the matrix E'_{12} to the upper triangular form E_{12}. The proof for (3.2b) is similar. □

Definition 3.1. The smallest nonnegative integer q is called the nilpotent index of the nilpotent matrix N if $N^q = 0$ and $N^{q-1} \neq 0$.

Lemma 3.2. If

$$\text{rank } E = r < \frac{n}{2} \tag{3.4}$$

then the nilpotent index q of the matrix E is

$$q = 2 \text{ for } r = 1,2,\dots,\frac{n}{2}-1. \tag{3.5}$$

Proof. If $r < \frac{n}{2}$ then by Definition 3.1 and (3.2a) we have

$$(P_1 E Q_1)^2 = \begin{bmatrix} 0 & E_{12} \\ 0 & 0 \end{bmatrix}^2 = \begin{bmatrix} 0 & 0 \\ 0 & 0 \end{bmatrix} \text{ for } r = 1,2,\dots,\frac{n}{2}-1. \tag{3.6}$$

Proof for (3.2b) is similar. □

Lemma 3.3. If the nilpotent matrix $N \in \Re^{n\times n}$ has the index $q = 2$ i.e. $N^2 = 0$ and

$$D = \det[d_1,\dots,d_n],\ d_k \neq 0,\ k = 1,2,\dots,n \tag{3.7}$$

then the solution $x(t)$ of the fractional differential equation

$$N\frac{d^\alpha x(t)}{dt^\alpha} = Dx,\ 0 < \alpha < 1 \tag{3.8}$$

satisfy the condition

$$x(t) = \begin{cases} -\sum_{k=0}^{q-2} \overline{N}^{(k+1)} \delta^{(k+1)\alpha-1}(t) & \text{for } t = 0 \\ 0 & \text{for } t > 0 \end{cases} \tag{3.9}$$

where $\delta^{(k)}(t)$ is the k-order derivative of the Dirac function $\delta(t)$.

Proof. Applying the Laplace transform to (3.8) and taking into account that

$$\mathcal{L}\left[\frac{d^\alpha x(t)}{dt^\alpha}\right] = \int_0^\infty \frac{d^\alpha x(t)}{dt^\alpha} e^{-st} dt = s^\alpha X(s) - s^{\alpha-1} x_0 \tag{3.10}$$

we obtain

$$Ns^{\alpha}X(s)-Ns^{\alpha-1}x_0=DX(s), \tag{3.11}$$

where $X(s)=\mathcal{L}[x(t)]=\int_0^\infty x(t)e^{-st}dt$ and $x_0=x(0)$.

Premultiplying (3.11) by the inverse matrix D^{-1} we obtain

$$X(s)=-[I_n-\overline{N}s^{\alpha}]^{-1}\overline{N}s^{\alpha-1}x_0, \tag{3.12}$$

where $\overline{N}=D^{-1}N$ and $\overline{N}^q=D^{-q}N^q=0$.

Taking into account that

$$[I_n-\overline{N}s^{\alpha}]^{-1}=\sum_{k=0}^{q-1}\overline{N}^k s^{k\alpha}, \tag{3.13}$$

from (3.12) w obtain

$$X(s)=-\sum_{k=0}^{q-2}\overline{N}^{(k+1)}s^{(k+1)\alpha-1}x_0. \tag{3.14}$$

Applying the inverse Laplace transform to (3.14) we obtain (3.9) since $\mathcal{L}^{-1}[s^{k\alpha}]=\delta^{(k\alpha)}(t)$. □

Let

$$e(t)=x(t)-\hat{x}(t). \tag{3.15}$$

Then using (2.1) and (2.4) we obtain

$$\begin{aligned}E\frac{d^{\alpha}e(t)}{dt^{\alpha}}&=E\frac{d^{\alpha}x(t)}{dt^{\alpha}}-E\frac{d^{\alpha}\hat{x}(t)}{dt^{\alpha}}\\&=Ax(t)+Bu(t)-(F\hat{x}(t)+Gu(t)+HCx(t))\\&=(A-HC)x(t)-F\hat{x}(t)+(B-G)u(t)\end{aligned} \tag{3.16}$$

and

$$E\frac{d^{\alpha}e(t)}{dt^{\alpha}}=Fe(t) \tag{3.17}$$

if

$$F=A-HC, \tag{3.18}$$

$$H=B. \tag{3.19}$$

By Lemma 3.1 using the elementary row and column operations the singular matrix E can be reduced to a suitable nilpotent matrix N and from (3.17) we obtain

$$N\frac{d^{\alpha}\bar{e}(t)}{dt^{\alpha}} = \bar{F}\bar{e}(t) \tag{3.20}$$

where

$$N = PEQ, \quad \bar{F} = PFQ, \quad \bar{e}(t) = Q^{-1}e(t) \tag{3.21}$$

and P and Q are matrices of elementary row and column operations.

If we choose the matrix H so that

$$\bar{F} = D \tag{3.22}$$

where D is given by (3.7) then by Lemma 3.3 $\bar{e}(t) = 0$ for $t > 0$ and the fractional descriptor observer (2.4) will be a perfect observer for the system (2.1).

Theorem 3.1. There exists the perfect fractional descriptor observer (2.4) of the fractional descriptor system (2.1) if and only if

$$\text{rank}\begin{bmatrix} \bar{A} - D \\ \bar{C} \end{bmatrix} = \text{rank}\,[\bar{C}] \tag{3.23}$$

where

$$\bar{A} = PAQ, \quad \bar{C} = CQ \tag{3.24}$$

and the matrices P, Q satisfy (3.21).

Proof. To design the perfect observer (2.4) for the system (2.1) with given matrices A, B, C we have to choose the matrices F, G, H of the observer so that the conditions (3.18), (3.19) and (3.22) are met. From (3.19) we have $H = B$ and the conditions (3.18) and (3.22) are met if and only if

$$\bar{A} - \overline{HC} = D \tag{3.25}$$

where $\bar{H} = PH$.

The equation (3.25) has a solution $\bar{H}$ (and $H = P^{-1}\bar{H}$) for given $\bar{C}$ and D if and only if the condition (3.23) is satisfied. Therefore, there exists the perfect observer (2.4) for the system (2.1) if and only if the condition (3.23) is satisfied. □

From the above considerations we have the following procedure for designing of the perfect observer (2.4) for the system (2.1).

Procedure 3.1

Step 1. Find the matrices P and Q of the elementary row and column operations reducing the matrix E to its nilpotent form $N = PEQ$.

Step 2. Knowing the matrices P, Q compute $\overline{A}$ and $\overline{C}$ defined by (3.24).

Step 3. Choose a diagonal matrix (3.7) and check the condition (3.23). If the condition is satisfied then there exists the perfect observer (2.4) for the system (2.1).

Step 4. Knowing the matrices $\overline{A}$ and $\overline{C}$ find the solution $\overline{H}$ of the equation (3.25).

Step 5. Compute the matrices of the perfect observer (2.4)

$$F = A - HC\,,\ G = B\,,\ H = P^{-1}\overline{H}\,. \tag{3.26}$$

4 Example

Consider the fractional descriptor system (2.1) with the matrices

$$E = \begin{bmatrix} 1 & 0 & 0 \\ 0 & 1 & 0 \\ 0 & 0 & 0 \end{bmatrix},\quad A = \begin{bmatrix} 2 & 0 & 1 \\ 3 & 0 & 2 \\ 0 & 2 & 0 \end{bmatrix},\quad B = \begin{bmatrix} 1 \\ 0 \\ 2 \end{bmatrix},\quad C = \begin{bmatrix} 0 & 0 & 1 \\ 1 & 0 & 0 \end{bmatrix}. \tag{4.1}$$

The descriptor system is regular since

$$\det[Es - A] = \begin{vmatrix} s-2 & 0 & -1 \\ -3 & s & -2 \\ 0 & -2 & 0 \end{vmatrix} = 2(1-2s) \neq 0\,. \tag{4.2}$$

To design the perfect fractional descriptor observer for the system we use Procedure 3.1 and we obtain the following:

Step 1. In this case we have

$$P = \begin{bmatrix} 1 & 0 & 0 \\ 0 & 1 & 0 \\ 0 & 0 & 1 \end{bmatrix},\quad Q = \begin{bmatrix} 0 & 1 & 0 \\ 0 & 0 & 1 \\ 1 & 0 & 0 \end{bmatrix} \tag{4.3}$$

and

$$N = PEQ = \begin{bmatrix} 0 & 1 & 0 \\ 0 & 0 & 1 \\ 0 & 0 & 0 \end{bmatrix}. \tag{4.4}$$

Step 2. Using (3.24) and (4.1) we obtain

$$\overline{A} = PAQ = \begin{bmatrix} 1 & 2 & 0 \\ 2 & 3 & 0 \\ 0 & 0 & 2 \end{bmatrix}, \quad \overline{C} = CQ = \begin{bmatrix} 1 & 0 & 0 \\ 0 & 1 & 0 \end{bmatrix}. \tag{4.5}$$

Step 3. In this case we choose

$$D = \begin{bmatrix} 3 & 0 & 0 \\ 0 & 2 & 0 \\ 0 & 0 & 2 \end{bmatrix} \tag{4.6}$$

and the condition (3.23) is satisfied since

$$\operatorname{rank}\begin{bmatrix} \overline{A} - D \\ \overline{C} \end{bmatrix} = \begin{bmatrix} -2 & 2 & 0 \\ 2 & 1 & 0 \\ 0 & 0 & 0 \\ 1 & 0 & 0 \\ 0 & 1 & 0 \end{bmatrix} = 2 = \operatorname{rank}[\overline{C}] = \operatorname{rank}\begin{bmatrix} 1 & 0 & 0 \\ 0 & 1 & 0 \end{bmatrix}. \tag{4.7}$$

Therefore, there exists the perfect observer (2.4) for the system (2.1) with (4.1).

Step 4. The equation

$$\overline{HC} = \begin{bmatrix} h_{11} & h_{12} \\ h_{21} & h_{22} \\ h_{31} & h_{32} \end{bmatrix}\begin{bmatrix} 1 & 0 & 0 \\ 0 & 1 & 0 \end{bmatrix} = \overline{A} - D = \begin{bmatrix} -2 & 2 & 0 \\ 2 & 1 & 0 \\ 0 & 0 & 0 \end{bmatrix} \tag{4.8}$$

has the solution

$$\overline{H} = \begin{bmatrix} -2 & 2 \\ 2 & 1 \\ 0 & 0 \end{bmatrix} = H \tag{4.9}$$

since $P = I_3$.

Step 5. Using (3.26), (4.1) and (4.9) we obtain

$$F = A - HC = \begin{bmatrix} 2 & 0 & 1 \\ 3 & 0 & 2 \\ 0 & 2 & 0 \end{bmatrix} - \begin{bmatrix} -2 & 2 \\ 2 & 1 \\ 0 & 0 \end{bmatrix} \begin{bmatrix} 0 & 0 & 1 \\ 1 & 0 & 0 \end{bmatrix} = \begin{bmatrix} 0 & 0 & 3 \\ 2 & 0 & 0 \\ 0 & 2 & 0 \end{bmatrix},$$
$$G = B = \begin{bmatrix} 1 \\ 0 \\ 2 \end{bmatrix}. \tag{4.10}$$

The perfect observer is described by the equation

$$\begin{bmatrix} 1 & 0 & 0 \\ 0 & 1 & 0 \\ 0 & 0 & 0 \end{bmatrix} \frac{d^\alpha \hat{x}(t)}{dt^\alpha} = \begin{bmatrix} 0 & 0 & 3 \\ 2 & 0 & 0 \\ 0 & 2 & 0 \end{bmatrix} \hat{x}(t) + \begin{bmatrix} 1 \\ 0 \\ 2 \end{bmatrix} u(t) + \begin{bmatrix} -2 & 2 \\ 2 & 1 \\ 0 & 0 \end{bmatrix} y(t). \tag{4.11}$$

5 Concluding Remarks

Perfect fractional descriptor observers for fractional descriptor continuous-time linear systems have been proposed. Necessary and sufficient conditions for the existence of perfect observers for the fractional descriptor linear systems have been established. Designing procedure of the fractional descriptor observers has been proposed and illustrated on a numerical example. The considerations can be easily extended to fractional descriptor discrete-time linear systems. An open problem is an extension for fractional descriptor 2D continuous-discrete linear systems.

Acknowledgment. This work was supported under work S/WE/1/11.

References

1. Dodig, M., Stosic, M.: Singular systems state feedbacks problems. Linear Algebra and its Applications 431(8), 1267–1292 (2009)
2. Cuihong, W.: New delay-dependent stability criteria for descriptor systems with interval time delay. Asian Journal of Control 14(1), 197–206 (2012)
3. Dai, L.: Singular Control Systems. LNCIS, vol. 118. Springer, Berlin (1989)
4. Fahmy, M.M., O'Reill, J.: Matrix pencil of closed-loop descriptor systems: Infinite-eigenvalue assignment. Int. J. Control 49(4), 1421–1431 (1989)
5. Gantmacher, F.R.: The Theory of Matrices. Chelsea Publishing Co., New York (1960)
6. Guang-Ren, D.: Analysis and Design of Descriptor Linear Systems. Springer, New York (2010)
7. Kaczorek, T.: Checking of the positivity of descriptor linear systems with singular pencils. Archives of Control Sciences 22(1), 77–86 (2012)

8. Kaczorek, T.: Positive fractional continuous-time linear systems with singular pencils. Bull. Pol. Acad. Sci. Techn. 60(1), 9–12 (2012)
9. Kaczorek, T.: Descriptor fractional linear systems with regular pencils. Asian Journal of Control 15(4), 1051–1064 (2013)
10. Kaczorek, T.: Full-order perfect observers for continuous-time linear systems. Bull. Pol. Acad. Sci. Techn. 49(4), 549–558 (2001)
11. Kaczorek, T.: Fractional positive continuous-time linear systems and their reachability. Int. J. Appl. Math. Comput. Sci. 18(2), 223–228 (2008)
12. Kaczorek, T.: Infinite eigenvalue assignment by output-feedback for singular systems. Int. J. Appl. Math. Comput. Sci. 14(1), 19–23 (2004)
13. Kaczorek, T.: Linear Control Systems, vol. 1. Research Studies Press, J. Wiley, New York (1992)
14. Kaczorek, T.: Positive linear systems consisting of n subsystems with different fractional orders. IEEE Trans. on Circuits and Systems 58(7), 1203–1210 (2011)
15. Kaczorek, T.: Selected Problems of Fractional Systems Theory. LNCIS, vol. 411. Springer, Berlin (2011)
16. Kaczorek, T.: Fractional descriptor observers for fractional descriptor continuous-time linear systems. Submitted to Archives of Control Sciences (2013)
17. Kociszewski, R.: Observer synthesis for linear discrete-time systems with different fractional orders. Pomiary Automatyka Robotyka (PAR) (2), 376–381 (CD-ROM) (2013)
18. Kucera, V., Zagalak, P.: Fundamental theorem of state feedback for singular systems. Automatica 24(5), 653–658 (1988)
19. Lewis, F.L.: Descriptor systems, expanded descriptor equation and Markov parameters. IEEE Trans. Autom. Contr. AC-28(5), 623–627 (1983)
20. Luenberger, D.G.: Time-invariant descriptor systems. Automatica 14(5), 473–480 (1978)
21. Luenberger, D.G.: Dynamical equations in descriptor form. IEEE Trans. Autom. Contr. AC-22(3), 312–321 (1977)
22. N'Doye, I., Darouach, M., Voos, H., Zasadzinski, M.: Design of unknown input fractional-order observers for fractional-order systems. Int. J. Appl. Math. Comput. Sci. 23(3), 491–500 (2013)
23. Podlubny, I.: Fractional Differential Equations. Academic Press, New York (1999)
24. Van Dooren, P.: The computation of Kronecker's canonical form of a singular pencil. Linear Algebra and its Applications 27, 103–140 (1979)
25. Virnik, E.: Stability analysis of positive descriptor systems. Linear Algebra and its Applications 429, 2640–2659 (2008)

Part II
Mathematical Fundamentals

Viable Solutions to Fractional Difference and Differential Equations

Ewa Girejko, Dorota Mozyrska, and Małgorzata Wyrwas

Bialystok University of Technology,
Wiejska 45A, 15-351 Białystok, Poland
{e.girejko,d.mozyrska,m.wyrwas}@pb.edu.pl

Abstract. The authors' purpose is to consider and formulate conditions providing the existence of viable solutions to a discrete fractional equation via viability properties of fractional differential equations. We show that the existence of viable solutions to a fractional differential equation suffices to get viable solutions to a difference fractional equation.

Keywords: viability, fractional differential equations, fractional discrete systems.

1 Introduction

Viability of fractional difference and differential equations is not so much exploited so far. Briefly, the problem of viability consists in finding at least one solution to the given equation that starts and remains in a certain constrained set. In the case of uniqueness of solutions viability coincides with the invariance idea, where all solutions have to stay in a constrained set. Our motivation for considering these kind of problems for fractional differential and difference equations was the value of viability theory. As the author in [2], the most important book on classical viability theory, says: "viability theory is a mathematical theory that offers mathematical metaphors of evolution of macrosystems arising in biology, economics, cognitive sciences, games, and similar areas, as well as in nonlinear systems of control theory". Although fractional calculus is an area in mathematics that develops very quickly and it is investigated by many mathematicians [1,4,13,17,19], the development of viability problem of solutions to fractional differential equations hasn't grown yet. Moreover, to our knowledge, discrete fractional systems haven't been investigated in this light yet. The area of our interest is to consider and formulate conditions providing existence of viable solutions to a discrete equation via viability properties of fractional differential equations. In this way we can employ the results on Nagumo Theorem that we presented in [9,15], namely necessary and sufficient conditions for solutions to be viable with respect to a constrained set but in continuous case. This idea allows us to use, indirectly, the classical tools as Bouligand cone or contingent vectors. The structure of the paper consists of four sections. Besides introduction we propose a preliminary section, where useful basic notions, definitions and facts

© Springer International Publishing Switzerland 2015
K.J. Latawiec et al. (eds.), *Advances in Modeling and Control of Non-integer Order Systems*,
Lecture Notes in Electrical Engineering 320, DOI: 10.1007/978-3-319-09900-2_2

are gathered. In Section 3 we present and formulate problems, for which viability properties are further examined. The main section is Section 4 that includes results on approximation and on viability of differential and difference equations.

2 Some Preliminaries

Throughout this section we present a set of notations, definitions, and some preliminary facts which are useful in the sequel of the paper.

2.1 Continuous Part

We start with definitions of fractional integrals of arbitrary order, the Caputo and Riemann–Liouville derivatives of order $\alpha \in (0,1)$.

Definition 1 ([11,17,18]). *Let $\varphi \in L_1([0,t_1],\mathbb{R})$. The integral*

$$(I_{0+}^{\alpha}\varphi)(t) = \frac{1}{\Gamma(\alpha)}\int_0^t \varphi(s)(t-s)^{\alpha-1}ds\,, \quad t>0,$$

where Γ is the gamma function and $\alpha > 0$, is called, the left-sided fractional integral of order α. *Additionally we define $I_{0+}^{0} := \boldsymbol{I}$ (identity operator).*

Remark 1 ([6]). Note that $I_{0+}^{\alpha}f(t) = (f * \varphi_\alpha)(t)$, where $\varphi_\alpha(t) = \frac{t^{\alpha-1}}{\Gamma(\alpha)}$ for $t>0$, $\varphi_\alpha(t) = 0$ for $t \le 0$, and $\varphi_\alpha \to \delta(t)$ as $\alpha \to 0$, with δ the delta Dirac pseudo function.

The fractional derivatives that we need are the Riemann–Liouville and the Caputo ones.

Definition 2 ([11,17]). *Let φ be defined on the interval $[0,t_1]$ and n be a natural number satisfying $n = [\alpha]+1$ with $[\alpha]$ denoting the integer part of α. The* left-sided Riemann–Liouville derivative of order α and the lower limit 0 *are defined through the following:*

$$\left(D_{0+}^{\alpha}\varphi\right)(t) = \frac{1}{\Gamma(n-\alpha)}\left(\frac{d}{dt}\right)^n \int_0^t \varphi(s)(t-s)^{n-\alpha-1}ds\,.$$

The left-sided Caputo derivative of order α and the lower limit 0 *are defined through the following:*

$$\left({}^{C}D_{0+}^{\alpha}\varphi\right)(t) = \frac{1}{\Gamma(n-\alpha)}\int_0^t \varphi^{(n)}(s)(t-s)^{n-\alpha-1}ds\,.$$

Remark 2. If $\alpha \in (0,1)$, then the left-sided Riemann–Liouville fractional derivative of order α takes the form:

$$\left(D_{0+}^{\alpha}\varphi\right)(t) = \frac{1}{\Gamma(1-\alpha)}\frac{d}{dt}\int_0^t \varphi(s)(t-s)^{-\alpha}ds = \frac{d}{dt}\left(\left(I_{0+}^{1-\alpha}\varphi\right)(t)\right)$$

and the left-sided Caputo fractional derivative of order α takes the form:

$$\left({}^C D^\alpha_{0+}\varphi\right)(t) = \frac{1}{\Gamma(1-\alpha)}\int_0^t \varphi'(s)(t-s)^{-\alpha}ds = \left(I^{1-\alpha}_{0+}\frac{d}{ds}(\varphi(s))\right)(t)\,.$$

If $\alpha \in (0,1]$ then the following comparison formula of the Caputo and Riemann–Liouville derivatives holds.

$$\left({}^C D^\alpha_{0+}\varphi\right)(t) = \left(D^\alpha_{0+}\varphi\right)(t) - \frac{t^{-\alpha}}{\Gamma(1-\alpha)}\varphi(0^+)\,, \tag{1}$$

where $\varphi(0^+) = \lim_{t\to 0^+}\varphi(t)$.

From [11, Theorem 2.4] we have the following property:

Proposition 1. *For $\alpha \in (0,1]$ we have*

$$\left(I^\alpha_{0+}D^\alpha_{0+}\varphi\right)(t) = \varphi(t) - \frac{t^{\alpha-1}}{\Gamma(\alpha)}\left(I^{1-\alpha}_{0+}\varphi\right)(t)\mid_a\ .$$

The following formulas are useful:

$$I^\alpha_{0+}t^p = \frac{\Gamma(p+1)}{\Gamma(p+\alpha+1)}t^{p+\alpha},\quad D^\alpha_{0+}t^p = \frac{\Gamma(p+1)}{\Gamma(p-\alpha+1)}t^{p-\alpha}\,.$$

2.2 Discrete Part

Let us denote by $\mathcal{F}_D$ the set of real valued functions defined on D. Let $h > 0, \alpha > 0$ and put $(h\mathbb{N})_a := \{a, a+h, a+2h, ...\}$ for $h > 0$ and $a \in \mathbb{R}$. Let $\sigma : (h\mathbb{N})_a \to (h\mathbb{N})_a$ be the operator defined by the formula $\sigma(t) := t + h$, $t \in (h\mathbb{N})_a$.

Definition 3 ([8]). *For a function $x \in \mathcal{F}_{(h\mathbb{N})_a}$ the* forward h-difference operator *is defined as*

$$(\Delta_h x)(t) := \frac{x(\sigma(t)) - x(t)}{h}, \quad t = a + nh,\ n \in \mathbb{N}_0\,,$$

while the h-difference sum *is given by*

$$\left({}_a\Delta_h^{-1}x\right)(t) := h\sum_{k=0}^{n} x(a+kh)\,,$$

where $t = a + (n+1)h$, $n \in \mathbb{N}_0$ and $\left({}_a\Delta_h^{-1}x\right)(a) = 0$.

The next definition comes from [5].

Definition 4. *For arbitrary $t, \alpha \in \mathbb{R}$ the* h-factorial function *is defined by*

$$t_h^{(\alpha)} := h^\alpha\frac{\Gamma(\frac{t}{h}+1)}{\Gamma(\frac{t}{h}+1-\alpha)}\,, \tag{2}$$

where $\frac{t}{h} \notin \mathbb{Z}_- := \{-1,-2,-3,\ldots\}$, and we use the convention that division at a pole yields zero.

Notice that if we use the general binomial coefficient $\binom{a}{b} := \frac{\Gamma(a+1)}{\Gamma(b+1)\Gamma(a-b+1)}$, then (2) can be rewritten as

$$t_h^{(\alpha)} = h^\alpha \Gamma(\alpha+1)\binom{\frac{t}{h}}{\alpha}.$$

In our consideration the crucial role plays the power rule formula presented in [8], i.e.

$$\left({}_a\Delta_h^{-\alpha}\psi\right)(t) = \frac{\Gamma(\mu+1)}{\Gamma(\mu+\alpha+1)}(t-a+\mu h)_h^{(\mu+\alpha)}, \tag{3}$$

where $\psi(r) = (r-a+\mu h)_h^{(\mu)}$, $r \in (h\mathbb{N})_a$, $t \in (h\mathbb{N})_{a+\alpha h}$. Note that using the general binomial coefficient one can write (3) as

$$\left({}_a\Delta_h^{-\alpha}\psi\right)(t) = \Gamma(\mu+1)\binom{n+\alpha+\mu}{n}h^{\mu+\alpha}.$$

Then if $\psi \equiv 1$, we have for $\mu = 0$, $a = (1-\alpha)h$ and $t = nh + a + \alpha h$

$$\left({}_a\Delta_h^{-\alpha}1\right)(t) = \frac{1}{\Gamma(\alpha+1)}(t-a)_h^{(\alpha)} = \frac{\Gamma(n+\alpha+1)}{\Gamma(\alpha+1)\Gamma(n+1)}h^\alpha = \binom{n+\alpha}{n}h^\alpha.$$

The next definition was stated in [5].

Definition 5. *For a function $x \in \mathcal{F}_{(h\mathbb{N})_a}$ the* fractional h-sum of order $\alpha > 0$ *is given by*

$$\left({}_a\Delta_h^{-\alpha}x\right)(t) := \frac{h}{\Gamma(\alpha)}\sum_{k=0}^{n}(t-\sigma(a+kh))_h^{(\alpha-1)}x(a+kh), \tag{4}$$

where $t = a + (\alpha+n)h$, $n \in \mathbb{N}_0$. Moreover we define $\left({}_a\Delta_h^0 x\right)(t) := x(t)$.

Remark 3. Note that ${}_a\Delta_h^{-\alpha} : \mathcal{F}_{(h\mathbb{N})_a} \to \mathcal{F}_{(h\mathbb{N})_{a+\alpha h}}$.

Accordingly to the definition of h-factorial function formula (4) can be rewritten as:

$$\begin{aligned}\left({}_a\Delta_h^{-\alpha}x\right)(t) &= h^\alpha\sum_{k=0}^{n}\frac{\Gamma(\alpha+n-k)}{\Gamma(\alpha)\Gamma(n-k+1)}x(a+kh)\\ &= h^\alpha\sum_{k=0}^{n}\binom{n-k+\alpha-1}{n-k}x(a+kh)\\ &= h^\alpha\sum_{j=0}^{n}(-1)^j\binom{-\alpha}{j}x(a-jh)\end{aligned}$$

for $t = a + (\alpha+n)h$, $n \in \mathbb{N}_0$.

Remark 4. In [10] one can find the following form of the fractional h-sum of order $\alpha > 0$:

$$\left({}_a\Delta_h^{-\alpha} x\right)(t) = \frac{h^\alpha}{\Gamma(\alpha)} \sum_{k=a}^{t-\alpha h} \left(\frac{t-\sigma(k)}{h}\right)^{(\alpha-1)} x(k)$$

that can be useful in implementation.

Definition 6. *Let* $\alpha \in (0,1]$. *The* Riemann–Liouville–type fractional h-difference operator ${}_a\Delta_h^\alpha x$ *of order* α *for a function* $x \in \mathcal{F}_{(h\mathbb{N})_a}$ *is defined by*

$$\left({}_a\Delta_h^\alpha x\right)(t) = \left(\Delta_h \left({}_a\Delta_h^{-(1-\alpha)} x\right)\right)(t), \quad t \in (h\mathbb{N})_{a+(1-\alpha)h}. \tag{5}$$

Remark 5. Note that ${}_a\Delta_h^\alpha : \mathcal{F}_{(h\mathbb{N})_a} \to \mathcal{F}_{(h\mathbb{N})_{a+(1-\alpha)h}}$, where $\alpha \in (0,1]$.

Definition of the Caputo–type h-difference operator was stated by the authors in [14].

Definition 7. *Let* $\alpha \in (0,1]$. *The* Caputo–type h-difference operator ${}_a\Delta_{h,*}^\alpha$ *of order* α *for a function* $x \in \mathcal{F}_{(h\mathbb{N})_a}$ *is defined by*

$$\left({}_a\Delta_{h,*}^\alpha x\right)(t) := \left({}_a\Delta_h^{-(1-\alpha)} (\Delta_h x)\right)(t), \quad t \in (h\mathbb{N})_{a+(1-\alpha)h}.$$

Remark 6. Note that: ${}_a\Delta_{h,*}^\alpha : \mathcal{F}_{(h\mathbb{N})_a} \to \mathcal{F}_{(h\mathbb{N})_{a+(1-\alpha)h}}$, where $\alpha \in (0,1]$.

The property of the composition of h-sums was proved in [14].

Proposition 2. *Let x be a real valued function defined on $(h\mathbb{N})_a$, where $a, h \in \mathbb{R}, h > 0$. For $\alpha, \beta > 0$ the following equalities hold:*

$$\left({}_{a+\beta h}\Delta_h^{-\alpha}\left({}_a\Delta_h^{-\beta} x\right)\right)(t) = \left({}_a\Delta_h^{-(\alpha+\beta)} x\right)(t) = \left({}_{a+\alpha h}\Delta_h^{-\beta}\left({}_a\Delta_h^{-\alpha} x\right)\right)(t),$$

where $t \in (h\mathbb{N})_{a+(\alpha+\beta)h}$.

The next proposition gives a useful identity of transforming Caputo fractional difference equations into fractional summations for the case when an order is from the interval $(0,1]$.

Proposition 3 ([14]). *Let $\alpha \in (0,1]$, $h > 0$, $a = (\alpha-1)h$ and x be a real valued function defined on $(h\mathbb{N})_a$. The following formula holds*

$$\left({}_0\Delta_h^{-\alpha}\left({}_a\Delta_{h,*}^\alpha x\right)\right)(nh+a) = x(nh+a) - x(a), \quad n \in \mathbb{N}_1.$$

The operators presented in this section can be extended to vector valued functions in a componentwise manner.

3 Differential and Difference Systems

Let us consider the initial value problem stated by the system with the Caputo fractional derivative with order $\alpha \in (0,1]$ as follows:

$$ {}_a^C D^\alpha x(t) = f(t, x(t)), \quad t \in (a, T],\ a > 0\,, \tag{6}$$

satisfying the initial condition

$$x(a) = x_a \in \mathbb{R}^n\,, \tag{7}$$

where $t \in (a, T]$, $x : (a, T] \to \mathbb{R}^n$ and x_a is a constant vector from $\mathbb{R}^n$.

Nextly, we consider an initial value discrete problem stated by the system:

$$\left({}_a\Delta_{h,*}^\alpha y\right)(nh) = f(nh, y(nh + a)) \tag{8}$$

with initial value:

$$y(a) = y_a\,, \tag{9}$$

where $h > 0$, $\alpha \in (0,1]$, $a = (\alpha - 1)h$, $n \in \mathbb{N}_0$, $y : (a, T]_{(h\mathbb{N})_a} \to \mathbb{R}^n$, $(a, T]_{(h\mathbb{N})_a} := (a, T] \cap (h\mathbb{N})_{a+(1-\alpha)h}$, and y_a is a constant vector from $\mathbb{R}^n$. Hereinafter we mean $(a, T]_{(h\mathbb{N})_a} = (a, T] \cap (h\mathbb{N})_a$.

Note that $y_a = x_a$. Then from [12] we know that the solution of (6) has the following form

$$x(t) = x_a + \frac{1}{\Gamma(\alpha)} \int_a^t (t - \tau)^{\alpha - 1} f(\tau, x(\tau)) d\tau\,. \tag{10}$$

Moreover, we have the Caputo recurrence formula of solution of the following form

$$y(a + (n+1)h) = y_a + h^\alpha \sum_{j=0}^{n} \binom{n - j + \alpha - 1}{n - j} f(jh, y(a + jh))\,, \tag{11}$$

for $n \in \mathbb{N}_0$.

4 Viability via Approximation

Similarly as for the ordinary differential equations (see [7]) one can define the viability of a subset with respect to the fractional differential equation (6). Before formulating viability problems we state the following results which has been proved in [16].

Proposition 4. *Let $\alpha \in (0,1]$, $h > 0$, $a \in \mathbb{R}$ and $t \geq a$. Then let $\bar{t}_h := a + (1 - \alpha)h + nh$, where $n = \left[\frac{t-a}{h}\right] + 1$ and $\left[\frac{t-a}{h}\right]$ is the greatest integer less than or equal to $\frac{t-a}{h}$. Then $\lim_{h\to 0} \bar{t}_h = t$ and*

$$\lim_{h\to 0} (\bar{t}_h - a)_h^{(-\alpha)} = (t - a)^{-\alpha}\,.$$

Proposition 5. *Let $\alpha \in (0,1]$, $h > 0$, f being continuous with integrable f' on some $[a,T]$ with $T > 0$, and $\bar{t}_h := a + (1-\alpha)h + nh$, where $n = \left[\frac{t-a}{h}\right] + 1$. Then*

$$\left({}_a^C D^\alpha f\right)(t) = \lim_{h\to 0} \left({}_a\Delta_{h,*}^\alpha f\right)(\bar{t}_h).$$

Proposition 6. *Let $\alpha \in (0,1]$, $h > 0$, f being continuous with integrable f' on some $[a,T]$ with $T > 0$, and $\bar{t}_h := a + (1-\alpha)h + nh$, where $n = \left[\frac{t-a}{h}\right] + 1$. Then*

$$\left({}_a^{RL} D^\alpha f\right)(t) = \lim_{h\to 0} \left({}_a\Delta_h^\alpha f\right)(\bar{t}_h). \tag{12}$$

Proof. Since the following formula holds:

$$\left({}_a\Delta_h^\alpha x\right)(t) = \left({}_a\Delta_{h,*}^\alpha x\right)(t) + \frac{x(a)(t-a)_h^{-\alpha}}{\Gamma(1-\alpha)}$$

and equality (12) holds true for Caputo–type operator, then applying Proposition 4 we get

$$\lim_{h\to 0} \left({}_a\Delta_h^\alpha f\right)(\bar{t}_h) = \left({}_a^{RL} D^\alpha f\right)(t).$$

The next propositions are direct consequences of Propositions 5 and 6, respectively.

Proposition 7. *The solution x of system*

$$\left({}_0^C D^\alpha x\right)(t) = f(t, x(t)),$$
$$x(0) = x_0.$$

is approximated by the solution of system

$$\left({}_a\Delta_{h,*}^\alpha \overline{x}\right)(t) = f(t, \overline{x}(t)),$$
$$\overline{x}(a) = x_0.$$

in values via the following limit:

$$\lim_{h\to 0} \overline{x}(t_h) = x(t),$$

where $\bar{t}_h := a + (1-\alpha)h + nh$ with $n = \left[\frac{t-a}{h}\right] + 1$ and $a = (\alpha-1)h$, $\alpha \in (0,1]$.

Proposition 8. *The solution x of system*

$$\left({}_0^{RL} D^\alpha x\right)(t) = f(t, x(t)),$$
$$x(a) = x_a.$$

is approximated by the solution of system

$$\left({}_a\Delta_h^\alpha \overline{x}\right)(t) = f(t, \overline{x}(t)),$$
$$\overline{x}(a) = x_a.$$

in values via the following limit:

$$\lim_{h\to 0} \overline{x}(t_h) = x(t),$$

where $\bar{t}_h := a + (1-\alpha)h + nh$ with $n = \left[\frac{t-a}{h}\right] + 1$ and $a = (\alpha-1)h$, $\alpha \in (0,1]$.

In order to formulate results on viability we need the following definitions. Let us denote by I an open interval in $\mathbb{R}$.

Definition 8. *Let $K \subset \mathbb{R}^n$ be nonempty and $f : I \times \mathbb{R}^n \to \mathbb{R}^n$. The subset K is* fractionally viable with respect to f *if for any $x_0 \in \mathbb{R}^n$ there exists $[t_0, T] \subseteq \mathbb{R}$, such that* (6) *has at least one solution such that $x : [t_0, T] \to K$, for $t \in [t_0, T]$, where $t_0 > 0$. We say then that x is a viable solution of* (6) *in K.*

The idea of viability of fractional differential equations can be expressed by using the concept of tangent cone. There are many notions of tangency of a vector to a set, see for example [7, Section 2.3]. We will follow the concept of the contingent vectors (see [3]).

Let us recall that for $K \subseteq \mathbb{R}^n$ and $x_0 \in K$ one can define the vector tangent to the set K as follows:

Definition 9. *The vector $\eta \in \mathbb{R}^n$ is* contingent *to the set K at the point x_0 if*

$$\liminf_{h \downarrow 0} \frac{1}{h} \operatorname{dist}(x_0 + h\eta; K) = 0\,. \tag{13}$$

The set of all vectors that are contingent to the set K at point x_0 is a closed cone, see [7, Proposition 2.3.1]. This cone, denoted by $\mathcal{T}_K(x_0)$, is called *contingent cone (Bouligand cone)* to the set K at $x_0 \in K$. From [7, Proposition 2.3.2] we know that $\eta \in \mathcal{T}_K(x_0)$ if and only if for every $\varepsilon > 0$ there exist $h \in (0, \varepsilon)$ and $p_h \in B(0, \varepsilon)$ such that $x_0 + h(\eta + p_h) \in K$, where $B(0, \varepsilon)$ denotes the closed ball in $\mathbb{R}^n$ centered at 0 and of radius $\varepsilon > 0$.

For $\varepsilon > 0$, by ε-neighborhood of a set $K \subset \mathbb{R}^n$ we mean the following:

$$K^\varepsilon := \{x \in \mathbb{R}^n : dist(x, K) < \varepsilon\}\,.$$

Proposition 9. *If $f(t_0, x_0) \in T_K(x_0)$, where f is the right hand side of system* (6), *then for every $\varepsilon > 0$ there exists $h > 0$ such that there exists a solution y to $h - difference$ system* (8) *that is viable in K_ε.*

Proof. By assumption, since (6) has a viable solution in K it means that the subset K is fractionally viable with respect to f. Then for any $x_0 \in \mathbb{R}^n$ there exists $[t_0, T] \subseteq \mathbb{R}$, such that (6) has at least one solution $x : [t_0, T] \to \mathbb{R}^n$, satisfying $x(a) \in K$ for $t \in [a, T]$, where $a > 0$. But since

$$\lim_{h \to 0} \overline{x}(t_h) = x(t)\,,$$

where $\bar{t}_h := a + (1 - \alpha)h + nh$ with $n = \left[\frac{t-a}{h}\right] + 1$ and $a = (\alpha - 1)h$, $\alpha \in (0, 1]$, this implies that for every $\varepsilon > 0$ there exists $\delta > 0$ such that the fact $|t_h - t| < \delta$ implies that $||y(\bar{t}_h) - x(t)|| < \varepsilon$. Now if we put $\delta = h$ we get that $y(\bar{t}_h) \in K_\varepsilon$, for every $\bar{t}_h \in [a, T]_{(h\mathbb{N})_a}$.

If one takes $x(a) = 0$ then Caputo and Riemann–Liouville–type operators in both, discrete and continuous cases, coincide. Thus the following statements is true.

Remark 7. If the fractional differential equation

$$ {}_{0}^{RL}D^{\alpha}x(t) = f(t, x(t)), \quad 0 < \alpha < 1, \quad t \in (0, T], \tag{14}$$

has a viable solution in K, then for every $\varepsilon > 0$ there exists $h > 0$ such that there exists a solution y to system

$$ \left({}_{a}\Delta_{h}^{\alpha} y\right)(nh) = f(nh, y(nh + a)) \tag{15}$$

that is viable in K_{ε},

Remark 8. Let us notice that if $K = \mathbb{R}_{+}^{n}$ one can consider all above results as positivity ones.

5 Conclusions

The problem of viability that consists in finding conditions under which there exists at least one solution to a system, which maintains in a certain set of constrains is considered. In the case of uniqueness of solutions, as it concerns the paper, viability coincides with the invariance idea, where all solutions have to stay in a constrained set. In the paper, the conditions for the existence of viable (invariant) solutions to a discrete fractional equation are formulated via viability properties of fractional differential equations. Our next step will be to examine the obtain results under potential applications. Going further, we want to consider the problem of the existence of viable solutions to fractional differential and difference inclusions.

Acknowledgments. The project was supported by the founds of National Science Centre granted on the bases of the decision number DEC-2011/03/B/ST7/03476. The work was supported by Bialystok University of Technology grant G/WM/3/12.

References

1. Almeida, R., Torres, D.F.M.: Necessary and sufficient conditions for the fractional calculus of variations with Caputo derivatives. Commun. Nonlinear Sci. Numer. Simul. 16(3), 1490–1500 (2011)
2. Aubin, J.P.: Viability Theory. Birkhäuser, Berlin (1991)
3. Aubin, J.P., Frankowska, H.: Set-Valued Analysis. Birkhäuser Boston Inc., Boston (1990)
4. Bai, Z., Lü, H.: Positive solutions for boundary value problems of nonlinear fractional differential equations. Journal of Mathematical Analysis and Applications 311(2), 495–505 (2005)
5. Bastos, N.R.O., Ferreira, R.A.C., Torres, D.F.M.: Necessary optimality conditions for fractional difference problems of the calculus of variations. Discrete Contin. Dyn. Syst. 29(2), 417–437 (2011)

6. Benchora, M., Hamani, S., Ntouyas, S.K.: Boundary value problems for differential equations with fractional order. Surveys in Mathematics and its Applications 3, 1843–7265 (2008) (print)
7. Cañada, A., Drábek, P., Fonda, A.: Handbook of Differential Equations Ordinary Differential Equations. Elsevier, North Holland (2005)
8. Ferreira, R.A.C., Torres, D.F.M.: Fractional h-difference equations arising from the calculus of variations. Appl. Anal. Discrete Math. 5(1), 110–121 (2011)
9. Girejko, E., Mozyrska, D., Wyrwas, M.: A sufficient condition of viability for fractional differential equations with the Caputo derivative. Journal of Mathematical Analysis and Applications 381 (2011)
10. Holm, M.T.: The theory of discrete fractional calculus: Development and application. PhD thesis, University of Nebraska - Lincoln (2011)
11. Kilbas, A.A., Srivastava, H.M., Trujillo, J.J.: Theory and applications of fractional differential equations. North–Holland Mathematics Studies, vol. 204. Elsevier Science B. V., Amsterdam (2006)
12. Lakshmikantham, V., Vatsala, A.S.: General uniqueness and monotone iterative technique for fractional differential equations. Applied Mathematics Letters 21, 828–834 (2008)
13. Miller, K.S., Ross, B.: Fractional difference calculus. In: Proceedings of the International Symposium on Univalent Functions, Fractional Calculus and their Applications, pp. 139–152. Nihon University, Kōriyama (1988)
14. Mozyrska, D., Girejko, E.: Overview of the fractional h-difference operators. In: Advances in Harmonic Analysis and Operator Theory: The Stefan Samko Anniversary Volume, vol. 229, pp. 253–267. Springer (2013)
15. Mozyrska, D., Girejko, E., Wyrwas, M.: A necessary condition of viability for fractional differential equations with initialization. Computers and Mathematics with Applications 62(9), 3642–3647 (2011)
16. Mozyrska, D., Girejko, E., Wyrwas, M.: Fractional nonlinear systems with sequential operators. Central European Journal of Physics, 9 (2013), doi:10.2478/s11534-013-0223-3
17. Podlubny, I.: Fractional Differential Equations. Mathematics in Sciences and Engineering, vol. 198. Academic Press, San Diego (1999)
18. Samko, S.G., Kilbas, A.A., Marichev, O.I.: Fractional Integrals and Derivatives: Theory and Applications. Gordon and Breach Science Publishers S.A., Yverdon (1993)
19. Zhang, S.: Existence of a positive solution for some class of nonlinear fractional differential equations. Journal of Mathematical Analysis and Applications 278, 136–148 (2003)

Regular Sturm-Liouville Problem with Riemann-Liouville Derivatives of Order in (1,2): Discrete Spectrum, Solutions and Applications

Malgorzata Klimek and Marek Blasik

Institute of Mathematics, Czestochowa University of Technology
Armii Krajowej 21, 42-201 Czestochowa, Poland
{mklimek,marek.blasik}@im.pcz.pl

Abstract. We study a regular fractional Sturm-Liouville problem formulated using left and right Riemann-Liouville derivatives of order in the range (1,2). We prove a theorem describing the eigenvalues and eigenfunctions of such a problem considered on the space of functions continuously differentiable in a finite interval and obeying vanishing Dirichlet and fractional Neumann boundary conditions. It appears that the spectrum of eigenvalues is discrete and that the eigenfunctions form a basis in the space of square-integrable functions. We also show applications of the derived eigenfunctions in the theory of partial fractional differential equations.

Keywords: Fractional Sturm-Liouville problem, Riemann-Liouville derivatives, eigenvalues and eigenfunctions, fractional differential equations.

1 Introduction

In the paper, we consider a regular fractional Sturm-Liouville problem (FSLP) with left and right Riemann-Liouville dervatives. Fractional differential equations, mixing the left and the right derivatives, arise in fractional calculus of variations (FVC) [20, 21] and lead to many meaningful and applicable problems which, however, are difficult to solve explicitly. For results on the minimum action principle and the derivation of Euler-Lagrange equations, we refer readers to monographs [6, 15] and the references given therein. FSLPs are eigenvalue problems extending the classical Sturm-Liouville theory. They were earlier formulated as fractional deformations of classical problems [1–3, 13, 16, 19], recently they have emerged within the framework of FVC. In the theory of differential equations, the solutions of Sturm-Liouville problems (SLPs) are important and useful tools in solving many equations appearing in the mathematical modelling of real-world phenomena [23, 26]. We introduced regular and singular FSLPs of a variational type in [7], then they were also studied in [8–11, 22, 25]. Some singular FSLPs were solved explicitly in [9, 10, 22, 25] and solutions were applied in anomalous diffusions theory which extends Pearson's diffusions [4] to the fractional version (compare also [14]). The variational formulation of the fractional

© Springer International Publishing Switzerland 2015

K.J. Latawiec et al. (eds.), *Advances in Modeling and Control of Non-integer Order Systems*,
Lecture Notes in Electrical Engineering 320, DOI: 10.1007/978-3-319-09900-2_3

version of SLPs yields solutions which correspond to real eigenvalues and are orthogonal eigenfunctions.

The aim of this paper is to propose a simple method of proof of the existence result for a discrete spectrum of a regular FSLP with vanishing Dirichlet and fractional Neumann boundary conditions. We extend here an approach developed in [12] in the case of a regular FSLP with derivatives in the range (0,1). By using the methods of functional analysis we also prove that the respective infinite countable set of eigenfunctions yields a basis in the space of square-integrable functions.

Let us begin with some basic definitions and properties of fractional operators [5, 24].

2 Preliminaries

Definition 1. *Left and the right Riemann–Liouville fractional integrals are defined as follows*

$$I^{\alpha}_{a+} f(x) := \frac{1}{\Gamma(\alpha)} \int_a^x \frac{f(t)dt}{(x-t)^{1-\alpha}}, \quad x \in (a,b], \tag{1}$$

$$I^{\alpha}_{b-} f(x) := \frac{1}{\Gamma(\alpha)} \int_x^b \frac{f(x)dt}{(t-x)^{1-\alpha}}, \quad x \in [a,b), \tag{2}$$

Here $\Gamma(\alpha)$ denotes Euler's gamma function.

Riemann–Liouville fractional integrals satisfy the semigroup property.

Property 1 (cf. Lemma 2.3 [5]). Let $\alpha, \beta > 0$ and $f \in L^p(a,b)$, $(1 \leq p \leq \infty)$. Then, relations

$$I^{\alpha}_{a+} I^{\beta}_{a+} f(x) = I^{\alpha+\beta}_{a+} f(x), \quad I^{\alpha}_{b-} I^{\beta}_{b-} f(x) = I^{\alpha+\beta}_{b-} f(x) \tag{3}$$

are fulfilled.

Definition 2. *The left Riemann–Liouville fractional derivative of order $\alpha \in (1,2)$ of a function f, denoted by $D^{\alpha}_{a+} f$, is given by*

$$\forall x \in (a,b], \quad D^{\alpha}_{a+} f(x) := D^2 \, I^{2-\alpha}_{a+} f(x). \tag{4}$$

Similarly, the right Riemann–Liouville fractional derivative of order $\alpha \in (1,2)$ of a function f, denoted by $D^{\alpha}_{b-} f$, is given by

$$\forall x \in [a,b), \quad D^{\alpha}_{b-} f(x) := D^2 \, I^{2-\alpha}_{b-} f(x), \tag{5}$$

where we denoted $D = \frac{d}{dx}$.

The following Caputo derivatives will also be applied in further considerations.

Definition 3. *The left and the right Caputo fractional derivatives of order $\alpha \in (1, 2)$ are given by*

$$\forall x \in (a, b], \quad {}^cD^\alpha_{a+} f(x) := D^\alpha_{a+} \left[f(x) - f(a) - f'(a)(x - a) \right], \tag{6}$$

$$\forall x \in [a, b), \quad {}^cD^\alpha_{b-} f(x) := D^\alpha_{b-} \left[f(x) - f(b) - f'(b)(b - x) \right]. \tag{7}$$

When order $\alpha \in (1, 2)$ and $f \in AC^2[a, b]$, the Caputo fractional derivatives satisfy the following relations:

$${}^cD^\alpha_{a+} f(x) = I^{2-\alpha}_{a+} D^2 f(x), \quad {}^cD^\alpha_{b-} f(x) = I^{2-\alpha}_{b-} D^2 f(x), \tag{8}$$

respectively.

Property 2 (cf. Lemma 2.4 [5]). If $\alpha > 0$ and $f \in L^p(a, b)$, $(1 \leq p \leq \infty)$, then the following composition rules are valid:

$$D^\alpha_{a+} I^\alpha_{a+} f(x) = f(x), \quad D^\alpha_{b-} I^\alpha_{b-} f(x) = f(x), \tag{9}$$

for almost all $x \in [a, b]$. If function f is continuous, then the composition rules hold for all $x \in [a, b]$.

In addition, in certain classes of functions the Caputo derivatives are the left inverse operators of Riemann–Liouville fractional integrals.

Property 3 (cf. Lemma 2.21 [5]). Let $\alpha > 0$ and $\alpha \in \mathbb{N}$ or $\alpha \notin \mathbb{N}$. If f is continuous on interval $[a, b]$, then

$${}^cD^\alpha_{a+} I^\alpha_{a+} f(x) = f(x), \quad {}^cD^\alpha_{b-} I^\alpha_{b-} f(x) = f(x). \tag{10}$$

3 Eigenvalues and Eigenfunctions of a Regular Fractional Sturm-Liouville Problem

We shall prove that a regular fractional Sturm-Liouville problem:

$$\mathcal{L}y(x) = D^\alpha_{b-} p(x) D^\alpha_{a+} y(x) = \lambda y(x), \tag{11}$$

$$y(a) = 0 \quad y(b) = 0, \tag{12}$$

$$y'(a) = 0 \quad {}^cD^\alpha_{a+} y(x)|_{x=b} = 0, \tag{13}$$

where order $\alpha \in \left(\frac{3}{2}, 2\right)$ and p is an arbitrary positive function from the $C[a, b]$-space, has an infinite countable set of simple positive eigenvalues and corresponding continuously differentiable eigenfunctions. We denote the space of continuously differentiable complex-valued functions obeying (12)–(13) as

$$C_{B2}[a, b] := \{g \in C^1[a, b];\ g(a) = g(b) = 0,\ g'(a) = {}^cD^\alpha_{a+} g(x)|_{x=b} = 0\}. \tag{14}$$

Remark 1. We note that similar to the classical case, the eigenvalues will be simple as the dimension of the space of solutions (11), obeying initial conditions $y(a) = y'(a) = 0$, is one-dimensional.

Now, we describe the stationary functions of Sturm-Liouville operator $\mathcal{L}$ in the $C^1[a,b]$ and $C_{B2}[a,b]$ spaces.

Lemma 1. *For order $\frac{3}{2} < \alpha < 2$ the solutions of equation*

$$D^{\alpha}_{b-}\, p(x)^c D^{\alpha}_{a+} u(x) = 0 \tag{15}$$

continuously differentiable in $[a,b]$, are of the form:

$$u(x) = C_1 + C_2 x + C_3\, I^{\alpha}_{a+} \frac{(b-x)^{\alpha-1}}{p(x)} + C_4\, I^{\alpha}_{a+} \frac{(b-x)^{\alpha-2}}{p(x)}, \tag{16}$$

where $C_j,\ j = 1, ..., 4$ are arbitrary constants.

Corollary 1. *Let $\frac{3}{2} < \alpha < 2$. If function $u \in C_{B2}[a,b]$ fulfills (15), then $u \equiv 0$.*

Lemma 2. *If λ is an eigenvalue of FSLP (11)–(13) corresponding to the non-trivial eigenfunction $y \in C_{B2}[a,b]$, then $\lambda > 0$.*

Proof. First, $\lambda \neq 0$ by Corollary 1. From the results of papers [7, 8], it follows that $\lambda \in \mathbb{R}$. Next, we multiply (11) by $\bar{y}$, integrate and obtain:

$$\int_a^b \overline{y(x)}\, D^{\alpha}_{b-} p(x)\, D^{\alpha}_{a+} y(x)\, dx = \int_a^b \overline{y(x)}\, D^{\alpha}_{b-} p(x)\, {}^c D^{\alpha}_{a+} y(x)\, dx$$

$$= -\overline{y(x)}\, D^{\alpha-1}_{b-} p(x)\, {}^c D^{\alpha}_{a+} y(x)|^{x=b}_{x=a} + \int_a^b \overline{y'(x)}\, D^{\alpha-1}_{b-} p(x)\, {}^c D^{\alpha}_{a+} y(x)\, dx =$$

$$= -\overline{y'(x)}\, I^{2-\alpha}_{b-} p(x)\, {}^c D^{\alpha}_{a+} y(x)|^{x=b}_{x=a} + \int_a^b \overline{y''(x)}\, I^{2-\alpha}_{b-} p(x)\, {}^c D^{\alpha}_{a+} y(x)\, dx =$$

$$= \int_a^b p(x) \left| {}^c D^{\alpha}_{a+} y(x) \right|^2 dx = \lambda \cdot ||y||^2_{L^2}$$

which implies

$$\lambda = \left(\frac{||\sqrt{p}\, {}^c D^{\alpha}_{a+} y||_{L^2}}{||y||_{L^2}} \right)^2 \quad \Longrightarrow \lambda > 0.$$

Similar to the classical Sturm-Liouville theory and the case of order $\alpha \in (\frac{1}{2}, 1)$ from paper [12], we shall replace the unbounded Sturm-Liouville operator from (11) (denoted as $\mathcal{L}$) with the right inverse integral and bounded operator denoted as T

$$\mathcal{L}u(x) = D^{\alpha}_{b-} p(x)\, {}^c D^{\alpha}_{a+} u(x) = f(x), \tag{17}$$

$$u(x) = I^{\alpha}_{a+} \frac{1}{p(x)} I^{\alpha}_{b-} f(x) - \frac{I^{\alpha}_{a+} \frac{(b-x)^{\alpha-1}}{p(x)}}{I^{\alpha}_{a+} \frac{(b-x)^{\alpha-1}}{p(x)}|_{x=b}} I^{\alpha}_{a+} \frac{1}{p(x)} I^{\alpha}_{b-} f(x)|_{x=b} = Tf(x). \tag{18}$$

We observe that the property:

$$\mathcal{L}Tu(x) = u(x), \tag{19}$$

resulting from the composition rules, implies the following equivalence on the $C_{B2}[a,b]$- space:

$$\mathcal{L}u(x) = \lambda u(x) \iff Tu(x) = \frac{1}{\lambda}u(x). \tag{20}$$

Let us note that operator T can be expressed as the following integral operator with kernel $K = K_1 + K_2$

$$u(x) = \int_a^b \left(K_1(x,s) + K_2(x,s)\right) f(s)ds, \tag{21}$$

where the K_1 and K_2 parts of the kernel look as follows (compare [12])

$$K_1(x,s) = \begin{cases} \frac{1}{[\Gamma(\alpha)]^2}\int_a^s \frac{(x-t)^{\alpha-1}(s-t)^{\alpha-1}}{p(t)}dt & s \le x \\ \frac{1}{[\Gamma(\alpha)]^2}\int_a^x \frac{(x-t)^{\alpha-1}(s-t)^{\alpha-1}}{p(t)}dt & s > x, \end{cases} \tag{22}$$

$$K_2(x,s) = -\frac{1}{\Gamma(\alpha)} \frac{I^\alpha_{a+}\frac{(b-x)^{\alpha-1}}{p(x)}}{I^\alpha_{a+}\frac{(b-x)^{\alpha-1}}{p(x)}|_{x=b}} \cdot I^\alpha_{a+}\frac{(b-s)^{\alpha-1}}{p(s)}. \tag{23}$$

Fractional integral operator T, defined for a positive continuous function p, is bounded on the $L^2(a,b)$- space since for real order $\alpha > 0$ we have relations

$$||I^\alpha_{a+}f||_{L^2} \le K_\alpha ||f||_{L^2} \quad ||I^\alpha_{a+}f||_{L^2} \le K_\alpha ||f||_{L^2} \quad K_\alpha = \frac{(b-a)^\alpha}{\Gamma(\alpha+1)} \tag{24}$$

which follow from Lemma 2.1 [5]. Now, we shall show that T is a compact and self-adjoint operator.

Lemma 3. *Let $2 > \alpha > \frac{3}{2}$ and $p \in C[a,b]$ be an arbitrary positive function. Then, operator T defined by (18) on the $L^2(a,b)$ - space is a compact operator.*

Proof. According to (24), operator T is correctly defined as a bounded operator mapping $L^2(a,b) \longrightarrow L^2(a,b)$. To prove its compactness, we show that

$$\int_a^b \int_a^b K^2(x,s)dxds < \infty. \tag{25}$$

This integral fulfills inequality

$$\int_a^b \int_a^b \left(K_1(x,s) + K_2(x,s)\right)^2 dxds \le \tag{26}$$

$$\le 2\int_a^b \int_a^b \left[\left(K_1(x,s)\right)^2 + \left(K_2(x,s)\right)^2\right] dxds.$$

Both, the K_1 and K_2 - kernels are continuous in $[a,b] \times [a,b]$, thence condition (25) is fulfilled and operator T, defined by kernel K, is compact.

Lemma 4. *Let $2 > \alpha > \frac{3}{2}$ and $p \in C[a,b]$ be an arbitrary positive function. Operator T defined by (18) on the $L^2(a,b)$ - space is a self-adjoint operator.*

Proof. Operator T is defined on a Hilbert space, namely on the $L^2(a,b)$-space of complex-valued functions, endowed with the $||\cdot||_{L^2}$ norm. It is self-adjoint, provided $\langle f, Tf\rangle \in \mathbb{R}$ for any function from this Hilbert space. Let function $f \in L^2(a,b)$, $f \neq 0$ be an arbitrary function. Denote

$$\psi(x) := \frac{I^{\alpha}_{a+}\frac{(b-x)^{\alpha-1}}{p(x)}}{I^{\alpha}_{a+}\frac{(b-x)^{\alpha-1}}{p(x)}|_{x=b}}, \tag{27}$$

$$\psi(a) = 0 \quad \psi(b) = 1, \tag{28}$$

$$\psi'(a) = 0 \quad {}^{c}D^{\alpha}_{a+}\psi(x)|_{x=b} = \frac{\frac{(b-b)^{\alpha-1}}{p(x)}}{I^{\alpha}_{a+}\frac{(b-x)^{\alpha-1}}{p(x)}|_{x=b}} = 0. \tag{29}$$

For inner product $\langle f, Tf\rangle$, we obtain:

$$\langle f, Tf\rangle = \tag{30}$$

$$= \int_a^b f(x)\overline{I^{\alpha}_{a+}\frac{1}{p}I^{\alpha}_{b-}\, f(x)}\, dx - \int_a^b f(x)\overline{\psi(x)I^{\alpha}_{a+}\frac{1}{p}I^{\alpha}_{b-}\, f(x)|_{x=b}}\, dx =$$

$$= \int_a^b f(x)I^{\alpha}_{a+}\frac{1}{p}I^{\alpha}_{b-}\, \overline{f(x)}\, dx - \int_a^b f(x)\psi(x)I^{\alpha}_{a+}\frac{1}{p}I^{\alpha}_{b-}\, \overline{f(x)}|_{x=b}\, dx =$$

$$= \int_a^b I^{\alpha}_{b-}f(x)\frac{1}{p(x)}I^{\alpha}_{b-}\, \overline{f(x)}\, dx - \frac{I^{\alpha}_{a+}\frac{1}{p}I^{\alpha}_{b-}\, \overline{f(x)}|_{x=b}}{I^{\alpha}_{a+}\frac{(b-x)^{\alpha-1}}{p(x)}|_{x=b}} \cdot \int_a^b f(x)I^{\alpha}_{a+}\frac{(b-x)^{\alpha-1}}{p(x)}dx =$$

$$= \int_a^b \frac{1}{p(x)}I^{\alpha}_{b-}f(x)\cdot\overline{I^{\alpha}_{b-}\, f(x)}\, dx - \frac{I^{\alpha}_{a+}\frac{1}{p}I^{\alpha}_{b-}\, \overline{f(x)}|_{x=b}}{I^{\alpha}_{a+}\frac{(b-x)^{\alpha-1}}{p(x)}|_{x=b}} \cdot \int_a^b \frac{(b-x)^{\alpha-1}}{p(x)}I^{\alpha}_{b-}f(x)dx =$$

$$= \int_a^b \frac{1}{p(x)}|I^{\alpha}_{b-}f(x)|^2\, dx - \frac{I^{\alpha}_{a+}\frac{1}{p}I^{\alpha}_{b-}\, \overline{f(x)}|_{x=b}\cdot I^{\alpha}_{a+}\frac{1}{p}I^{\alpha}_{b-}\, f(x)|_{x=b}}{I^{\alpha}_{a+}\frac{(b-x)^{\alpha-1}}{p(x)}|_{x=b}} \cdot \Gamma(\alpha) =$$

$$= \int_a^b \frac{1}{p(x)}|I^{\alpha}_{b-}f(x)|^2\, dx - \frac{|I^{\alpha}_{a+}\frac{1}{p}I^{\alpha}_{b-}\, f(x)|_{x=b}|^2}{I^{\alpha}_{a+}\frac{(b-x)^{\alpha-1}}{p(x)}|_{x=b}} \cdot \Gamma(\alpha) \in \mathbb{R},$$

where $|\cdot|$ denotes the complex modulus. From the above property it follows that operator T is a self-adjoint operator.

Corollary 2. *Let $2 > \alpha > \frac{3}{2}$ and $p \in C[a,b]$ be an arbitrary positive function. In the $L^2(a,b)$ - space an infinite, countable and orthonormal basis of continuously differentiable eigenfunctions of operator T: $\{\phi_k;\ \ k \in \mathbb{N}\}$ corresponding to* *simple, positive eigenvalues* $\frac{1}{\lambda_1} > \frac{1}{\lambda_2} > ... > \frac{1}{\lambda_k}...;\quad k \in \mathbb{N}$ *exists. Moreover, the following series are convergent:*

$$\sum_{k=1}^{\infty} | \langle f, \phi_k \rangle |^2 < \infty, \quad \langle f, \phi_k \rangle = \int_a^b f(x)\overline{\phi_k(x)}dx, \quad \forall f \in L^2(a,b), \tag{31}$$

$$\sum_{k=1}^{\infty} \frac{1}{(\lambda_k)^2} < \infty \tag{32}$$

and relation

$$\frac{|\phi_k(x)|}{|\lambda_k|} < M_+ \tag{33}$$

holds for certain $M_+ \in \mathbb{R}$ *and any* $k \in \mathbb{N}, x \in [a,b]$.

Proof. According to Lemma 3 and 4, operator T is a self-adjoint and compact operator. Thus, from the Hilbert-Schmidt theorem, it follows that the infinite, countable and orthonormal basis of eigenvectors exists in $L^2(a,b)$. Moreover, 0 is the only accumulation point of eigenvalues sequence: $\left(\frac{1}{\lambda_k}\right)_{k\in\mathbb{N}}$. Each eigenvector ϕ_k corresponds to eigenvalue $\frac{1}{\lambda_k}$ and fulfills equation:

$$\frac{1}{\lambda_k}\phi_k(x) = T\phi_k(x) = I^{\alpha}_{a+}\frac{1}{p(x)}I^{\alpha}_{b-}\phi_k(x) - \psi(x)I^{\alpha}_{a+}\frac{1}{p(x)}I^{\alpha}_{b-}\phi_k(x)|_{x=b}. \tag{34}$$

Let us note that for any positive function $p \in C[a,b]$, function $\frac{(b-x)^{\alpha-1}}{p(x)} \in C[a,b]$, therefore function ψ given in (27), obtained after integration with I^{α}_{a+} is continuously differentiable. This remark also applies to arbitrary function $f \in L^2(a,b)$, provided $\frac{3}{2} < \alpha < 2$, which yields $I^{\alpha}_{b-}f \in C^1[a,b]$ and $I^{\alpha}_{a+}f \in C^1[a,b]$. Thus, $\phi_k \in C^1[a,b]$ for any $k \in \mathbb{N}$. Next, each function $f \in L^2(a,b)$ can be expanded with regard to the basis $\{\phi_k;\ \ k \in \mathbb{N}\}$ as follows:

$$f(x) = \sum_{k=1}^{\infty}\langle f, \phi_k\rangle\phi_k(x).$$

From the fact that this series is convergent in $L^2(a,b)$, we obtain (31). As kernel $K \in L^2([a,b]\times[a,b])$, then we expand it by using the basis of eigenvectors $\{\phi_k \otimes \overline{\phi_l};\ \ k,l \in \mathbb{N}\}$:

$$K(x,s) = \sum_{k,l=1}^{\infty} a_{k,l}\phi_k(x)\overline{\phi_l(s)}.$$

We calculate coefficients $a_{k,l}$:

$$\frac{1}{\lambda_m}\phi_m(x) = T\phi_m(x) = \int_a^b K(x,s)\phi_m(s)ds = \sum_{k=1}^{\infty} a_{k,m}\phi_k(x)$$

and obtain

$$a_{k,m} = \frac{1}{\lambda_m}\delta_{k,m}.$$

Thus, integral kernel K is of the form:

$$K(x,s) = \sum_{k=1}^{\infty} \frac{1}{\lambda_k} \phi_k(x) \overline{\phi_k(s)}.$$

From property (25) we obtain (32).

Finally, relation (33) follows from (34) and the Schwarz-Bunyakovsky inequality for integrals. We can estimate complex modulus $|\phi_k(x)|$ as follows

$$\frac{|\phi_k(x)|}{|\lambda_k|} \leq \left(I^{\alpha}_{a+} \frac{1}{p(x)} I^{\alpha}_{b-} |\phi_k(x)| + |\psi(x)| \cdot I^{\alpha}_{a+} \frac{1}{p(x)} I^{\alpha}_{b-} |\phi_k(x)| \,|_{x=b} \right) \leq$$

$$\leq \frac{1}{\Gamma(\alpha)} \left(\int_a^b |x-s|^{\alpha-1} \frac{1}{p(s)} \left(I^{\alpha}_{b-} |\phi_k(s)| \right) ds + ||\psi|| \int_a^b (b-s)^{\alpha-1} \frac{1}{p(s)} \left(I^{\alpha}_{b-} |\phi_k(s)| \right) ds \right) \leq$$

$$\leq \frac{||\frac{1}{p}||}{\Gamma(\alpha)} \cdot ||I^{\alpha}_{b-} \phi_k||_{L^2} \cdot \frac{(b-a)^{\alpha-\frac{1}{2}}}{\sqrt{\alpha - \frac{1}{2}}} \left(1 + ||\psi||/\sqrt{2}\right) \leq$$

$$\leq \frac{||\frac{1}{p}||}{\Gamma(\alpha)} \cdot K_\alpha \cdot \frac{(b-a)^{\alpha-\frac{1}{2}}}{\sqrt{\alpha - \frac{1}{2}}} \left(1 + ||\psi||/\sqrt{2}\right) = M_+ < \infty,$$

where $||\cdot||$ denotes the supremum norm in space $C[a,b]$, ψ is given in (27) and constant K_α in (24).

Theorem 1. *Fractional Sturm-Liouville problem (11)–(12) has an infinite, countable set of positive, simple eigenvalues: $\lambda_1 < \lambda_2 < ...$ and corresponding orthonormal continuously differentiable eigenfunctions, provided $\frac{3}{2} < \alpha < 2$ and function p is an arbitrary positive and continuous function.*

Proof. We consider mapping $\mathcal{L}: \quad C_{B2}[a,b] \longrightarrow L^2(a,b)$ determined by the fractional Sturm-Liouville operator:

$$\mathcal{L}u(x) = D^{\alpha}_{b-}\, p(x) D^{\alpha}_{a+} u(x) = f(x)$$

defined on the subspace of continuously differentiable functions $C_{B2}[a,b]$ obeying boundary conditions (12)–(13). We note that on the $C_{B2}[a,b]$- space this mapping is linear and injective as by using Lemma 1 and Corollary 1 we infer that the only solution for

$$D^{\alpha}_{b-}\, p(x)^c D^{\alpha}_{a+} u(x) = 0$$

is $u \equiv 0$. By applying the composition rules, we previously derived the right inverse mapping (18) which can be extended to the Hilbert space $L^2(a,b)$. From Lemma 3 and 4, it follows that T is a self-adjoint and compact operator, defined on the Hilbert space, therefore by applying Corollary 2, it has an infinite countable set of eigenvalues and its continuously differentiable eigenfunctions form an orthonormal basis in $L^2(a,b)$:

$$\frac{1}{\lambda_k} \phi_k(x) = T\phi_k.$$

We recall that $I^{\alpha}_{b-} f \in C^1[a,b]$ and $I^{\alpha}_{a+} f \in C^1[a,b]$ for any $f \in L^2(a,b)$, provided $\frac{3}{2} < \alpha < 2$. Thus, $\phi_k \in C^1[a,b]$ for any $k \in \mathbb{N}$ and all eigenvectors obey boundary conditions (12)–(13), as by applying (27)–(29) and the fact that $\phi_k \in C^1[a,b]$, we have

$$\frac{1}{\lambda_k}\phi_k(a) = I^{\alpha}_{a+}\frac{1}{p}I^{\alpha}_{b-}\phi_k(x)|_{x=a} - \psi(a)I^{\alpha}_{a+}\frac{1}{p(x)}I^{\alpha}_{b-}\phi_k(x)|_{x=b} = 0, \tag{35}$$

$$\frac{1}{\lambda_k}\phi_k(b) = I^{\alpha}_{a+}\frac{1}{p}I^{\alpha}_{b-}\phi_k(x)|_{x=b} - \psi(b)I^{\alpha}_{a+}\frac{1}{p(x)}I^{\alpha}_{b-}\phi_k(x)|_{x=b} = 0, \tag{36}$$

$$\frac{1}{\lambda_k}\phi_k'(a) = I^{\alpha-1}_{a+}\frac{1}{p}I^{\alpha}_{b-}\phi_k(x)|_{x=a} - \psi'(a)I^{\alpha}_{a+}\frac{1}{p(x)}I^{\alpha}_{b-}\phi_k(x)|_{x=b} = 0, \tag{37}$$

$$\frac{1}{\lambda_k}{}^cD^{\alpha}_{a+}\phi_k(x)|_{x=b} = \frac{1}{p}I^{\alpha}_{b-}\phi_k(x)|_{x=b} - {}^cD^{\alpha}_{a+}\psi(x)|_{x=b}I^{\alpha}_{a+}\frac{1}{p(x)}I^{\alpha}_{b-}\phi_k(x)|_{x=b} = 0. \tag{38}$$

Now, we differentiate relation (34) for eigenvectors by applying the composition rules described in Properties 2 and 3. We obtain:

$$D^{\alpha}_{b-}p(x)\ {}^cD^{\alpha}_{a+}\phi_k(x) = \lambda_k\phi_k(x) \tag{39}$$

therefore

$$D^{\alpha}_{b-}p(x)\ D^{\alpha}_{a+}\phi_k(x) = \lambda_k\phi_k(x) \tag{40}$$

because $\phi_k \in C_{B2}[a,b]$, $k \in \mathbb{N}$. Thus, we conclude that regular fractional Sturm-Liouville problem (11)–(13) has in the case of order $\frac{3}{2} < \alpha < 2$ an infinite countable set of simple positive eigenvalues and corresponding continuously differentiable eigenfunctions. The eigenfunctions form an orthonormal countable basis in $L^2(a,b)$.

4 Applications - Solving Partial Fractional Differential Equations

We shall apply the obtained results to solve a time- and space-fractional partial differential equation in a finite space domain. The solution will be subjected to boundary conditions (12)–(13). We denote the partial version of FSLO (11) as follows:

$$\mathcal{L}_x := D^{\alpha}_{b-,x}p(x)\ D^{\alpha}_{a+,x}. \tag{41}$$

The following theorem describes a weak solution of the partial differential fractional equation, derived using the expansion with regard to the eigenfunctions basis.

Theorem 2. *The time- and space-fractional differential equation with the initial and boundary conditions given below:*

$$ {}^{c}D^{\beta}_{0+,t}u(t,x) = -\mathcal{L}_x u(t,x), \tag{42} $$

$$ u(0,x) = g(x) \quad g \in C_{B2}[a,b] \quad \mathcal{L}_x g \in L^2(a,b), \quad x \in [a,b], \tag{43} $$

$$ u(t,a) = u(t,b) = 0 \quad t \in [0,\infty) \tag{44} $$

$$ \frac{\partial u(t,x)}{\partial x}|_{x=a} = 0 \quad {}^{c}D^{\alpha}_{a+,x}u(t,x)|_{x=b} = 0, \tag{45} $$

has a unique weak solution continuous in $(0,\infty) \times [a,b]$*:*

$$ u(t,x) = \sum_{k=1}^{\infty} \langle g, \phi_k \rangle \, E_{\beta}(-\lambda_k t^{\beta}) \phi_k(x), \tag{46} $$

where coefficients $\langle g, \phi_k \rangle = \int_a^b g(x) \cdot \overline{\phi_k(x)} \, dx$*,* E_{β} *is the Mittag-Leffler function and* $\{\phi_k \in C_{B2}[a,b]; \quad k \in \mathbb{N}\}$ *is the orthonormal basis of eigenvectors in the* $L^2(a,b)$*-space corresponding to fractional Sturm-Liouville operator* $\mathcal{L}_x$*.*

Proof. Let us expand the solution with regard to the eigenfunctions basis:

$$ u(t,x) = \sum_{k=1}^{\infty} b_k(t) \phi_k(x). \tag{47} $$

By using the orthonormality of eigenvectors we derive the following set of fractional differential equations for variable coefficients b_k, $k \in \mathbb{N}$

$$ {}^{c}D^{\beta}_{0+,t} b_k(t) = -\lambda_k b_k(t). \tag{48} $$

Solutions of the above equation are proportional to the Mittag-Leffler function:

$$ b_k(t) = c_k E_{\beta}(-\lambda_k t^{\beta}), \tag{49} $$

where constants c_k can be easily derived from initial condition (43). By using the orthonormality of basis ϕ_k, $k \in \mathbb{N}$, we obtain: $c_k = \int_a^b g(x) \cdot \overline{\phi_k(x)} \, dx = \langle g, \phi_k \rangle$. Thus, solution u is of the form (46) and it also obeys boundary conditions (44)–(45) as each eigenfunction ϕ_k fulfills (12)–(13). Now, we test the uniform convergence of the above series. For any $t > t_0 > 0$ the following estimation is valid:

$$ |\langle g, \phi_k \rangle \, E_{\beta}(-\lambda_k t^{\beta}) \phi_k(x)| \leq |\langle g, \phi_k \rangle \cdot \frac{C}{1 + \lambda_k t_0^{\beta}} \phi_k(x)| \leq \frac{C M_+}{t_0^{\beta}} \frac{|\langle \mathcal{L}_x g, \phi_k \rangle|}{|\lambda_k|}. $$

In the above estimations, we applied the following inequality for the one-parameter Mittag-Leffler function, valid for $\lambda, t \in \mathbb{R}_+$ [18]:

$$ E_{\beta}(-\lambda t^{\beta}) \leq \frac{C}{1 + \lambda t^{\beta}}. $$

Next, we observe that the dominating number series is convergent, namely from the Schwarz-Bunyakovsky inequality for series we have:

$$ \sum_{k=1}^{\infty} \frac{|\langle \mathcal{L}_x g, \phi_k \rangle|}{|\lambda_k|} \leq \sqrt{\sum_{k=1}^{\infty} |\langle \mathcal{L}_x g, \phi_k \rangle|^2} \sqrt{\sum_{k=1}^{\infty} \frac{1}{(\lambda_k)^2}} < \infty. $$

Thus, the series defining solution u is uniformly convergent in any set $[t_0, \infty) \times [a, b]$. We infer from the Weierstrass' Majorant theorem that u is continuous in $(0, \infty) \times [a, b]$. Finally, we check that the obtained series is a weak solution to equation (42). To prove this fact, we show that for each eigenfunction ϕ_k from the countable basis of the $L^2(a, b)$- space, we have

$$\left\langle \left({}^c D^{\beta}_{0+,t} + \mathcal{L}_x \right) u, \phi_k \right\rangle = 0. \tag{50}$$

By using the series form of solution, we have for any $k \in \mathbb{N}$

$$\left\langle \left({}^c D^{\beta}_{0+,t} + \mathcal{L}_x \right) u, \phi_k \right\rangle = {}^c D^{\beta}_{0+,t} \langle u, \phi_k \rangle + \langle \mathcal{L}_x u, \phi_k \rangle =$$

$$= {}^c D^{\beta}_{0+,t} \langle g, \phi_k \rangle E_\beta \left(-\lambda_k t^\beta \right) + \langle u, \mathcal{L}_x \phi_k \rangle = -\lambda_k \, \langle g, \phi_k \rangle E_\beta \left(-\lambda_k t^\beta \right) + \lambda_k \langle u, \phi_k \rangle =$$

$$= -\lambda_k \, \langle g, \phi_k \rangle E_\beta \left(-\lambda_k t^\beta \right) + \lambda_k \, \langle g, \phi_k \rangle E_\beta \left(-\lambda_k t^\beta \right) = 0.$$

5 Conclusion

We considered a regular FSLP with Riemann-Liouville derivatives of order in the range (1,2), subjected to a set of boundary conditions including vanishing Dirichlet and fractional Neumann conditions. We extended the classical method of deriving the existence result on a discrete spectrum and the corresponding eigenvectors to FSLO. By replacing the analysis of an unbounded FSLO with a study of its inverse integral and bounded operator and by using the Hilbert-Schmidt operators' properties, we proved that the inverse, therefore also FSL operator, have a discrete spectrum and their eigenfunctions, continuously differentiable, form a basis in the space of square-integrable functions. Thus, they can be applied in solving certain partial fractional differential equations. Indeed, we constructed for the considered example a continuous weak solution. Further investigations should extend the simple regular FSLO to a more general case. We also shall study other 1D and 2D partial differential equations.

References

1. Al-Mdallal, Q.M.: An efficient method for solving fractional Sturm–Liouville problems. Chaos Solitons and Fractals 40, 183–189 (2009)
2. Al–Mdallal, Q.M.: On the numerical solution of fractional Sturm–Liouville problem. Int. J. of Comput. Math. 87, 2837–2845 (2010)
3. Erturk, V.S.: Computing eigenelements of Sturm-Liouville Problems of fractional order via fractional differential transform method. Mathematical and Computational Applications 16, 712–720 (2011)
4. Forman, J.L., Soerensen, M.: The Pearson diffusions: a class of statistically tractable diffusion processes. Scandinavian J. Stat. 35, 438–465 (2008)
5. Kilbas, A.A., Srivastava, H.M., Trujillo, J.J.: Theory and applications of fractional differential equations. Elsevier, Amsterdam (2006)

6. Klimek, M.: On solutions of linear fractional differential equations of a variational type. The Publishing Office of Czestochowa University of Technology, Czestochowa (2009)
7. Klimek, M., Agrawal, O.P.: On a regular fractional Sturm–Liouville problem with derivatives of order in $(0, 1)$. In: Proceedings of the 13th International Carpathian Control Conference, Vysoke Tatry (Podbanske), Slovakia, May 28-31 (2012), doi:dx.doi.org/10.1109/CarpathianCC.2012.6228655
8. Klimek, M., Agrawal, O.P.: Regular fractional Sturm–Liouville problem with generalized derivatives of order in (0,1). In: Proceedings of the IFAC Joint Conference: 5th SSSC, 11th WTDA, 5th WFDA, Grenoble, France, February 4-6 (2013), doi:dx.doi.org/10.3182/20130204-3-FR-4032.00170
9. Klimek, M., Agrawal, O.P.: Fractional Sturm–Liouville problem. Comput. Math. Appl. 66, 795–812 (2013)
10. Klimek, M., Agrawal, O.P.: Space- and time-fractional Legendre-Pearson diffusion equation. In: Proceedings of the ASME International Design Engineering Technical Conferences & Computers and Information in Engineering Conference (DETC 2013), August 4-7. Oregon USA Paper DETC2013-12604, Portland (2013)
11. Klimek, M., Odzijewicz, T., Malinowska, A.: Variational methods for the fractional Sturm-Liouville problems. J. Math. Anal. Appl. 416, 402–426 (2014)
12. Klimek, M., Blasik, M.: Regular fractional Sturm-Liouville problem with discrete spectrum: solutions and applications. In: Proceedings of the 2014 International Conference on Fractional Differentiaton and Its Applications, Catania, Italy, June 23-25 (2014)
13. Lin, Y., He, T., Shi, H.: Existence of positive solutions for Sturm–Liouville BVPs of singular fractional differential equations. U. P. B. Sci. Bull. Series A, 74 (2012)
14. Leonenko, N.N., Meerschaert, M.M., Sikorskii, A.A.: Fractional Pearson diffusion. J. Math Anal. Appl. 403, 532–546 (2013)
15. Malinowska, A.B., Torres, D.F.M.: Introduction to the Fractional Calculus of Variations. Imperial College Press, London (2012)
16. Neamaty, A., Darzi, R., Dabbaghian, A., Golipoor, J.: Introducing an Iterative Method for Solving a Special FDE. International Mathematical Forum 4, 1449–1456 (2009)
17. d'Ovidio, M.: From Sturm–Liouville problems to fractional and anomalous diffusions. Stochastic Processes and their Applications 122, 3513–3544 (2012)
18. Podlubny, I.: Fractional differential equations. Academic Press, San Diego (1999)
19. Qi, J., Chen, S.: Eigenvalue problems of the model from nonlocal continuum mechanics. J. Math. Phys. 52, 073516 (2011)
20. Riewe, F.: Nonconservative Lagrangian and Hamiltonian mechanics. Phys. Rev. E 53, 1890–1899 (1996)
21. Riewe, F.: Mechanics with fractional derivatives. Phys. Rev. E 55, 3581–3592 (1997)
22. Rivero, M., Trujillo, J.J., Velasco, M.P.: A fractional approach to the Sturm-Liouville problem. Cent. Eur. J. Phys. (2013), doi: 10.2478/s11534-013-0216-2
23. Sagan, H.: Boundary and Eigenvalue Problems in Mathematical Physics. J. Wiley & Sons, Inc., New York (1961)
24. Samko, S.G., Kilbas, A.A., Marichev, O.I.: Fractional integrals and derivatives. Gordon and Breach, Yverdon (1993)
25. Zayernouri, M., Karniadakis, G.E.: Fractional Sturm-Liouville eigen-problems: Theory and numerical approximation. J. Comput. Phys. 252, 495–517 (2013)
26. Zettl, A.: Sturm-Liouville Theory. Mathematical Surveys and Monographs, vol. 121. American Mathematical Society (2005)

Noether's Second Theorem for Variable Order Fractional Variational Problems

Agnieszka B. Malinowska[1] and Tatiana Odzijewicz[2]

[1] Faculty of Computer Science
Bialystok University of Technology
15-351 Białystok, Poland
[2] Center for Research and Development in Mathematics and Applications
Department of Mathematics, University of Aveiro
3810-193 Aveiro, Portugal

Abstract. We prove an analog of the second Noether theorem for variable order fractional variational problems. From this theorem, we get identities between Euler–Lagrange expressions and their variable order fractional derivatives.

Keywords: variable order fractional integrals, variable order fractional derivatives, fractional variational analysis, Euler–Lagrange equations, Noether's second theorem.

1 Introduction

In 1993, Samko and Ross [20] proposed an interesting generalization of fractional operators. Namely, they introduced the study of fractional integration and differentiation when the order is not a constant but a function. With time several works were dedicated to variable order fractional operators [1, 3, 8], and many interesting applications of those fractional operators were proposed, e.g., in mechanics and in the theory of viscous flows [3, 4, 8, 17–19]. Recently, also the study of the calculus of variations with variable order fractional operators has been introduced. In [2, 15] generalizations of the fractional Hamilton principle are proposed and appropriated Euler–Lagrange equations were proved. The works [13, 14] are devoted to problems of the calculus of variations with functionals given by multi-dimensional definite integrals involving partial derivatives of variable fractional order. An analog of the first Noether theorem for variable order fractional variational problem is proved in [16].

The aim of the current work is to prove an analog of the second Noether theorem, asserting that if a variational integral is invariant under transformations parameterized linearly by r arbitrary functions and their derivatives up to a given order m, then there are r identities between Euler–Lagrange expressions and their derivatives up to order m (see, e.g., [9, 11]. These identities Noether called "dependencies". For example, the Bianchi identities, in the general theory of relativity, are examples of such "dependencies". Noether's second theorem has applications in general relativity, electrodynamics, hydromechanics, quantum

© Springer International Publishing Switzerland 2015

K.J. Latawiec et al. (eds.), *Advances in Modeling and Control of Non-integer Order Systems*,
Lecture Notes in Electrical Engineering 320, DOI: 10.1007/978-3-319-09900-2_4

chromodynamics and other gauge field theories. It should be mentioned that a generalize of the second Noether theorem to fractional problems was already proved in [10]. However, in contrast with this work, where fractional operators with a constant non-integer order α were considered, here we study more general fractional variational problems with variable order derivatives.

The article is organized as follows. In Section 2 we give the definitions and basic properties of both ordinary and partial integrals and derivatives of variable fractional order. An analog of the second Noether theorem is stated and proved in Section 3. We finish with a simple illustrative example in Section 4.

2 Preliminaries

Let us introduce the following triangle:

$$\Delta := \left\{(t,\tau) \in \mathbb{R}^2 :\ a \leq \tau < t \leq b\right\},$$

and let $\alpha(t,\tau) : \bar{\Delta} \to [0,1]$ be such that $\alpha \in C^1\left(\bar{\Delta};\mathbb{R}\right)$.

Definition 1 (Left and right Riemann–Liouville integrals of variable order). *Operator*

$${}_aI_t^{\alpha(\cdot,\cdot)}[f](t) := \int\limits_a^t \frac{1}{\Gamma(\alpha(t,\tau))}(t-\tau)^{\alpha(t,\tau)-1} f(\tau)d\tau \quad (t > a)$$

is the left Riemann–Liouville integral of variable fractional order $\alpha(\cdot,\cdot)$, while

$${}_tI_b^{\alpha(\cdot,\cdot)}[f](t) := \int\limits_t^b \frac{1}{\Gamma(\alpha(\tau,t))}(\tau-t)^{\alpha(\tau,t)-1} f(\tau)d\tau \quad (t < b)$$

is the right Riemann–Liouville integral of variable fractional order $\alpha(\cdot,\cdot)$.

Next we define two types of variable order fractional derivatives.

Definition 2 (Left and right Riemann–Liouville derivatives of variable order). *The left Riemann–Liouville derivative of variable fractional order $\alpha(\cdot,\cdot)$ of a function f is defined by*

$$\forall t \in (a,b],\quad {}_aD_t^{\alpha(\cdot,\cdot)}[f](t) := \frac{d}{dt}{}_aI_t^{1-\alpha(\cdot,\cdot)}[f](t),$$

while the right Riemann–Liouville derivative of variable fractional order $\alpha(\cdot,\cdot)$ is defined by

$$\forall t \in [a,b),\quad {}_tD_b^{\alpha(\cdot,\cdot)}[f](t) := -\frac{d}{dt}{}_tI_b^{1-\alpha(\cdot,\cdot)}[f](t).$$

Definition 3 (Left and right Caputo derivatives of variable fractional order). *The left Caputo derivative of variable fractional order* $\alpha(\cdot,\cdot)$ *is defined by*

$$\forall t \in (a,b], \quad {}_a^C D_t^{\alpha(\cdot,\cdot)}[f](t) := {}_a I_t^{1-\alpha(\cdot,\cdot)} \left[\frac{d}{dt} f\right](t),$$

while the right Caputo derivative of variable fractional order $\alpha(\cdot,\cdot)$ *is given by*

$$\forall t \in [a,b), \quad {}_t^C D_b^{\alpha(\cdot,\cdot)}[f](t) := -{}_t I_b^{1-\alpha(\cdot,\cdot)} \left[\frac{d}{dt} f\right](t).$$

Partial variable order fractional integrals and derivatives are natural generalizations of the corresponding fractional operators of one variable. Along the work, for $i = 1, \ldots, n$, let a_i, b_i be numbers in $\mathbb{R}$ and $t = (t_1, \ldots, t_n)$ be such that $t \in \Omega_n$, where $\Omega_n = (a_1, b_1) \times \cdots \times (a_n, b_n)$ is a subset of $\mathbb{R}^n$. Moreover, let us define the following sets:

$$\Delta_i := \left\{(t_i, \tau) \in \mathbb{R}^2 : \; a_i \leq \tau < t_i \leq b_i\right\}, \quad i = 1 \ldots, n.$$

In the following we assume that $\alpha_i : \bar{\Delta}_i \to [0,1]$, $\alpha_i \in C^1\left(\bar{\Delta}_i; \mathbb{R}\right)$, $i = 1, \ldots, n$, $t \in \Omega_n$ and $f : \Omega_n \to \mathbb{R}$.

Definition 4. *The left Riemann–Liouville partial integral of variable fractional order* $\alpha_i(\cdot,\cdot)$ *with respect to the ith variable* t_i*, is given by*

$${}_{a_i} I_{t_i}^{\alpha_i(\cdot,\cdot)}[f](t) := \int\limits_{a_i}^{t_i} \frac{1}{\Gamma(\alpha_i(t_i,\tau))} (t_i - \tau)^{\alpha_i(t_i,\tau)-1} f(t_1, \ldots, t_{i-1}, \tau, t_{i+1}, \ldots, t_n) d\tau,$$

$t_i > a_i$*, while*

$${}_{t_i} I_{b_i}^{\alpha_i(\cdot,\cdot)}[f](t) := \int\limits_{t_i}^{b_i} \frac{1}{\Gamma(\alpha_i(\tau,t_i))} (\tau - t_i)^{\alpha_i(\tau,t_i)-1} f(t_1, \ldots, t_{i-1}, \tau, t_{i+1}, \ldots, t_n) d\tau,$$

$t_i < b_i$*, is the right Riemann–Liouville partial integral of variable fractional order* $\alpha_i(\cdot,\cdot)$ *with respect to variable* t_i.

Definition 5. *The left Riemann–Liouville partial derivative of variable fractional order* $\alpha_i(\cdot,\cdot)$*, with respect to the ith variable* t_i*, is given by*

$$\forall t_i \in (a_i, b_i], \quad {}_{a_i} D_{t_i}^{\alpha_i(\cdot,\cdot)}[f](t) = \frac{\partial}{\partial t_i} {}_{a_i} I_{t_i}^{1-\alpha_i(\cdot,\cdot)}[f](t)$$

while the right Riemann–Liouville partial derivative of variable fractional order $\alpha_i(\cdot,\cdot)$*, with respect to the ith variable* t_i*, is defined by*

$$\forall t_i \in [a_i, b_i), \quad {}_{t_i} D_{b_i}^{\alpha_i(\cdot,\cdot)}[f](t) = -\frac{\partial}{\partial t_i} {}_{t_i} I_{b_i}^{1-\alpha_i(\cdot,\cdot)}[f](t)$$

Definition 6. *The left Caputo partial derivative of variable fractional order $\alpha_i(\cdot,\cdot)$, with respect to the ith variable t_i, is defined by*

$$\forall t_i \in (a_i, b_i],\ \ {}^{C}_{a_i}D^{\alpha_i(\cdot,\cdot)}_{t_i}[f](t) = {}_{a_i}I^{1-\alpha_i(\cdot,\cdot)}_{t_i}\left[\frac{\partial}{\partial t_i}f\right](t),$$

while the right Caputo partial derivative of variable fractional order $\alpha_i(\cdot,\cdot)$, with respect to the ith variable t_i, is given by

$$\forall t_i \in [a_i, b_i),\ \ {}^{C}_{t_i}D^{\alpha_i(\cdot,\cdot)}_{b_i}[f](t) = -{}_{t_i}I^{1-\alpha_i(\cdot,\cdot)}_{b_i}\left[\frac{\partial}{\partial t_i}f\right](t).$$

From now on we make the assumption:

(*H*) $\alpha_i(t,\tau) = \alpha_i(t_i - \tau)$ such that $\alpha_i \in C_1([0, b_i - a_i]; (0,1))$, $i = 1, \ldots, n$.

Corollary 1. *[Cf. [12]] If functions $f : \mathbb{R}^n \to \mathbb{R}$ and $\eta : \mathbb{R}^n \to \mathbb{R}$ are such that $f, \eta \in C^1(\bar{\Omega}_n; \mathbb{R})$ and ${}_{t_i}I^{1-\alpha_i(\cdot)}_{b_i}[f] \in C^1(\bar{\Omega}_n; \mathbb{R})$, then*

$$\int\limits_{\Omega_n} f(t)\cdot {}^{C}_{a_i}D^{\alpha_i(\cdot)}_{t_i}[\eta](t)\, dt = \int\limits_{\partial\Omega_n} \eta(t)\cdot {}_{t_i}I^{1-\alpha_i(\cdot)}_{b_i}[f](t)\cdot \nu^i\, d(\partial\Omega_n) + \int\limits_{\Omega_n} \eta(t)\cdot {}_{t_i}D^{\alpha_i(\cdot)}_{b_i}[f](t)\, dt,$$

where ν^i is the outward pointing unit normal to $\partial\Omega_n$.

3 Main Result

For $n, m \in \mathbb{N}$ let us assume that $y : \mathbb{R}^n \to \mathbb{R}^m$, and $\zeta : \partial\Omega_n \to \mathbb{R}^m$ is a given function. Consider the following functional:

$$\begin{aligned} \mathcal{I} : \mathcal{A}(\zeta) &\longrightarrow \mathbb{R} \\ y &\longmapsto \int\limits_{\Omega_n} F(y(t), \nabla_D[y](t), t)\, dt \end{aligned} \tag{1}$$

where $dt = dt_1 \ldots dt_n$,

$$\mathcal{A}(\zeta) := \left\{ y \in C^1(\bar{\Omega}_n; \mathbb{R}^m) : y|_{\partial\Omega_n} = \zeta, {}^{C}_{a_i}D^{\alpha_i(\cdot)}_{t_i}[y_j] \in C(\bar{\Omega}_n; \mathbb{R}), j = 1, \ldots, m \right\},$$

and $\nabla_D[y] = (\nabla_D[y_1], \ldots, \nabla_D[y_m])$, $\nabla_D = \sum\limits_{i=1}^{n} e_i \cdot {}^{C}_{a_i}D^{\alpha_i(\cdot)}_{t_i}$, $\{e_i : i = 1, \ldots, n\}$ is the standard basis in $\mathbb{R}^n$. We assume that F is a Lagrangian of class C^1:

$$\begin{aligned} F : \mathbb{R}^m \times \mathbb{R}^{mn} \times \bar{\Omega}_n &\longrightarrow \mathbb{R} \\ (x_1, x_2, t) &\longmapsto F(x_1, x_2, t), \end{aligned}$$

and ${}_{t_i}I^{1-\alpha_i(\cdot)}_{b_i}\left[\frac{\partial F(y(\tau), \nabla_D[y](\tau), \tau)}{\partial {}^{C}_{a_i}D^{\alpha_i(\cdot)}_{t_i}[y_j]}\right](t)$ is continuously differentiable on $\bar{\Omega}_n$, $i = 1, \ldots, n$, $j = 1, \ldots, m$.

Following theorem states that if a function minimizes (or maximizes) functional (1), then it necessarily must satisfy (2).

Theorem 1 (Cf. [14]). *If $\tilde{y} \in \mathcal{A}(\zeta)$ minimizes (or maximizes) the functional* (1) *then $\tilde{y}$ satisfies the following equations*

$$\frac{\partial F}{\partial y_j}(y(t), \nabla_D[y](t), t) + \sum_{i=1}^{n} {}_{t_i}D_{b_i}^{\alpha_i(\cdot)} \left[\frac{\partial F(y(\tau), \nabla_D[y](\tau), \tau)}{\partial {}_{a_i}^{C}D_{t_i}^{\alpha_i(\cdot)}[y_j]} \right](t) = 0, \tag{2}$$

for $t \in \Omega_n$, $j = 1, \ldots, m$.

Define

$$E_j^f(F) := \frac{\partial F}{\partial y_j} + \sum_{i=1}^{n} {}_{t_i}D_{b_i}^{\alpha_i(\cdot)} \left[\frac{\partial F}{\partial {}_{a_i}^{C}D_{t_i}^{\alpha_i(\cdot)}[y_j]} \right], \quad j = 1, \ldots, m.$$

We shall call $E_j^f(F)$ the fractional Euler–Lagrange expressions.
Let us consider infinitesimal transformations that depend upon arbitrary functions of independent variables and their partial variable order fractional derivatives. Namely, let

$$\begin{cases} \bar{t} = t, \\ \bar{y}_j(t) = y_j(t) + T^{j1}[p_1](t) + \cdots + T^{jr}[p_r](t), \quad j = 1, \ldots, n, \end{cases} \tag{3}$$

where

$$T^{js} := c^{js}(t) + \sum_{i=1}^{n} c_i^{js}(t) {}_{a_i}^{C}D_{t_i}^{\beta_{jis}(\cdot)}, \quad 0 < \beta_{jsi}(\cdot) \leq 1,$$

are formal linear fractional differential operators and p_s, $s = 1, \ldots, r$ are the r arbitrary, independent C^1 functions defined on $\bar{\Omega}_n$. We assume that functions c_i^{js} and ${}_{a_i}^{C}D_{t_i}^{\beta_{jis}(\cdot)}[p_s]$ are of the C^1 class on $\bar{\Omega}_n$. To formulate the second Noether theorem we make use of the formal adjoint operator $\tilde{T}^{js}$ of T^{js}:

$$\int_{\Omega_n} f(t) T^{js}[p_s](t)\, dt = \int_{\Omega_n} p_s(t) \left[c^{js}(t) f(t) + \sum_{i=1}^{n} {}_{t_i}D_{b_i}^{\alpha_i(\cdot)} [c_i^{js} \cdot f](t) \right] dt + [\cdot]$$
$$= \int_{\Omega_n} p_s(t) \tilde{T}^{js}[f](t)\, dt + [\cdot] \tag{4}$$

for $f \in C^1(\bar{\Omega}_n; \mathbb{R})$, ${}_{t_i}I_{b_i}^{1-\alpha_i(\cdot)}[c_i^{js} \cdot f] \in C^1(\bar{\Omega}_n; \mathbb{R})$, and where $[\cdot]$ represents boundary terms; this can be obtained by the integration by parts (see Corollary 1).

Now, let us introduce the notion of invariance.

Definition 7. *Functional* (1) *is invariant under transformations* (3) *if and only if for all $y \in C^1(\bar{\Omega}_n, \mathbb{R}^m)$, such that ${}_{a_i}^{C}D_{t_i}^{\alpha_i(\cdot)}[y_j] \in C(\bar{\Omega}_n; \mathbb{R})$, we have*

$$\int_{\Omega_n} F(\bar{y}(t), \nabla_D[\bar{y}](t), t) dt = \int_{\Omega_n} F(y(t), \nabla_D[y](t), t) dt.$$

Theorem 2 (The second Noether theorem). *If functional* (1) *is invariant under transformations* (3), *then there exist r identities of the form*

$$\sum_{j=1}^{n} \tilde{T}^{js}\left[E_j^f(F)\right] = 0, \quad s = 1, \ldots, r,$$

where $\tilde{T}^{js}$ is the formal adjoint of T^{js}.

Proof. Using the definition of invariance and noting that the family of transformations (3) depend upon arbitrary functions $p_1, \ldots, p_r$, we conclude that, for any real number ε,

$$\begin{aligned}\int_{\Omega_n} & F(y_1(t), \ldots, y_m(t), \nabla_D[y_1](t), \ldots \nabla_D[y_m](t), t)dt \\ &= \int_{\Omega_n} F(y_1(t) + \varepsilon T^{1s}[p_s](t), \ldots, y_m(t) + \varepsilon T^{ms}[p_s](t), \\ &\nabla_D[y_1](t) + \nabla_D[\varepsilon T^{1s}[p_s]](t), \ldots, \nabla_D[y_m](t) + \nabla_D[\varepsilon T^{ms}[p_s]](t), t)dt,\end{aligned}$$

where $T^{js}[p_s] = \sum_{s=1}^{r} T^{js}[p_s]$, $j = 1, \ldots, m$. Differentiating with respect to ε and taking $\varepsilon = 0$, we get

$$\begin{aligned}0 = \sum_{j=1}^{m} \int_{\Omega_n} &\left[\frac{\partial F}{\partial y_j}(y(t), \nabla_D[y](t), t) T^{js}[p_s](t)\right. \\ &\left. + \sum_{i=1}^{n} \frac{\partial F}{\partial\, {}_{a_i}^{C}D_{t_i}^{\alpha_i(\cdot)}[y_j]}(y(t), \nabla_D[y](t), t)\, {}_{a_i}^{C}D_{t_i}^{\alpha_i(\cdot)}\left[T^{js}[p_s]\right](t)\right] dt. \quad (5)\end{aligned}$$

Applying Corollary 1 we obtain

$$\begin{aligned}\sum_{i=1}^{n} \int_{\Omega_n} & \frac{\partial F}{\partial\, {}_{a_i}^{C}D_{t_i}^{\alpha_i(\cdot)}[y_j]}(y(t), \nabla_D[y](t), t)\, {}_{a_i}^{C}D_{t_i}^{\alpha_i(\cdot)}\left[T^{js}[p_s]\right](t)dt \\ &= \sum_{i=1}^{n} \int_{\Omega_n} {}_{t_i}D_{b_i}^{\alpha_i(\cdot)}\left[\frac{\partial F(y(\tau), \nabla_D[y](\tau), \tau)}{\partial\, {}_{a_i}^{C}D_{t_i}^{\alpha_i(\cdot)}[y_j]}\right](t) T^{js}[p_s](t)dt \\ &+ \sum_{i=1}^{n} \int_{\partial\Omega_n} T^{js}[p_s](t) \cdot {}_{t_i}I_{b_i}^{1-\alpha_i(\cdot)}\left[\frac{\partial F(y(\tau), \nabla_D[y](\tau), \tau)}{\partial\, {}_{a_i}^{C}D_{t_i}^{\alpha_i(\cdot)}[y_j]}\right](t) \cdot \nu^i \, d(\partial\Omega_n). \quad (6)\end{aligned}$$

Since p_s are arbitrary, we may choose p_s such that:

$$p_s(x)|_{\partial\Omega_n} = 0, \quad {}_{t_i}^{C}D_{b_i}^{\beta_{jis}(\cdot)}[p_s](t)|_{\partial\Omega_n} = 0.$$

Therefore, the boundary term in (6) vanishes and combining (6) with (5) we get

$$\begin{aligned}0 = \sum_{j=1}^{m} \int_{\Omega_n} &\left\{\frac{\partial F}{\partial y_j}(y(t), \nabla_D[y](t), t)\right. \\ &\left. + \sum_{i=1}^{n} {}_{t_i}D_{b_i}^{\alpha_i(\cdot)}\left[\frac{\partial F(y(\tau), \nabla_D[y](\tau), \tau)}{\partial\, {}_{a_i}^{C}D_{t_i}^{\alpha_i(\cdot)}[y_j]}\right](t)\right\} T^{js}[p_s](t)dt. \quad (7)\end{aligned}$$

Using the definition of the formal adjoint operator (see (4)) we may write (7) as

$$0 = \sum_{j=1}^{m} \int_{\Omega_n} p_s(t)\tilde{T}^{js} \left[E_j^f(F) \right] (t) \, dt + [\cdot]$$

Again appealing to the arbitrariness of p_s we can force the boundary term to vanish. Finally, by the fundamental lemma of the calculus of variations and by arbitrariness of p_s, we get $\sum_{j=1}^{m} \tilde{T}^{js} \left[E_j^f(F) \right] = 0$ for all $s = 1, \ldots, r$.

Remark 1. Observe that in the special case, when α_i and β_{jsi} are constant, Theorem 2 gives Theorem 3.6 that was proved in [10].

4 Example

In electrodynamics, if the Lagrangian represents a charged particle interacting with a electromagnetic field, the Lagrangian density for the electromagnetic field (see [5, 7]) has the following form:

$$\mathcal{L} = \frac{1}{8\pi}(\mathbf{E}^2 - \mathbf{H}^2), \tag{8}$$

where $\mathbf{E}$ and $\mathbf{H}$ are the electric field vector and the magnetic field vector, respectively. Let $t = (t_0, t_1, t_2, t_3, t_4)$ and $\mathbf{A}(t) = (A_1(t), A_2(t), A_3(t))$, $A_0(t)$ be a vector potential and a scalar potential, respectively. They are defined by setting

$$\mathbf{E} = \nabla A_0 - \frac{\partial \mathbf{A}}{\partial t_0}, \quad \mathbf{H} = curl \mathbf{A}. \tag{9}$$

Replacing $\mathbf{E}$ and $\mathbf{H}$ in (8) by their expressions (9) we obtain the Lagrangian density in the form

$$\mathcal{L} = \frac{1}{8\pi}\left[(\nabla A_0 - \frac{\partial \mathbf{A}}{\partial t_0})^2 - (curl \mathbf{A})^2 \right]. \tag{10}$$

Note that, the four-potential $(A_0, \mathbf{A})$ is not uniquely determined by the vectors $\mathbf{E}$ and $\mathbf{H}$. Namely, $\mathbf{E}$ and $\mathbf{H}$ do not change if we make a gauge transformation:

$$\tilde{A}_j(t) = A_j(x) + \frac{\partial f}{\partial t_j}(x), \quad j = 0, \ldots, 3, \tag{11}$$

where $f : \Omega \to \mathbb{R}$, $\Omega \subset \mathbb{R}^4$, is an arbitrary function of class C^2 in all of its argument. Therefore, the Lagrangian density (10), and hence the action functional, is invariant under transformation (11). By the second Noether theorem (see [6, 7]), we conclude that

$$\sum_{j=0}^{3} \frac{\partial}{\partial t_j} (E_j(\mathcal{L})) = 0,$$

where

$$E_j(\mathcal{L}) = \frac{\partial \mathcal{L}}{\partial A_j} - \sum_{i=0}^{3} \frac{\partial \mathcal{L}}{\partial \left(\frac{\partial A_j}{\partial t_i}\right)}, \quad j = 0, 1, 2, 3,$$

are Lagrange expressions corresponding to (10). Therefore, equations $E_j(\mathcal{L}) = 0$ do not uniquely determine the potential $(A_0, \mathbf{A})$ and to avoid this lack of uniqueness, e.g., the Lorentz condition $div\mathbf{A} - \frac{\partial A_0}{\partial t_0} = 0$ can be imposed on $(A_0, \mathbf{A})$.

We will show that result known for the Lagrangian density for the electromagnetic field (8) can be generalized to the fractional Lagrangian density by changing classical partial derivatives by fractional partial derivatives of variable order. In order to do this we use the functional:

$$\begin{aligned} \mathcal{I} : \mathcal{A} &\longrightarrow \mathbb{R} \\ y &\longmapsto \int_{\Omega_4} F(y(t), \nabla_D[y](t), t)\, dt \end{aligned} \tag{12}$$

where $dt = dt_1 \ldots dt_4$,

$$\mathcal{A} = \left\{ y \in C^1(\bar{\Omega}_4; \mathbb{R}^4) : {}^C_{a_i}D^{\alpha_i(\cdot)}_{t_i}[y_j] \in C(\bar{\Omega}_4; \mathbb{R}), j = 1, \ldots, 4 \right\},$$

and

$$\begin{aligned} F(y(t), \nabla_D[y](t), t) &= \left({}^C_{a_2}D^{\alpha_2(\cdot)}_{t_2}[y_1](t) - {}^C_{a_1}D^{\alpha_1(\cdot)}_{t_1}[y_2](t)\right)^2 \\ &+ \left({}^C_{a_3}D^{\alpha_3(\cdot)}_{t_3}[y_1](t) - {}^C_{a_1}D^{\alpha_1(\cdot)}_{t_1}[y_3](t)\right)^2 + \left({}^C_{a_4}D^{\alpha_4(\cdot)}_{t_4}[y_1](t) - {}^C_{a_1}D^{\alpha_1(\cdot)}_{t_1}[y_4](t)\right)^2 \\ &+ \left({}^C_{a_3}D^{\alpha_3(\cdot)}_{t_3}[y_4](t) - {}^C_{a_4}D^{\alpha_4(\cdot)}_{t_4}[y_3](t)\right)^2 + \left({}^C_{a_4}D^{\alpha_4(\cdot)}_{t_4}[y_2](t) - {}^C_{a_2}D^{\alpha_2(\cdot)}_{t_2}[y_4](t)\right)^2 \\ &+ \left({}^C_{a_2}D^{\alpha_2(\cdot)}_{t_2}[y_3](t) - {}^C_{a_3}D^{\alpha_3(\cdot)}_{t_3}[y_2](t)\right)^2. \end{aligned} \tag{13}$$

For Lagrangian density F we have the the following fractional Euler–Lagrange expressions:

$$E^f_j(F) := \frac{\partial F}{\partial y_j} + \sum_{i=1}^{4} {}_{t_i}D^{\alpha_i(\cdot)}_{b_i} \left[\frac{\partial F}{\partial {}^C_{a_i}D^{\alpha_i(\cdot)}_{t_i}[y_j]} \right], \quad j = 1, \ldots, 4. \tag{14}$$

Observe that functional (12) is invariant under a gauge transformation:

$$\tilde{y}_j(t) = y_j(t) + {}^C_{a_j}D^{\alpha_j(\cdot)}_{t_j} f(t), \quad j = 1, \ldots, 4,$$

where $f : \Omega \to \mathbb{R}$ is an arbitrary function of class C^2 in all of its argument. Therefore, by Theorem (2), we conclude that

$$\sum_{j=1}^{4} {}_{t_j}D^{\alpha_j(\cdot)}_{b_j} \left(E^f_j(F) \right) = 0,$$

where $E^f_j(F)$ are Euler–Lagrange expressions (14). Equations $E_j(F) = 0$ do not uniquely determine the vector function y. Therefore, as it is in the case of the classical electromagnetic field, to avoid this lack of uniqueness, the additional condition should be imposed on $y = (y_1, y_2, y_3, y_4)$.

Acknowledgement. A.B. Malinowska is supported by Bialystok University of Technology grant S/WI/02/2011, T. Odzijewicz is supported by Portuguese funds through the CIDMA - Center for Research and Development in Mathematics and Applications, and the Portuguese Foundation for Science and Technology (FCTFundao para a Cincia e a Tecnologia), within project PEst-OE/MAT/UI4106/2014, and also by the FCT postdoctoral fellowship SFRH/BPD/92691/2013.

References

1. Almeida, A., Samko, S.G.: Fractional and hypersingular operators in variable exponent spaces on metric measure spaces. Mediterr. J. Math. 6, 215–232 (2009)
2. Atanackovic, T.M., Pilipovic, S.: Hamilton's principle with variable order fractional derivatives. Fract. Calc. Appl. Anal. 14, 94–109 (2011)
3. Coimbra, C.F.M.: Mechanics with variable-order differential operators. Ann. Phys. 12, 692–703 (2003)
4. Diaz, G., Coimbra, C.F.M.: Nonlinear dynamics and control of a variable order oscillator with application to the van der Pol equation. Nonlinear Dynam. 56, 145–157 (2009)
5. Gelfand, I.M., Fomin, S.V.: Calculus of variations. Dover Publications, Mineola (2000)
6. Logan, J.D.: On variational problems which admit an infinite continuous group. Yokohama Mathematical Journal 22, 31–42 (1974)
7. Logan, J.D.: Invariant variational principles. Academic Press (Harcourt Brace Jovanovich Publishers), New York (1977)
8. Lorenzo, C.F., Hartley, T.T.: Variable order and distributed order fractional operators. Nonlinear Dynam. 29, 57–98 (2002)
9. Kosmann-Schwarzbach, Y.: The Noether theorems: Invariance and conservation laws in the twentieth century. Springer, New York (2011)
10. Malinowska, A.B.: On fractional variational problems which admit local transformations. J. Vib. Control 19, 1161–1169 (2013)
11. Noether, E.: Invariante Variationsprobleme. Nachr. v. d. Ges. d. Wiss. zu Gttingen, 235–257 (1918)
12. Odzijewicz, T.: Generalized fractional calculus of variations, PhD thesis, Doctoral Programme in Mathematics and Applications (PDMA). University of Aveiro and University of Minho (2013), `http://arxiv.org/abs/1403.4379`
13. Odziehjewicz, T.: Variable order fractional isoperimetric problem of several variables. In: Mitkowski, W., Kacprzyk, J., Baranowski, J. (eds.) Theory & Appl. of Non-integer Order Syst. LNEE, vol. 257, pp. 133–139. Springer, Heidelberg (2013)
14. Odzijewicz, T., Malinowska, A.B., Torres, D.F.M.: Variable order fractional variational calculus for double integrals. In: 51st IEEE Conference on Decision and Control, Maui, Hawaii, USA, December 10-13, pp. 6873–6878 (2012), Article Source/Identifier: PLZ-CDC12.1240.d4462b33

15. Odzijewicz, T., Malinowska, A.B., Torres, D.F.M.: Fractional variational calculus of variable order. In: Almeida, A., Castro, L., Speck, F.-O. (eds.) Advances in Harmonic Analysis and Operator Theory, The Stefan Samko Anniversary Volume. Operator Theory: Advances and Applications, vol. 229, pp. 291–301. Birkhäuser Verlag (2013)
16. Odzijewicz, T., Malinowska, A.B., Torres, D.F.M.: Noether's theorem for fractional variational problems of variable order. Cent. Eur. J. Phys. 11, 691–701 (2013)
17. Pedro, H.T.C., Kobayashi, M.H., Pereira, J.M.C., Coimbra, C.F.M.: Variable order modeling of diffusive-convective effects on the oscillatory flow past a sphere. J. Vib. Control 14, 1569–1672 (2008)
18. Ramirez, L.E.S., Coimbra, C.F.M.: On the selection and meaning of variable order operators for dynamic modeling. Int. J. Differ. Equ. 2010, 16, Art. ID 846107 (2010)
19. Ramirez, L.E.S., Coimbra, C.F.M.: On the variable order dynamics of the nonlinear wake caused by a sedimenting particle. Phys. D 240, 1111–1118 (2011)
20. Samko, S.G., Ross, B.: Integration and differentiation to a variable fractional order. Integral Transform. Spec. Funct. 1, 277–300 (1993)

Fractional Linear Equations with Discrete Operators of Positive Order

Dorota Mozyrska and Małgorzata Wyrwas

Faculty of Computer Science
Bialystok University of Technology
Wiejska 45A, 15-351 Białystok, Poland
{m.wyrwas,d.mozyrska}@pb.edu.pl

Abstract. The Caputo- and Riemann–Liouville–type fractional order difference initial value problems for linear and semilinear equations are discussed. We take under our consideration the possible solution via the classical $\mathcal{Z}$-transform method for any positive order. We stress the formulas that used the concept of discrete Mittag–Leffler fractional function.

Keywords: fractional difference operator, linear fractional order system, $\mathcal{Z}$-transform method.

1 Introduction

Recently, in many papers systems with fractional derivatives and differences are widely discussed and their properties are presented usually for fractional orders from the interval $(0, 1]$. In the paper the possible solutions of linear and semilinear systems with the Caputo– and Riemann–Liouville–type (difference) operators are studied for any positive order $\alpha > 0$. However we use notation that $\alpha \in (q-1, q]$, where $q \in \mathbb{N}_1$. The possible solution via the classical $\mathcal{Z}$-transform method for any positive order are taken under our consideration. We stress the formulas that used the concept of discrete Mittag–Leffler fractional function.

Basic properties of fractional sums and difference operators were developed firstly in [11] and continued by Atici and Eloe in [5,6], Baleanu and Abdeljawad in [2,3]. Another concept of the fractional sum/difference was introduced in [8,10,7]. In the cited literature there are usually presented the methods of solutions via recurrence or transform methods but not so often via the $\mathcal{Z}$-transform. The problem of stability properties for fractional difference systems with higher orders authors studied in [4,19,20]. In the presented paper we only state formulas for solutions to initial value problems without studying the stability property of the considered systems. As we stressed in paper for example [13] for commensurate case and in [18] for multi–order case, the conversion of the Grünwald–Letnikov–type operator to the Riemann–Liouville–type gives the same result for the first mentioned operator.

Fractional differences used in models of control systems and could be understand as an approximation of continuous operators (see [14]) and a possibility of

© Springer International Publishing Switzerland 2015
K.J. Latawiec et al. (eds.), *Advances in Modeling and Control of Non-integer Order Systems*,
Lecture Notes in Electrical Engineering 320, DOI: 10.1007/978-3-319-09900-2_5

involving some memory to difference systems, i.e. systems in which the current state depends on the full history of systems' states. The main advantage of the use of the $\mathcal{Z}$-transform is to introduce the natural language for discrete systems, it means to work with sequences instead of discrete functions defined on various domains.

The structure of the paper is the following. In Section 2 the preliminary material is presented. Section 3 gives the formula for the $\mathcal{Z}$-transform of the Caputo–type operator, then in Section 4 we investigate Riemann-Liouville–type operator with positive order.

2 Preliminaries

In this section, we make a review of notations, definitions, and some preliminary facts which are useful for the paper. The necessary definitions and technical propositions that are used in the sequel therein the paper are recalled.

Let $h > 0, a \in \mathbb{R}$ and $(h\mathbb{N})_a := \{a, a+h, a+2h, ...\}$.

For a function $x : (h\mathbb{N})_a \to \mathbb{R}$ the forward h-difference operator is defined as (see [10]) $(\Delta_h x)(t) = \frac{x(t+h)-x(t)}{h}$, where $t \in (h\mathbb{N})_a$ and $(\Delta_h^0 x)(t) := x(t)$. Let $q \in \mathbb{N}_0$ and $\Delta_h^q := \Delta_h \circ \cdots \circ \Delta_h$ is q-fold application of operator Δ_h. Then $(\Delta_h^q x)(t) = h^{-q} \sum_{k=0}^{q} (-1)^{q-k} \binom{q}{k} x(t+kh)$.

Let us introduce the family of binomial functions on $\mathbb{Z}$ parameterized by $\mu > 0$ and given by the values: $\tilde{\varphi}_\mu(n) = \binom{n+\mu-1}{n}$ for $n \in \mathbb{N}_0$ and $\tilde{\varphi}_\mu(n) = 0$ for $n < 0$.

Definition 1. *For a function $x : (h\mathbb{N})_a \to \mathbb{R}$ the* fractional h-sum of order $\alpha > 0$ *is given by* $\left({}_a\Delta_h^{-\alpha} x\right)(t) := h^\alpha \left(\tilde{\varphi}_\alpha * \overline{x}\right)(n)$, *where* $t = a+(\alpha+n)h$, $\overline{x}(s) := x(a+sh)$, $n \in \mathbb{N}_0$ *and "$*$" denotes a convolution operator, i.e.* $\left(\tilde{\varphi}_\alpha * \overline{x}\right)(n) := \sum_{s=0}^{n} \binom{n-s+\alpha-1}{n-s} \overline{x}(s) q$. *Additionally, we define* $\left({}_a\Delta_h^0 x\right)(t) := x(t)$.

For $a = 0$ we will write shortly $\Delta_h^{-\alpha}$ instead of ${}_0\Delta_h^{-\alpha}$. Note that ${}_a\Delta_h^{-\alpha} x : (h\mathbb{N})_{a+\alpha h} \to \mathbb{R}$. Let us recall that the $\mathcal{Z}$-transform of a sequence $\{y(n)\}_{n\in\mathbb{N}_0}$ is a complex function given by $Y(z) := \mathcal{Z}[y](z) = \sum_{k=0}^{\infty} \frac{y(k)}{z^k}$, where $z \in \mathbb{C}$ is a complex number for which the series $\sum_{k=0}^{\infty} y(k) z^{-k}$ converges absolutely. Note that since $\binom{k+\alpha-1}{k} = (-1)^k \binom{-\alpha}{k}$, then for $|z| > 1$ we have

$$\mathcal{Z}\left[\tilde{\varphi}_\alpha\right](z) = \sum_{k=0}^{\infty} \frac{1}{z^k} \binom{k+\alpha-1}{k} = \left(\frac{z}{z-1}\right)^\alpha . \tag{1}$$

Proposition 1 ([17]). *For* $t = a + \alpha h + nh \in (h\mathbb{Z})_{a+\alpha h}$ *let us define* $y(n) := \left({}_a\Delta_h^{-\alpha} x\right)(t)$ *and* $\overline{x}(n) = x(a+nh)$. *Then*

$$\mathcal{Z}[y](z) = \left(\frac{hz}{z-1}\right)^\alpha X(z), \tag{2}$$

where $X(z) := \mathcal{Z}[\overline{x}](z)$.

In [10] the authors prove the following lemma that gives transition between fractional summation operators for any $h > 0$ and $h = 1$.

Lemma 1. *Let $x : (h\mathbb{N})_a \to \mathbb{R}$ and $\alpha > 0$. Then, $\left({}_a\Delta_h^{-\alpha}x\right)(t) = h^\alpha \left({}_{\frac{a}{h}}\Delta_1^{-\alpha}\tilde{x}\right)\left(\frac{t}{h}\right)$, where $t \in (h\mathbb{N})_{a+\alpha h}$ and $\tilde{x}(s) = x(sh)$.*

We define the discrete version of Mittag-Leffler function and prove that with some values of parameters it is an eigenfunction of difference equation with Caputo– or Riemann–Liouville–type difference operator with order $\alpha \in (q-1, q]$, where $q \in \mathbb{N}_1$. In [17] we use such a function but for orders from $(0, 1]$ or also cite from [3]. Here let us define the *discrete Mittag-Leffler two-parameter function* as follows:

$$E_{(\alpha,\beta)}(\lambda, n) := \sum_{k=0}^{\infty} \lambda^k \widetilde{\varphi}_{k\alpha+\beta}(n - qk) = \sum_{k=0}^{n} \lambda^k \widetilde{\varphi}_{k\alpha+\beta}(n - qk)\,, \tag{3}$$

where the second equation only claims that for $n < qk$ we have values of $\widetilde{\varphi}_{k\alpha+\beta}(n - qk) = 0$. This is not in contradiction with the definition of Mittag–Leffler discrete type functions stated in [1] or used in [17]. Later on we will show that for $\beta = 1$ and $\beta = \alpha$ the formula (3) gives an eigenfunction of difference equation with Caputo– or Riemann–Liouville–type difference operator, respectively. In fact in the paper we will use the following discrete Mittag–Leffler–type functions

$$E_{(\alpha,\alpha)}(\lambda, n) = \sum_{k=0}^{\infty} \lambda^k \widetilde{\varphi}_{k\alpha+\alpha}(n - qk) = \sum_{k=0}^{\infty} \lambda^k \binom{n - qk + (k+1)\alpha - 1}{n - qk}, \tag{4}$$

$$E_{(\alpha)}(\lambda, n) := E_{(\alpha,1)}(\lambda, n) = \sum_{k=0}^{\infty} \lambda^k \widetilde{\varphi}_{k\alpha+1}(n - qk) = \sum_{k=0}^{\infty} \lambda^k \binom{n - qk + k\alpha}{n - qk}, \tag{5}$$

$$E_{(\alpha,0)}(\lambda, n) = \sum_{k=0}^{\infty} \lambda^k \widetilde{\varphi}_{k\alpha}(n - qk) = \sum_{k=0}^{\infty} \lambda^k \binom{n - qk + k\alpha - 1}{n - qk}. \tag{6}$$

Based on (1) for family of functions $\widetilde{\varphi}_{k\alpha+\beta}$ we can state the following result for discrete Mittag-Leffler function.

Proposition 2. *Let $\alpha \in (q-1, q]$ and $\nu = \alpha - q$, $q = \nu h$, $q \in \mathbb{N}_1$. Then*

1. *$E_{(\alpha,\beta)}(\lambda, 0) = 1$.*
2. *For z such that $|z| > 1$ we have*

$$\mathcal{Z}\left[E_{(\alpha,\beta)}(\lambda, \cdot)\right](z) = \left(\frac{z}{z-1}\right)^\beta \left(1 - \frac{\lambda}{z^q}\left(\frac{z}{z-1}\right)^\alpha\right)^{-1},$$

where $|z| > 1$ and $|z-1|^\alpha |z|^{q-\alpha} > |\lambda|$.

Proof. The item 1. is obvious, we prove part 2. By basic calculations we have

$$\begin{aligned}\mathcal{Z}\left[E_{(\alpha,\beta)}(\lambda,\cdot)\right](z) &= \sum_{k=0}^{\infty}\sum_{n=0}^{\infty}\lambda^k\binom{n-qk+k\alpha+\beta-1}{n-qk}z^{-n}\\ &= \sum_{k=0}^{\infty}\lambda^k z^{-qk}\sum_{s=0}^{\infty}\binom{s+k\alpha+\beta-1}{s}z^{-s}\\ &= \sum_{k=0}^{\infty}\left(\frac{\lambda}{z^q}\right)^k\sum_{s=0}^{\infty}(-1)^s\binom{-k\alpha-\beta}{s}z^{-s}\\ &= \left(\frac{z}{z-1}\right)^{\beta}\sum_{k=0}^{\infty}\left(\frac{\lambda}{z^q}\right)^k\left(\frac{z}{z-1}\right)^{k\alpha}\\ &= \left(\frac{z}{z-1}\right)^{\beta}\left(1-\frac{\lambda}{z^q}\left(\frac{z}{z-1}\right)^{\alpha}\right)^{-1},\end{aligned}$$

where the summation exists for $|z|>1$ and $|z-1|^{\alpha}|z|^{q-\alpha}>|\lambda|$.

3 Caputo-Type Operator with Positive Order

Let us define, like in [15], the family of functions $\varphi_{k,\alpha}:\mathbb{Z}\to\mathbb{R}$ parameterized by $k\in\mathbb{N}_0$ and $\alpha\in(q-1,q]$, $q\in\mathbb{N}_1$, with the following values

$$\varphi^*_{k,\alpha}(n) := \begin{cases}\binom{n-qk+k\alpha}{n-qk}, & \text{for } n\in\mathbb{N}_{qk}\\ 0, & \text{for } n<qk\end{cases}. \tag{7}$$

Proposition 3 ([15]). *Let function $\varphi^*_{k,\alpha}$ be defined by (7). Then*

$$\mathcal{Z}\left[\varphi^*_{k,\alpha}\right](z) = \frac{1}{z^{qk}}\left(\frac{z}{z-1}\right)^{k\alpha+1} \tag{8}$$

for z such that $|z|>1$.

For family of functions $\varphi^*_{k,\alpha}$ we can state the following proposition.

Proposition 4 ([15]). *Let $\alpha\in(q-1,q]$, $q\in\mathbb{N}_1$, and $\nu=\alpha-q$. Then for $n\in\mathbb{N}_0$ one has $\left(\Delta^{-\alpha}\varphi^*_{k,\alpha}\right)(n+\nu)=\varphi^*_{k+1,\alpha}(n)$.*

In this section we recall the definition of Caputo–type operator and give the properties of this operator, in particular the formula for its $\mathcal{Z}$-transform is proved.

The definition of the Caputo–type fractional h-difference operator can be found, for example, in [9] (for $h=1$) or in [12] (for any $h>0$).

Definition 2. *Let $\alpha\in(q-1,q]$, $q\in\mathbb{N}_1$ and $a\in\mathbb{R}$. The* Caputo–type fractional h-difference operator *${}_a\Delta_h^{\alpha}x$ of order α for a function $x:(h\mathbb{N})_a\to\mathbb{R}$ is defined by*

$$\left({}_a\Delta^{\alpha}_{h,*}x\right)(t) = \left({}_a\Delta_h^{-(q-\alpha)}\left(\Delta_h^q x\right)\right)(t), \tag{9}$$

where $t\in(h\mathbb{N})_{a+(q-\alpha)h}$.

Moreover, for $\alpha = q \in \mathbb{N}_1$ we have $\left({}_a\Delta^q_{h,*}x\right)(t) = \left(\Delta^q_h x\right)(t)$.

There exists the transition formula for the Caputo–type operator between the cases for any $h > 0$ and $h = 1$. Let $x : (h\mathbb{N})_a \to \mathbb{R}$ and $\alpha \in (q-1, q]$, $q \in \mathbb{N}_1$. Then, $\left({}_a\Delta^\alpha_{h,*}x\right)(t) = h^{-\alpha}\left({}_{\frac{a}{h}}\Delta^\alpha_{1,*}\tilde{x}\right)\left(\frac{t}{h}\right)$, where $t \in (h\mathbb{N})_{a+(q-\alpha)h}$ and $\tilde{x}(s) = x(sh)$. For the case $h = 1$ we will write: ${}_{\frac{a}{h}}\Delta^\alpha_* :={}_{\frac{a}{h}}\Delta^\alpha_{1,*}$ and $\Delta^q := \Delta^q_1$.

Proposition 5 ([15]). *For $a \in \mathbb{R}$, $\alpha \in (q-1, q]$, $q \in \mathbb{N}_1$ let us define $y(n) := \left({}_a\Delta^\alpha_{h,*}x\right)(t)$, where $t \in (h\mathbb{N})_{a+(q-\alpha)h}$ and $t = a + (q-\alpha)h + nh$. Then*

$$\mathcal{Z}\left[y\right](z) = h^{-\alpha}z^q\left(\frac{z}{z-1}\right)^{-\alpha}\left(X(z) - \frac{z}{z-1}\sum_{k=0}^{q-1}(z-1)^{-k}\left(\Delta^k_h x\right)(a)\right), \quad (10)$$

where $X(z) = \mathcal{Z}[\overline{x}](z)$ and $\overline{x}(n) := x(a+nh)$.

The $\mathcal{Z}$-transform can be used to show some properties of the Caputo–type operator of functions related with the solutions of initial value problems. In [15] we proved that for $k \in \mathbb{N}_1$

$$\left({}_0\Delta^\alpha\varphi^*_{k,\alpha}\right)(n+q-\alpha) = \varphi^*_{k-1,\alpha}(n). \quad (11)$$

Using the notation of the family of functions $\varphi^*_{k,\alpha}$ we can write the formula for Mittag–Leffler function defined by (5) as

$$E_{(\alpha)}(\lambda, n) := E_{(\alpha,1)}(\lambda, n) = \sum_{k=0}^{\infty}\lambda^k\varphi^*_{k,\alpha}(n).$$

Moreover, by equations: (5) and (11) we easily see the following property.

Proposition 6. *Let $\alpha \in (q-1, q]$ and $\nu = q - \alpha$, $q = \nu h$, $q \in \mathbb{N}_1$.*

1. $\left(\Delta^{-\alpha}E_{(\alpha)}(\lambda,\cdot)\right)(n-\nu) = \frac{1}{\lambda}E_{(\alpha)}(\lambda, n-1)$,
2. $\left(\Delta^\alpha_* E_{(\alpha)}(\lambda,\cdot)\right)(n+\nu) = \lambda E_{(\alpha)}(\lambda, n)$.

The next proposition states that the function $E_{(\alpha)}(\lambda h^\alpha, \cdot)$ is an eigenfunction of fractional difference equation with the Caputo–type operator.

Proposition 7. *Let $\alpha \in (q-1, q]$ and $a = (\alpha - q)h$. The initial value problem*

$$\left({}_a\Delta^\alpha_{h,*}x\right)(nh) = \lambda x(nh+a), \quad n \in \mathbb{N}_q \quad (12)$$

$$\left(\Delta^i_h x\right)(a) = b_i, \quad i = 0, \ldots, q-1 \quad (13)$$

has the unique solution given by the formula

$$x(a+nh) = \overline{x}(n) = \sum_{i=0}^{q-1}\left(\Delta^{q-i-1}E_{(\alpha)}(\lambda h^\alpha,\cdot)\right)(n-(q-1))b_i. \quad (14)$$

Proof. Let $y(n) = \left({}_a\Delta^{\alpha}_{h,*}x\right)(nh)$ and $\overline{x}(n) = x(nh+a)$. Then using the $\mathcal{Z}$-transform of both sides of (12) we get the algebraic equation

$$\left(z^q\left(\frac{z}{z-1}\right)^{-\alpha} - h^{\alpha}\lambda\right)X(z) = z^q\left(\frac{z}{z-1}\right)^{-\alpha}\frac{z}{z-1}\sum_{i=0}^{q-1}(z-1)^{-i}b_i\,.$$

Then we calculate that

$$X(z) = \frac{z}{z-1}\left(1-\frac{1}{z^q}\left(\frac{z}{z-1}\right)^{\alpha}\lambda h^{\alpha}\right)^{-1}\sum_{i=0}^{q-1}(z-1)^{-i}b_i\,.$$

Taking

$$W(z) = \frac{z}{z-1}\left(1-\frac{1}{z^q}\left(\frac{z}{z-1}\right)^{\alpha}\lambda h^{\alpha}\right)^{-1}$$

and

$$\begin{aligned}G(z) &= \frac{1}{z^{q-1}}\sum_{i=0}^{q-1}(z-1)^{q-i-1}b_i\\ &= \frac{1}{z^{q-1}}\sum_{i=0}^{q-1}\sum_{s=0}^{q-i-1}\binom{q-i-1}{s}(-1)^{q-i-s-1}z^s b_i\\ &= \sum_{i=0}^{q-1}\sum_{s=0}^{q-i-1}\binom{q-i-1}{s}(-1)^{q-i-s-1}\frac{1}{z^{q-1-s}}b_i\end{aligned}$$

we get

$$X(z) = W(z)G(z)\,.$$

For $W(z)$ from Proposition 2 we see that

$$\mathcal{Z}^{-1}[W](n) = E_{(\alpha)}(\lambda h^{\alpha}, n)\,.$$

Then

$$\begin{aligned}x(a+nh) =&\overline{x}(n) = \mathcal{Z}^{-1}[X](n) = \mathcal{Z}^{-1}[W(z)G(z)](n)\\ =&\sum_{i=0}^{q-1}\sum_{s=0}^{q-i-1}\binom{q-i-1}{s}(-1)^{q-i-s-1}E_{(\alpha)}(\lambda h^{\alpha}, n-(q-s-1))b_i\\ =&\sum_{i=0}^{q-1}\left(\Delta^{q-i-1}E_{(\alpha)}(\lambda h^{\alpha},\cdot)\right)(n-(q-1))b_i\,.\end{aligned}$$

The immediate consequence of Proposition 7 is the formula for solution of semilinear equation.

Proposition 8. *Let* $\alpha \in (q-1, q]$ *and* $a = (\alpha - q)h$. *The initial value problem*

$$\left({}_a\Delta^{\alpha}_{h,*}x\right)(nh) = \lambda x(nh+a) + f(nh)\,, \tag{15}$$

$$\left(\Delta^{i}_{h}(x)\right)(a) = b_i\,,\quad i = 0,\ldots,q-1 \tag{16}$$

where $n \in \mathbb{N}_q$ *has the unique solution given by the formula*

$$x(a+nh) = \sum_{i=0}^{q-1} \left(\Delta^{q-i-1} E_{(\alpha)}(\lambda h^{\alpha}, \cdot)\right)(n-(q-1))b_i + E_{(\alpha)}(\lambda h^{\alpha}, n-q) * \overline{f}(n)$$

and $\overline{f}(n) = f(nh)$.

Example 1. Let us consider the following initial value problem

$$\left({}_a\Delta^{\alpha}_{h,*}x\right)(nh) = 0.1 \cdot x(nh+a)\,, \tag{17a}$$

$$x(a) = 1 \tag{17b}$$

$$\left(\Delta_h x\right)(a) = 0.1 \tag{17c}$$

Then Figure 1 presents the graphs of solutions to initial value problems (17) with the Caputo–type operator with orders: 1.1, 1.5, 1.9.

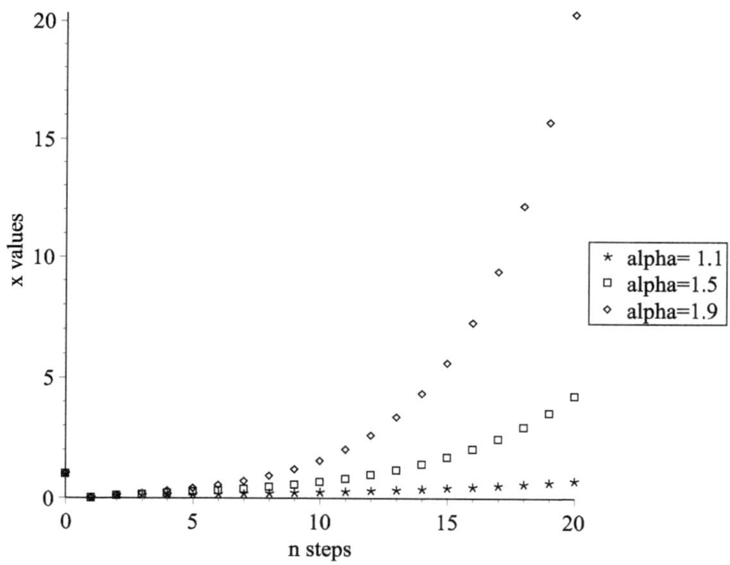

Fig. 1. Trajectories of equation $\left({}_a\Delta^{\alpha}_{h,*}x\right)(nh) = \lambda \cdot x(nh+a)$ with the Caputo–type operator for $\lambda = 0.1$ and the initial conditions $x(a) = 1$, $\left(\Delta_h x\right)(a) = \lambda$, see Example 1

4 Riemann–Liouville–Type Operator with Positive Order

Let us define, like in [16], the family of functions $\varphi_{k,\alpha} : \mathbb{Z} \to \mathbb{R}$ parameterized by $k \in \mathbb{N}_0$ and $\alpha \in (q-1, q]$, $q \in \mathbb{N}_1$, with the following values

$$\varphi_{k,\alpha}(n) := \begin{cases} \binom{n-qk+k\alpha+\alpha-1}{n-qk}\,, & \text{for } n \in \mathbb{N}_{qk} \\ 0, & \text{for } n < qk \end{cases}\,. \tag{18}$$

Observe that $\varphi_{k,\alpha}(n) = \widetilde{\varphi}_{k\alpha+\alpha}(n-qk)$.

Proposition 9 ([16]). *Let function $\varphi_{k,\alpha}$ be defined by* (18). *Then*

$$\mathcal{Z}\left[\varphi_{k,\alpha}\right](z) = \frac{1}{z^{qk}}\left(\frac{z}{z-1}\right)^{(k+1)\alpha} \tag{19}$$

for z such that $|z| > 1$.

For family of functions $\varphi_{k,\alpha}$, $k \in \mathbb{N}_0$, $\alpha > 0$, we can state the following proposition.

Proposition 10 ([16]). *Let $\alpha \in (q-1, q]$, $q \in \mathbb{N}_1$, and $\nu = \alpha - q$. Then for $n \in \mathbb{N}_0$ one has*

$$\left(\Delta^{-\alpha}\varphi_{k,\alpha}\right)(n+\nu) = \varphi_{k+1,\alpha}(n)\,. \tag{20}$$

The definition of the fractional h-difference Riemann-Liouville–type operator can be found, for example, in [5] (for $h = 1$) or in [8] (for any $h > 0$).

Definition 3. *Let $\alpha \in (q-1, q]$, $q \in \mathbb{N}_1$ and $a \in \mathbb{R}$. The* Riemann-Liouville–type fractional h-difference operator *${}_a\Delta_h^\alpha x$ of order α for a function $x : (h\mathbb{N})_a \to \mathbb{R}$ is defined by*

$$\left({}_a\Delta_h^\alpha x\right)(t) = \left(\Delta_h^q\left({}_a\Delta_h^{-(q-\alpha)}x\right)\right)(t)\,, \tag{21}$$

where $t \in (h\mathbb{N})_{a+(q-\alpha)h}$.

Moreover, for $\alpha = q \in \mathbb{N}_1$ we have: $\left({}_a\Delta_h^q x\right)(t) = \left(\Delta_h^q x\right)(t)$.

There exists the transition formula for the Riemann-Liouville–type operator between the cases for any $h > 0$ and $h = 1$. Let $x : (h\mathbb{N})_a \to \mathbb{R}$ and $\alpha \in (q-1, q]$, $q \in \mathbb{N}_1$. Then, $\left({}_a\Delta_h^\alpha x\right)(t) = h^{-\alpha}\left({}_{\frac{a}{h}}\Delta_1^\alpha \tilde{x}\right)\left(\frac{t}{h}\right)$, where $t \in (h\mathbb{N})_{a+(q-\alpha)h}$ and $\tilde{x}(s) = x(sh)$. For the case $h = 1$ we will write: ${}_{\frac{a}{h}}\Delta^\alpha :={}_{\frac{a}{h}}\Delta_1^\alpha$ and $\Delta^q := \Delta_1^q$.

Proposition 11 ([16]). *For $a \in \mathbb{R}$, $\alpha \in (q-1, q]$, $q \in \mathbb{N}_1$ let us define $y(n) := \left({}_a\Delta_h^\alpha x\right)(t)$, where $t \in (h\mathbb{N})_{a+(q-\alpha)h}$ and $t = a + (q-\alpha)h + nh$, $t_0 = a + (q-\alpha)h$. Then*

$$\begin{aligned}\mathcal{Z}[y](z) =& h^{-\alpha}\left(z^q\left(\frac{z}{z-1}\right)^{-\alpha}X(z)\right.\\ &\left.-z\sum_{k=0}^{q-1}(z-1)^{q-k-1}\left(\Delta_h^k\left({}_a\Delta^{-(q-\alpha)}x\right)\right)(t_0)\right),\end{aligned} \tag{22}$$

where $X(z) = \mathcal{Z}[\overline{x}](z)$ and $\overline{x}(n) := x(a+nh)$.

Using the $\mathcal{Z}$-transform of the Riemann-Liouville–type operator we get

$$\left({}_0\Delta^\alpha\varphi_{k,\alpha}\right)(n+q-\alpha) = \varphi_{k-1,\alpha}(n)\,, \tag{23}$$

for $k \in \mathbb{N}_q$.

Using the notation of the family of functions $\varphi_{k,\alpha}$ we can write the formula for Mittag–Leffler function defined by (4) as $E_{(\alpha,\alpha)}(\lambda, n) = \sum_{k=0}^{\infty}\lambda^k\varphi_{k,\alpha}(n)\,.$

Moreover, by equations: (4) and (23) we easily see the property of $E_{(\alpha,\alpha)}(\lambda,\cdot)$ analogous to that one for $E_{(\alpha)}(\lambda,\cdot)$ stated in Proposition 6

The next proposition states that the function $E_{(\alpha,\alpha)}(\lambda h^\alpha,\cdot)$ is an eigenfunction of fractional difference equation with the Riemann–Liouville–type operator.

Proposition 12. *Let $\alpha \in (q-1,q]$ and $a=(\alpha-q)h$. The initial value problem*

$$\left({}_a\Delta_h^\alpha x\right)(nh) = \lambda x(nh+a)\,, \quad n\in\mathbb{N}_q \tag{24}$$

$$\left(\Delta_h^i\left({}_a\Delta_h^{-(q-\alpha)}x\right)\right)(a) = b_i\,, \quad i=0,\dots,q-1 \tag{25}$$

has the unique solution given by the formula

$$x(a+nh) = \overline{x}(n) = \sum_{i=0}^{q-1}\left(\Delta^{q-i-1}E_{(\alpha,\alpha)}(\lambda h^\alpha,\cdot)\right)(n-(q-1))b_i\,. \tag{26}$$

Proof. Let $y(n) = \left({}_a\Delta_h^\alpha x\right)(nh)$ and $\overline{x}(n) = x(nh+a)$. Then using the $\mathcal{Z}$-transform of both sides of (24) we get the algebraic equation

$$\left(z^q\left(\frac{z}{z-1}\right)^{-\alpha} - h^\alpha\lambda\right)X(z) = z\sum_{i=0}^{q-1}(z-1)^{q-i-1}b_i\,.$$

Then we calculate that

$$X(z) = W(z)G(z)\,,$$

where

$$W(z) = \left(\frac{z}{z-1}\right)^\alpha\left(1-\frac{1}{z^q}\left(\frac{z}{z-1}\right)^\alpha\lambda h^\alpha\right)^{-1}$$

and

$$\begin{aligned}G(z) &= \frac{1}{z^{q-1}}\sum_{i=0}^{q-1}(z-1)^{q-i-1}b_i = \frac{1}{z^{q-1}}\sum_{i=0}^{q-1}\sum_{s=0}^{q-i-1}\binom{q-i-1}{s}(-1)^{q-i-s-1}z^s b_i\\ &=\sum_{i=0}^{q-1}\sum_{s=0}^{q-i-1}\binom{q-i-1}{s}(-1)^{q-i-s-1}\frac{1}{z^{q-1-s}}b_i\,.\end{aligned}$$

For $W(z)$ from Proposition 2 we see that

$$\mathcal{Z}^{-1}[W](n) = E_{(\alpha,\alpha)}(\lambda h^\alpha,n)\,.$$

Then

$$\begin{aligned}x(a+nh) &=\overline{x}(n) = \mathcal{Z}^{-1}[X](n) = \mathcal{Z}^{-1}[W(z)G(z)](n)\\ &=\sum_{i=0}^{q-1}\sum_{s=0}^{q-i-1}\binom{q-i-1}{s}(-1)^{q-i-s-1}E_{(\alpha,\alpha)}(\lambda h^\alpha,n-(q-s-1))b_i\\ &=\sum_{i=0}^{q-1}\left(\Delta^{q-i-1}E_{(\alpha,\alpha)}(\lambda h^\alpha,\cdot)\right)(n-(q-1))b_i\,.\end{aligned}$$

The immediate consequence of Proposition 12 is the formula for solution of semilinear equation.

Proposition 13. *Let $\alpha \in (q-1, q]$ and $a = (\alpha - q)h$. The initial value problem*

$$\left({}_a\Delta_h^\alpha x\right)(nh) = \lambda x(nh + a) + f(nh)\,, \tag{27}$$

$$\left(\Delta_h^i \left({}_a\Delta_h^{-(q-\alpha)} x\right)\right)(a) = b_i\,, \quad i = 0, \ldots, q-1 \tag{28}$$

where $n \in \mathbb{N}_q$ has the unique solution given by the formula

$$x(a+nh) = \sum_{i=0}^{q-1} \left(\Delta^{q-i-1} E_{(\alpha,\alpha)}(\lambda h^\alpha, \cdot)\right)(n-(q-1))b_i + E_{(\alpha,\alpha)}(\lambda h^\alpha, n-q) * \overline{f}(n)$$

and $\overline{f}(n) = f(nh)$.

Example 2. Let us consider the following equations

$$\left({}_a\Delta_{h,*}^{1.5} x\right)(nh) = 0.1 \cdot x(nh + a)\,, \tag{29}$$

$$\left({}_a\Delta_h^{1.5} x\right)(nh) = 0.1 \cdot x(nh + a) \tag{30}$$

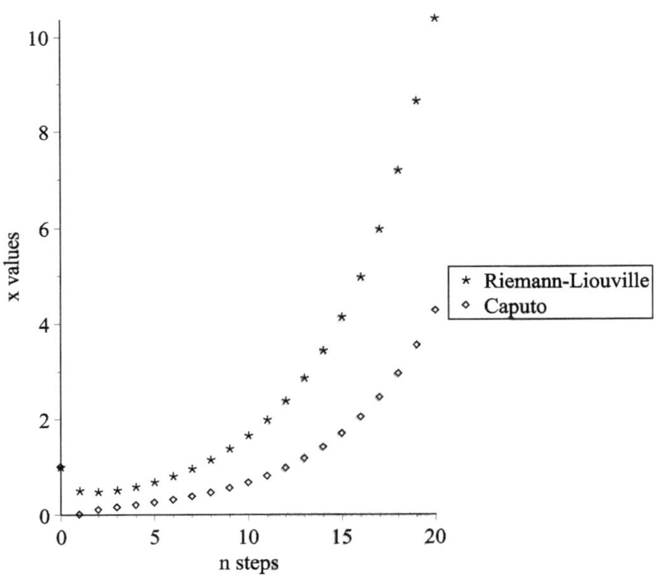

Fig. 2. Trajectories of equations $\left({}_a\Delta_{h,*}^\alpha x\right)(nh) = \lambda \cdot x(nh + a)$ and $\left({}_a\Delta_h^\alpha x\right)(nh) = \lambda \cdot x(nh+a)$ with the Caputo–type and Riemann-Liouville–type operators, respectively, for $\lambda = 0.1$ and order $\alpha = 1.5$, see Example 2

and initial conditions

$$b_0 = 1 \tag{31a}$$

$$b_1 = 0.1 \tag{31b}$$

Then Figure 2 presents the graphs of solutions to equations (29) and (30) with order 1.5 and the initial conditions (31) with the Caputo–type and Riemann-Liouville–type operators, respectively.

5 Conclusions

The Caputo– and Riemann-Liouville–type fractional order difference initial value problems for linear and semilinear systems with any positive order are discussed. The next step is to consider the semilinear control systems and develop conditions on controllability or stability properties via the complex domain.

Acknowledgment. The project was supported by the founds of National Science Centre granted on the bases of the decision number DEC-2011/03/B/ST7/03476. The work was supported by Bialystok University of Technology grant G/WM/3/2012.

References

1. Abdeljawad, T.: On Riemann and Caputo fractional differences. Computers and Mathematics with Applications 62(3), 1602–1611 (2011), doi:10.1016/j.camwa.2011.03.036.
2. Abdeljawad, T., Baleanu, D.: Fractional differences and integration by parts. Journal of Computational Analysis and Applications 13(3), 574–582 (2011)
3. Abdeljawad, T., Jarad, F., Baleanu, D.: A semigroup-like property for discrete Mittag-Leffler functions. Advances in Difference Equations 72 (2010)
4. Abu-Saris, R., Al-Mdallal, Q.: On the asymptotic stability of linear system of fractional-order differencce equations. An International Journal for Theory and Applications of Fractional Calculus and Applied Analysis 16(3), 613–629 (2013)
5. Atıcı, F.M., Eloe, P.W.: A transform method in discrete fractional calculus. International Journal of Difference Equations 2, 165–176 (2007)
6. Atıcı, F.M., Eloe, P.W.: Initial value problems in discrete fractional calculus. Proceedings of the American Mathematical Society 137(3), 981–989 (2009)
7. Bastos, N.R.O., Ferreira, R.A.C., Torres, D.F.M.: Discrete-time fractional variational problems. Signal Processing 91(3), 513–524 (2011)
8. Bastos, N.R.O., Ferreira, R.A.C., Torres, D.F.M.: Necessary optimality conditions for fractional difference problems of the calculus of variations. Discrete and Continuous Dynamical Systems 29(2), 417–437 (2011)
9. Chen, F., Luo, X., Zhou, Y.: Existence results for nonlinear fractional difference equation. Advances in Difference Equations 2011, 12 (2011), doi:10.1155/2011/713201

10. Ferreira, R.A.C., Torres, D.F.M.: Fractional h-difference equations arising from the calculus of variations. Applicable Analysis and Discrete Mathematics 5(1), 110–121 (2011)
11. Miller, K.S., Ross, B.: Fractional difference calculus. In: Proceedings of the International Symposium on Univalent Functions, Fractional Calculus and their Applications, pp. 139–152. Nihon University, Kōriyama (1988)
12. Mozyrska, D., Girejko, E.: Overview of the fractional h-difference operators. In: Operator Theory: Advances and Applications, vol. 229, pp. 253–267. Birkhäuser (2013), doi:10.1007/978-3-0348-0516-2.
13. Mozyrska, D., Girejko, E., Wyrwas, M.: Comparison of h-difference fractional operators. In: Mitkowski, W., Kacprzyk, J., Baranowski, J. (eds.) Theory & Appl. of Non-integer Order Syst. LNEE, vol. 257, pp. 191–197. Springer, Heidelberg (2013)
14. Mozyrska, D., Girejko, E., Wyrwas, M.: Fractional nonlinear systems with sequential operators. Central European Journal of Physics, 9 (2013), doi:10.2478/s11534-013-0223-3
15. Mozyrska, D., Wyrwas, M.: Solutions of fractional linear difference systems with Caputo–type operator via transform method. accepted to ICFDA 2014, p. 6 (2014)
16. Mozyrska, D., Wyrwas, M.: Solutions of fractional linear difference systems with Riemann-Liouville-type operator via transform method. accepted to ICFDA 2014, p. 6 (2014)
17. Mozyrska, D., Wyrwas, M.: Z-transforms of delta type fractional difference operators (submitted to publication, 2014)
18. Mozyrska, D.: Multiparameter fractional difference linear control systems. Discrete Dynamics in Nature and Society 8, article ID 183782 (2014)
19. Stanisławski, R., Latawiec, K.: Stability analysis for discrete-time fractional-order LTI state-space systems. Part I: New necessary and sufficient conditions for the asymptotic stability. Bulletin of the Polish Academy of Sciences. Technical Sciences 61(2), 353–361 (2013)
20. Stanisławski, R., Latawiec, K.: Stability analysis for discrete-time fractional-order LTI state-space systems. Part II: New stability criterion for FD-based systems. Bulletin of the Polish Academy of Sciences. Technical Sciences 61(2), 363–370 (2013)

The Fractional-Order Backward-Difference of a Product of Two Discrete-Variable Functions (Discrete Fractional Leibnitz Rule)

Piotr Ostalczyk

Institute of Applied Computer Science, Lodz University of Technology,
ul. Stefanowskiego 18/22, 90-924 Lodz, Poland
piotr.ostalczyk@p.lodz.pl

Abstract. In this paper a discrete fractional case of the so called Leibnitz Rule is presented The fractional-order backward difference of a product of two discrete-variable functions is derived. It is a generalisation to the first-order bacward difference of a product. The formula may be useful in the fractional-order backward differences of selected functions evaluation.

Keywords: Fractional Calculus, discrete-variable function.

1 Introduction

The Fractional Calculus [3],[6],[7] for discrete-variable functions [4] is getting more and more important in technical applications.One can mention here digital signal processing, image processing, fractional-order digital control [2] algorithms and real dynamical systems modelling [6].

The paper is orgainsed in a following order. First fundamental definition of the fractional-order backward difference [5] is given with some selected properties. The main result is presented in Section 2. Considerations are supported by numerical examples.

1.1 Mathematical Preliminaries

Consider a discrete-variable k real-valued bounded function $f(k)$ defined over interval $[k_0, k_0+1, \cdots, k-1, k]$. On a function $f(k)$ imposes a condition that $f(k) = 0$ for $k > k_0$. The Grünwald-Letnikov backward difference of $f(k)$ is defined as a following sum

$$ {}_{k_0}^{GL}\Delta_k^{(\nu)} f(k) = \sum_{i=k_0}^{k} a^{(\nu)}(i-k_0) f(k+k_0-i). \tag{1} $$

where a discrete-variable function $a^{(\nu)}(k)$ is defined as

$$ a^{(\nu)}(k) = \begin{cases} 0 & \text{for} \quad i < 0 \\ 1 & \text{for} \quad i = 0 \\ (-1)^i \frac{\nu(\nu-1)\cdots(\nu-k+1)}{k!} & \text{for } k = 1, 2, \cdots \end{cases} \tag{2} $$

© Springer International Publishing Switzerland 2015
K.J. Latawiec et al. (eds.), *Advances in Modeling and Control of Non-integer Order Systems*,
Lecture Notes in Electrical Engineering 320, DOI: 10.1007/978-3-319-09900-2_6

and $[k_0, k]$ is a difference calculation range, $\nu \in \mathbb{R}_+$ is an order which may be non-integer for the fractional-order backward-difference (FOBD), as well as integer one for a classical (integer-order) backward difference (IOBD). Further, to simplify notation ane assumes $k_0 = 0$ what does not reduce presented results generality. One should note that for $\nu \in \mathbb{R}_+$ one considers formula (1) as the fractional-order backward sum (FOBS) or n-th fold sum. The definition formula can be also presented in a vector form

$$ {}_0^{GL}\Delta_k^{(\nu)} f(k) = \left[\boldsymbol{a}^{(\nu)}(k)\right]^T \boldsymbol{N}_k \boldsymbol{f}(k) \tag{3}$$

where

$$\boldsymbol{f}(k) = \begin{bmatrix} f(0) \\ f(1) \\ \vdots \\ f(k-1) \\ f(k) \end{bmatrix}, \quad \boldsymbol{a}^{(\nu)}(k) = \begin{bmatrix} a(0) \\ a(1) \\ \vdots \\ a(k-1) \\ a(k) \end{bmatrix} \quad \boldsymbol{N}_k = \begin{bmatrix} 0\,0 \cdots 0\,1 \\ 0\,0 \cdots 1\,0 \\ \vdots\,\vdots \quad\; \vdots\,\vdots \\ 0\,1 \cdots 0\,0 \\ 1\,0 \cdots 0\,0 \end{bmatrix}. \tag{4}$$

It can be easily proved that for (1) and $\nu, \mu \in \mathbb{R}_+$

$$ {}_0^{GL}\Delta_k^{(\nu)} [{}_0^{GL}\Delta_k^{(\mu)} f(k)] = {}_0^{GL}\Delta_k^{(\nu+\mu)} f(k). \tag{5}$$

From (3) one immediately states that

$$ {}_0^{GL}\Delta_k^{(\nu)} f(k) \neq {}_0^{GL}\Delta_{k+1}^{(\nu)} f(k) \neq {}_0^{GL}\Delta_k^{(\nu)} f(k+1). \tag{6}$$

1.2 FOBD Vector Description

On a base of the FOBD definition formula (3) one gets a vector-matrix equation

$$ {}^{GL}\Delta^{(\nu)} \boldsymbol{f}(k) = \begin{bmatrix} {}_0^{GL}\Delta_k^{(\nu)} f(k) \\ {}_0^{GL}\Delta_{k-1}^{(\nu)} f(k-1) \\ \vdots \\ {}_0^{GL}\Delta_1^{(\nu)} f(1) \\ {}_0^{GL}\Delta_0^{(\nu)} f(0) \end{bmatrix} = \boldsymbol{A}_k^{(\nu)} \boldsymbol{N}_k \boldsymbol{f}(k) \tag{7}$$

where

$$\boldsymbol{A}_k^{(\nu)} = \begin{bmatrix} a_0^{(\nu)} & a_1^{(\nu)} & \cdots & a_{k-1}^{(\nu)} & a_k^{(\nu)} \\ 0 & a_0^{(\nu)} & \cdots & a_{k-2}^{(\nu)} & a_{k-1}^{(\nu)} \\ \vdots & \vdots & & \vdots & \vdots \\ 0 & 0 & \cdots & a_0^{(\nu)} & a_1^{(\nu)} \\ 0 & 0 & \cdots & 0 & a_0^{(\nu)} \end{bmatrix} \tag{8}$$

The matrix defined by (8) has following properties:

a) For $\nu, \mu \in \mathbb{R}_+$

$$\boldsymbol{A}_k^{(\nu)} \boldsymbol{A}_k^{(\mu)} = \boldsymbol{A}_k^{(\nu+\mu)}, \tag{9}$$

b) For $\nu \in \mathbb{R}_+$

$$[\boldsymbol{A}_k^{(\nu)}]^{-1} = \boldsymbol{A}_k^{(-\nu)}. \tag{10}$$

2 Main Result (FO Leibnitz Rule)

Now the FOBD of two discret-variable real functions $f(k)$ and $g(k)$ defined over interval k_0, k is evaluated. The main result - The Leibnitz Rule for the FOBD is stated in a form of a following theorem.

Theorem 2.1. *The BD of the order ν of a product of two functions $f(k)$ and $g(k)$ equals to*

$$ {}_0^{GL}\Delta_k^{(\nu)}[f(k)g(k)] = [{}^{GL}\Delta^{(\nu)}\boldsymbol{f}(k)]^T \boldsymbol{M}_k^{(\nu)} {}^{GL}\Delta^{(\nu)}\boldsymbol{g}(k), \tag{11}$$

where

$$\boldsymbol{g}(k) = \begin{bmatrix} g(0) \\ g(01) \\ \vdots \\ g(k-1) \\ g(k) \end{bmatrix}, \ {}^{GL}\Delta^{(\nu)}\boldsymbol{g}(k) = \begin{bmatrix} {}_0^{GL}\Delta_k^{(\nu)} g(k) \\ {}_0^{GL}\Delta_{k-1}^{(\nu)} g(k-1) \\ \vdots \\ {}_0^{GL}\Delta_1^{(\nu)} g(1) \\ {}_0^{GL}\Delta_0^{(\nu)} g(0) \end{bmatrix} = \boldsymbol{A}_k^{(\nu)} \boldsymbol{N}_k \boldsymbol{g}(k), \tag{12}$$

$$\boldsymbol{M}_k^{(\nu)} = [\boldsymbol{A}_k^{(-\nu)}]^T \boldsymbol{D}_k^{(\nu)} \boldsymbol{A}_k^{(\nu)}, \tag{13}$$

$$\boldsymbol{D}_k^{(\nu)} = \begin{bmatrix} a_0^{(\nu)} & 0 & \cdots & 0 & 0 \\ 0 & a_1^{(\nu)} & \cdots & 0 & 0 \\ \vdots & \vdots & & \vdots & \vdots \\ 0 & 0 & \cdots & a_{k-1}^{(\nu)} & 0 \\ 0 & 0 & \cdots & 0 & a_k^{(\nu)} \end{bmatrix}, \tag{14}$$

and ${}^{GL}\Delta^{(\nu)}\boldsymbol{f}(k)$ is defined by formula (7). A superscript T in (13) denotes a matrix transposition.

Proof. By Definition (1) formula (11) is expressed in a form

$$ {}_0^{GL}\Delta_k^{(\nu)}[f(k)g(k)] = \left[f(k)\ f(k-1)\ \cdots\ f(1)\ f(0)\right] \boldsymbol{D}_k^{(\nu)} \begin{bmatrix} g(k) \\ g(k-1) \\ \vdots \\ g(1) \\ g(0) \end{bmatrix} \tag{15}$$

$$= \left[f(k)\ f(k-1)\ \cdots\ f(1)\ f(0)\right] [[\boldsymbol{A}_k^{(\nu)}]^{-1} \boldsymbol{A}_k^{(\nu)}]^T \boldsymbol{D}_k^{(\nu)} [\boldsymbol{A}_k^{(\nu)}]^{-1} \boldsymbol{A}_k^{(\nu)} \begin{bmatrix} g(k) \\ g(k-1) \\ \vdots \\ g(1) \\ g(0) \end{bmatrix}$$

$$= \left[f(k)\ f(k-1)\ \cdots\ f(1)\ f(0)\right] [\boldsymbol{A}_k^{(\nu)}]^T [\boldsymbol{A}_k^{(-\nu)}]^T \boldsymbol{D}_k^{(\nu)} \boldsymbol{A}_k^{(-\nu)GL} \Delta^{(\nu)} \boldsymbol{g}(k)$$

$$= [\boldsymbol{A}_k^{(\nu)} \begin{bmatrix} f_k \\ f_{k-1} \\ \vdots \\ f_1 \\ f_0 \end{bmatrix}]^T [\boldsymbol{A}_k^{(-\nu)}]^T \boldsymbol{D}_k^{(\nu)} \boldsymbol{A}_k^{(-\nu)GL} \Delta^{(\nu)} \boldsymbol{g}(k)$$

$$= [{}^{GL}\Delta^{(\nu)} \boldsymbol{f}(k)]^T [\boldsymbol{A}_k^{(-\nu)}]^T \boldsymbol{D}_k^{(\nu)} \boldsymbol{A}_k^{(-\nu)GL} \Delta^{(\nu)} \boldsymbol{g}(k)$$

$$[{}^{GL}\Delta^{(\nu)} \boldsymbol{f}(k)]^T \boldsymbol{M}_k^{(\nu)GL} \Delta^{(\nu)} \boldsymbol{g}(k)$$

■

The result proved above is valid for integer orders $n \in \mathbb{Z}_+$. One should only realise that in this case

$$^{GL}\Delta^{(\nu)} \boldsymbol{f}(k) = \begin{bmatrix} {}^{GL}_{k-n}\Delta_k^{(n)} f(k) \\ {}^{GL}_{k-1-n}\Delta_{k-1}^{(n)} f(k-1) \\ \vdots \\ {}^{GL}_{1}\Delta_{n+1}^{(n)} f(n+1) \\ {}^{GL}_{0}\Delta_{n}^{(n)} f(n) \\ \vdots \\ {}^{GL}_{0}\Delta_{0}^{(n)} f(0) \end{bmatrix} = \boldsymbol{A}_k^{(n)} \boldsymbol{f}_k, \tag{16}$$

$$^{GL}\Delta^{(\nu)} \boldsymbol{g}(k) = \begin{bmatrix} {}^{GL}_{k-n}\Delta_k^{(n)} g(k) \\ {}^{GL}_{k-1-n}\Delta_{k-1}^{(n)} g(k-1) \\ \vdots \\ {}^{GL}_{1}\Delta_{n+1}^{(n)} g(n+1) \\ {}^{GL}_{0}\Delta_{n}^{(n)} g(n) \\ \vdots \\ {}^{GL}_{0}\Delta_{0}^{(n)} g(0) \end{bmatrix} = \boldsymbol{A}_k^{(n)} \boldsymbol{g}(k), \tag{17}$$

$$\boldsymbol{D}_k^{(n)} = \begin{bmatrix} a_0^{(n)} & \cdots & 0 & \cdots & 0 \\ \vdots & & \vdots & & \vdots \\ 0 & \cdots & a_n^{(n)} & \cdots & 0 \\ \vdots & & \vdots & & \vdots \\ 0 & \cdots & 0 & \cdots & 0 \end{bmatrix}, \tag{18}$$

with matrix $\boldsymbol{M}_k^{(\nu)}$ evaluated by formula (13). Below examplary matrices calculated for $n = 1, \cdots, 5$ are given

$$\boldsymbol{M}_k^{(1)} = \begin{bmatrix} 1 & 1 & 1 & 1 & \cdots \\ 1 & 0 & 0 & 0 & \cdots \\ 1 & 0 & 0 & 0 & \cdots \\ 1 & 0 & 0 & 0 & \cdots \\ \vdots & \vdots & \vdots & \vdots & \end{bmatrix}, \ \boldsymbol{M}_k^{(2)} = \begin{bmatrix} 1 & 2 & 3 & 4 & \cdots \\ 2 & 2 & 2 & 2 & \cdots \\ 3 & 2 & 2 & 2 & \cdots \\ 4 & 2 & 2 & 2 & \cdots \\ \vdots & \vdots & \vdots & \vdots & \end{bmatrix}, \ \boldsymbol{M}_k^{(3)} = \begin{bmatrix} 1 & 3 & 6 & 10 & 15 & \cdots \\ 3 & 6 & 9 & 12 & 15 & \cdots \\ 6 & 9 & 12 & 15 & 18 & \cdots \\ 10 & 12 & 15 & 18 & 21 & \cdots \\ \vdots & \vdots & \vdots & \vdots & \vdots & \end{bmatrix}, \tag{19}$$

$$\boldsymbol{M}_k^{(4)} = \begin{bmatrix} 1 & 4 & 10 & 20 & 35 & 56 & \cdots \\ 4 & 12 & 24 & 40 & 60 & 84 & \cdots \\ 10 & 24 & 42 & 64 & 90 & 120 & \cdots \\ 20 & 40 & 64 & 92 & 124 & 160 & \cdots \\ 35 & 60 & 90 & 124 & 162 & 204 & \cdots \\ 56 & 84 & 120 & 160 & 204 & 252 & \cdots \\ \vdots & \vdots & \vdots & \vdots & \vdots & \vdots & \end{bmatrix}, \ \boldsymbol{M}_k^{(5)} = \begin{bmatrix} 1 & 5 & 15 & 35 & 70 & 126 & \cdots \\ 5 & 20 & 50 & 100 & 60 & 84 & \cdots \\ 15 & 24 & 42 & 64 & 90 & 120 & \cdots \\ 35 & 40 & 64 & 92 & 124 & 160 & \cdots \\ 70 & 60 & 90 & 124 & 162 & 204 & \cdots \\ 126 & 84 & 120 & 160 & 204 & 252 & \cdots \\ \vdots & \vdots & \vdots & \vdots & \vdots & \vdots & \end{bmatrix}.$$

The correctness of the above formulae will be checked in a following numerical example.

Numerical Example 2.1. *Check the correctness of formula (13) for two integer orders $n = 1, 2$.*

Solution. For $n = 1$ by (13) one gets

$${}_{k_0}^{GL}\Delta_k^{(1)}[f(k)g(k)] = [{}^{GL}\Delta^{(1)}\boldsymbol{f}(k)]^T \boldsymbol{M}_k^{(1)}{}^{GL}\Delta^{(1)}\boldsymbol{g}(k) \tag{20}$$

$$= \begin{bmatrix} {}^{GL}_{k-1}\Delta^{(1)}_{k} f(k) \\ {}^{GL}_{k-2}\Delta^{(1)}_{k-1} f(k-1) \\ \vdots \\ {}^{GL}_{1}\Delta^{(1)}_{2} f(2) \\ {}^{GL}_{0}\Delta^{(1)}_{1} f(1) \\ \vdots \\ {}^{GL}_{0}\Delta^{(1)}_{0} f(0) \end{bmatrix}^T \begin{bmatrix} 1\,1\,1\,1 \cdots \\ 1\,0\,0\,0 \cdots \\ 1\,0\,0\,0 \cdots \\ 1\,0\,0\,0 \cdots \\ \vdots\;\vdots\;\vdots\;\vdots \end{bmatrix} \begin{bmatrix} {}^{GL}_{k-1}\Delta^{(1)}_{k} g(k) \\ {}^{GL}_{k-2}\Delta^{(1)}_{k-1} g(k-1) \\ \vdots \\ {}^{GL}_{1}\Delta^{(1)}_{2} g(2) \\ {}^{GL}_{0}\Delta^{(1)}_{1} g(1) \\ {}^{GL}_{0}\Delta^{(1)}_{0} g(0) \end{bmatrix}$$

$$= \begin{bmatrix} {}^{GL}_{k-1}\Delta^{(1)}_{k} f(k) \\ {}^{GL}_{k-2}\Delta^{(1)}_{k-1} f(k-1) \\ \vdots \\ {}^{GL}_{1}\Delta^{(1)}_{2} f(2) \\ {}^{GL}_{0}\Delta^{(1)}_{1} f(1) \\ {}^{GL}_{0}\Delta^{(1)}_{0} f(0) \end{bmatrix}^T \begin{bmatrix} \sum_{i=0}^{k} {}^{GL}_{i-1}\Delta^{(1)}_{i} g(i) \\ {}^{GL}_{k-1}\Delta^{(1)}_{k} g(k) \\ \vdots \\ {}^{GL}_{k-1}\Delta^{(1)}_{k} g(k) \\ {}^{GL}_{k-1}\Delta^{(1)}_{k} g(k)) \\ {}^{GL}_{k-1}\Delta^{(1)}_{k} g(k) \end{bmatrix}$$

$$= \begin{bmatrix} {}^{GL}_{k-1}\Delta^{(1)}_{k} f(k) \\ {}^{GL}_{k-2}\Delta^{(1)}_{k-1} f(k-1) \\ \vdots \\ {}^{GL}_{1}\Delta^{(1)}_{2} f(2) \\ {}^{GL}_{0}\Delta^{(1)}_{1} f(1) \\ {}^{GL}_{0}\Delta^{(1)}_{0} f(0) \end{bmatrix}^T \begin{bmatrix} g(k) \\ {}^{GL}_{k-1}\Delta^{(1)}_{k} g(k) \\ \vdots \\ {}^{GL}_{k-1}\Delta^{(1)}_{k} g(k) \\ {}^{GL}_{k-1}\Delta^{(1)}_{k} g(k)) \\ {}^{GL}_{k-1}\Delta^{(1)}_{k} g(k) \end{bmatrix}$$

$$= g(k) {}^{GL}_{k-1}\Delta^{(1)}_{k} f(k) + \begin{bmatrix} {}^{GL}_{k-2}\Delta^{(1)}_{k-1} f(k) \\ {}^{GL}_{k-3}\Delta^{(1)}_{k-2} f(k-1) \\ \vdots \\ {}^{GL}_{1}\Delta^{(1)}_{2} f(2) \\ {}^{GL}_{0}\Delta^{(1)}_{1} f(1) \\ {}^{GL}_{0}\Delta^{(1)}_{0} f(0) \end{bmatrix}^T \begin{bmatrix} 1 \\ \vdots \\ 1 \\ 1 \\ 1 \end{bmatrix} {}^{GL}_{k-1}\Delta^{(1)}_{k} g(k)$$

$$= g(k) {}^{GL}_{k-1}\Delta^{(1)}_{k} f(k) + f(k-1) {}^{GL}_{k-1}\Delta^{(1)}_{k} g(k)$$

$$= g(k)[f(k) - f(k-1)] + f(k-1)[g(k) - g(k-1)] = f(k)g(k) - f(k-1)g(k-1).$$

Comment 2.1. The last formula can be obtained by direct calculation of the first-order backwar differecne of the porduct of the two discrete-variable functions

$$ {}^{GL}_{k-1}\Delta^{(1)}_{k} f(k)g(k) = f(k)g(k) - f(k-1)g(k-1). \tag{21}$$

Moreover, the last but one formula in (20) is very close to the continuous function case known as [3]

$$\frac{d}{dt}[f(t)g(t)] = g(t)\frac{d}{dt}f(t) + f(t)\frac{d}{dt}g(t). \tag{22}$$

Similar calculations performed for $n = 2$ yield

$$ {}_{k_0}^{GL}\Delta_k^{(2)}[f(k)g(k)] = [{}^{GL}\Delta^{(2)}\boldsymbol{f}(k)]^T \boldsymbol{M}_k^{(2)}{}^{GL}\Delta^{(2)}\boldsymbol{g}(k) \tag{23}$$

$$= \begin{bmatrix} {}_{k-1}^{GL}\Delta_k^{(2)} f(k) \\ {}_{k-2}^{GL}\Delta_{k-1}^{(2)} f(k-1) \\ \vdots \\ {}_{1}^{GL}\Delta_2^{(2)} f(2) \\ {}_{0}^{GL}\Delta_1^{(2)} f(1) \\ \vdots \\ {}_{0}^{GL}\Delta_0^{(2)} f(0) \end{bmatrix}^T \begin{bmatrix} 1\,2\,3\,4\cdots \\ 2\,2\,2\,2\cdots \\ 3\,2\,2\,2\cdots \\ 4\,2\,2\,2\cdots \\ \vdots\;\vdots\;\vdots\;\vdots \end{bmatrix} \begin{bmatrix} {}_{k-1}^{GL}\Delta_k^{(2)} g(k) \\ {}_{k-2}^{GL}\Delta_{k-1}^{(2)} g(k-1) \\ \vdots \\ {}_{1}^{GL}\Delta_2^{(2)} g(2) \\ {}_{0}^{GL}\Delta_1^{(2)} g(1) \\ {}_{0}^{GL}\Delta_0^{(2)} g(0) \end{bmatrix}$$

$$= \begin{bmatrix} {}_{k-1}^{GL}\Delta_k^{(2)} f(k) \\ {}_{k-2}^{GL}\Delta_{k-1}^{(2)} f(k-1) \\ \vdots \\ {}_{1}^{GL}\Delta_2^{(2)} f(2) \\ {}_{0}^{GL}\Delta_1^{(2)} f(1) \\ {}_{0}^{GL}\Delta_0^{(2)} f(0) \end{bmatrix}^T \begin{bmatrix} {}_{0}^{GL}\Delta_k^{(-2)}[{}_{0}^{GL}\Delta_k^{(2)} g(k)] \\ 2\,{}_{0}^{GL}\Delta_k^{(-1)}[{}_{0}^{GL}\Delta_k^{(2)} g(k)] \\ 2\,{}_{k-1}^{GL}\Delta_k^{(2)} g(k) \\ 2\,{}_{k-1}^{GL}\Delta_k^{(2)} g(k) \\ 2\,{}_{k-1}^{GL}\Delta_k^{(2)} g(k) \\ \vdots \end{bmatrix}$$

$$= \begin{bmatrix} {}_{k-1}^{GL}\Delta_k^{(2)} f(k) \\ {}_{k-2}^{GL}\Delta_{k-1}^{(2)} f(k-1) \\ \vdots \\ {}_{1}^{GL}\Delta_2^{(2)} f(2) \\ {}_{0}^{GL}\Delta_1^{(2)} f(1) \\ {}_{0}^{GL}\Delta_0^{(2)} f(0) \end{bmatrix}^T \begin{bmatrix} g(k) \\ 2\,{}_{k-1}^{GL}\Delta_k^{(1)} g(k) \\ 1\,{}_{k-1}^{GL}\Delta_k^{(2)} g(k) + 2\,{}_{k-1}^{GL}\Delta_k^{(1)} g(k) \\ 2\,{}_{k-1}^{GL}\Delta_k^{(2)} g(k) + 2\,{}_{k-1}^{GL}\Delta_k^{(1)} g(k) \\ 3\,{}_{k-1}^{GL}\Delta_k^{(2)} g(k) + 2\,{}_{k-1}^{GL}\Delta_k^{(1)} g(k) \\ 4\,{}_{k-1}^{GL}\Delta_k^{(2)} g(k) + 2\,{}_{k-1}^{GL}\Delta_k^{(1)} g(k) \\ \vdots \end{bmatrix}$$

$$= \begin{bmatrix} {}_{k-1}^{GL}\Delta_k^{(2)} f(k) \\ {}_{k-2}^{GL}\Delta_{k-1}^{(2)} f(k-1) \\ \vdots \\ {}_{1}^{GL}\Delta_2^{(2)} f(2) \\ {}_{0}^{GL}\Delta_1^{(2)} f(1) \\ {}_{0}^{GL}\Delta_0^{(2)} f(0) \end{bmatrix}^T \left(\begin{bmatrix} g(k) - 2\,{}_{k-1}^{GL}\Delta_k^{(1)} g(k) \\ 0 \\ 1\,{}_{k-1}^{GL}\Delta_k^{(2)} g(k) \\ 2\,{}_{k-1}^{GL}\Delta_k^{(2)} g(k) \\ 3\,{}_{k-1}^{GL}\Delta_k^{(2)} g(k) \\ 4\,{}_{k-1}^{GL}\Delta_k^{(2)} g(k) \\ \vdots \end{bmatrix} + 2\,{}_{k-1}^{GL}\Delta_k^{(1)} g(k) \begin{bmatrix} 1 \\ 1 \\ 1 \\ 1 \\ \vdots \end{bmatrix} \right)$$

$$= ({}^{GL}_{k-1}\Delta^{(2)}_{k} f(k))[-g(k)+2g(k-1)] + {}^{GL}_{k-1}\Delta^{(2)}_{k} g(k) \begin{bmatrix} {}^{GL}_{k-3}\Delta^{(2)}_{k-2} f(k-2) \\ {}^{GL}_{k-4}\Delta^{(2)}_{k-3} f(k-3) \\ \vdots \\ {}^{GL}_{1}\Delta^{(2)}_{2} f(2) \\ {}^{GL}_{0}\Delta^{(2)}_{1} f(1) \\ {}^{GL}_{0}\Delta^{(2)}_{0} f(0) \end{bmatrix}^T \begin{bmatrix} 1 \\ 2 \\ 3 \\ 4 \\ \vdots \end{bmatrix}$$

$$+2{}^{GL}_{k-1}\Delta^{(1)}_{k} g(k) \begin{bmatrix} {}^{GL}_{k-2}\Delta^{(2)}_{k} f(k) \\ {}^{GL}_{k-3}\Delta^{(2)}_{k-1} f(k-1) \\ \vdots \\ {}^{GL}_{1}\Delta^{(2)}_{2} f(2) \\ {}^{GL}_{0}\Delta^{(2)}_{1} f(1) \\ {}^{GL}_{0}\Delta^{(2)}_{0} f(0) \end{bmatrix}^T \begin{bmatrix} 1 \\ 1 \\ 1 \\ 1 \\ \vdots \end{bmatrix}$$

$$= {}^{GL}_{k-2}\Delta^{(2)}_{k} f(k)[-g(k)+2g(k-1)] + {}^{GL}_{k-4}\Delta^{(2)}_{k-2} f(k-2) {}^{GL}_{k-2}\Delta^{(2)}_{k} g(k)$$

$$+2{}^{GL}_{k-1}\Delta^{(1)}_{k} g(k) {}^{GL}_{k-1}\Delta^{(1)}_{k} f(k)$$

$$= f(k)g(k) - 2f(k-1)g(k-1) + f(k-2)g(k-2).$$

Corollary 2.2. *The FOBD of a product $f(k)g(k)$ equals to the FOBD of a product $g(k)f(k)$.*

$${}^{GL}_{0}\Delta^{(\nu)}_{k} f(k)g(k) = {}^{GL}_{0}\Delta^{(\nu)}_{k} g(k)f(k). \tag{24}$$

Proof. By Theorem 2.1

$${}^{GL}_{k_0}\Delta^{(\nu)}_{k}[f(k)g(k)] = [{}^{GL}\Delta^{(\nu)}\boldsymbol{f}(k)]^T \boldsymbol{M}^{(\nu)}_{k} {}^{GL}\Delta^{(\nu)}\boldsymbol{g}(k) \tag{25}$$

$$= [[{}^{GL}\Delta^{(\nu)}\boldsymbol{f}(k)]^T \boldsymbol{M}^{(\nu)}_{k} {}^{GL}\Delta^{(\nu)}\boldsymbol{g}(k)]^T$$

$$= [{}^{GL}\Delta^{(\nu)}\boldsymbol{g}(k)]^T \boldsymbol{M}^{(\nu)}_{k}]^T [{}^{GL}\Delta^{(\nu)}\boldsymbol{f}(k) = [{}^{GL}\Delta^{(\nu)}\boldsymbol{g}(k)]^T \boldsymbol{M}^{(\nu)}_{k} {}^{GL}\Delta^{(\nu)}\boldsymbol{f}(k)$$

$$= {}^{GL}_{k_0}\Delta^{(\nu)}_{k}[g(k)f(k)].$$

■

One of the possible applications of the FO Leibnitz Rule demonstrates a numerical example presented below.

Numerical Example 2.2. *Consider a function* $f(k) = k\mathbf{1}(k)$. *Knowing that* ${}^{GL}_{k-1}\Delta^{(1)}_{k} f(k) = {}^{GL}_{k-1}\Delta^{(1)}_{k} k\mathbf{1}(k) = \mathbf{1}(k) = \mathbf{1}(k)$ *calculate* ${}^{GL}_{k-1}\Delta^{(1)}_{k} k^2\mathbf{1}(k)$ *using formula(11).*

Solution. Denote $f(k) = g(k) = k\mathbf{1}(k)$. Then for $\nu = n = 1$

$$ {}^{GL}_{k-1}\Delta^{(1)}_{k} k^2\mathbf{1}(k) \tag{26}$$

$$= \begin{bmatrix} {}^{GL}_{k-1}\Delta^{(1)}_{k} f(k) \\ {}^{GL}_{k-2}\Delta^{(1)}_{k-1} f(k-1) \\ \vdots \\ {}^{GL}_{1}\Delta^{(1)}_{2} f(2) \\ {}^{GL}_{0}\Delta^{(1)}_{1} f(1) \\ {}^{GL}_{0}\Delta^{(1)}_{0} f(0) \end{bmatrix}^T \begin{bmatrix} 1\,1\,1\,1 \cdots \\ 1\,0\,0\,0 \cdots \\ 1\,0\,0\,0 \cdots \\ 1\,0\,0\,0 \cdots \\ \vdots\,\vdots\,\vdots\,\vdots\,\vdots \end{bmatrix} \begin{bmatrix} {}^{GL}_{k-1}\Delta^{(1)}_{k} g(k) \\ {}^{GL}_{k-2}\Delta^{(1)}_{k-1} g(k-1) \\ \vdots \\ {}^{GL}_{1}\Delta^{(1)}_{2} g(2) \\ {}^{GL}_{0}\Delta^{(1)}_{1} g(1) \\ {}^{GL}_{0}\Delta^{(1)}_{0} g(0) \end{bmatrix}$$

$$= \begin{bmatrix} 1\,1\,1 \cdots 1\,1 \end{bmatrix} \begin{bmatrix} 1\,1\,1\,1 \cdots \\ 1\,0\,0\,0 \cdots \\ 1\,0\,0\,0 \cdots \\ 1\,0\,0\,0 \cdots \\ \vdots\,\vdots\,\vdots\,\vdots\,\vdots \end{bmatrix} \begin{bmatrix} 1 \\ 1 \\ \vdots \\ 1 \\ 1 \\ 1 \end{bmatrix}$$

$$= \begin{bmatrix} 1\,1\,1 \cdots 1\,1 \end{bmatrix} \begin{bmatrix} k\mathbf{1}(k) \\ 1 \\ \vdots \\ 1 \\ 1 \\ 1 \end{bmatrix} = k\mathbf{1}(k) + (k-1)\mathbf{1}(k-1) = (2k-1)\mathbf{1}(k-1)$$

3 Conclusions

The formula derived as a main resut is a discrete counterpart of the well-known Leibnitz Rule served in calcuation of the first-order derivative of a product of two continuous-variable functions. The FO Leibnitz Rule is a generalization of the commonly known first-order derivative of a product of two continuous-variable functions to the fractional case applied to discrete-variable functions.

References

1. Ortegueira, M.D., Coito, F.: From Differences to Derivatives. Fractional Calculus & Applied analysis 7, 459–471 (2004)

2. Kaczorek, T.: Selected Problems of Fractional Systems Theory. LNCIS, vol. 411. Springer, Berlin (2011)
3. Oldham, K.B., Spanier, J.: The Fractional Calculus. Academic Press, New York (1974)
4. Ostalczyk, P.: The non-integer difference of the discrete-time function and its application to the control system synthesis. I. J. System Science 31, 1551–1561 (2000)
5. Ostalczyk, P.: Discrete-variable Functions. A Series of Monographs Technical University of Lodz, Łódź (2001)
6. Ostaloup, A.: La dérivation non entière. Hermes, Paris (1994)
7. Podlubny, I.: Fractional Differential Equations. Academic Press, San Diego (1999)

Part III
Approximation, Modeling and Simulations

Matrix Approach and Analog Modeling for Solving Fractional Variable Order Differential Equations*

Wiktor Malesza, Michal Macias, and Dominik Sierociuk

Institute of Control and Industrial Electronics
Warsaw University of Technology
Koszykowa 75, 00-662 Warsaw, Poland
{wmalesza,michal.macias,dsieroci}@ee.pw.edu.pl

Abstract. In the paper, a matrix approach for solving fractional variable order linear differential equations of an additive-switching type will be presented. Introduced method is based on a duality property between additive and recursive type of variable order differential definitions. Obtained solutions will be validated by comparing them with analog model results.

Keywords: fractional calculus, differential equations, analog modeling.

1 Introduction

Fractional calculus is a natural generalization of traditional differential calculus when order of derivatives or integrals can be fractional (non-integer). The case, when order of the fractional derivative is changing in time, is much more complicated in description and analysis than in constant order case. One of the main problems is a variety types of definitions. In [1], nine different variable order derivative definitions are given, however, without clear interpretation of them. In papers [2–5], interpretation in the form for switching strategies, for two main types and two recursive types of derivatives, are given. This interpretation is very helpful in system analysis, and gives a chance to categorize definitions according to its switching scheme. Moreover, these switching schemes allow to build analog models of different variable order systems, that was also shown in mentioned papers. It is a very important feature, because numerical realization of variable order systems are much more complicated than in integer order case. Some other numerical realization methods are presented in [6, 7].

The matrix form of the fractional constant order ordinary and partial differential equations was presented in [8, 9]. In this paper, an extension of this method for numerical solving variable order differential equations will be proposed for two types of variable order definitions. Moreover, obtained numerical methods will be compared with results obtained by analog modeling method.

* This work was supported by the Polish National Science Center with the decision number DEC-2011/03/D/ST7/00260.

© Springer International Publishing Switzerland 2015

K.J. Latawiec et al. (eds.), *Advances in Modeling and Control of Non-integer Order Systems*,
Lecture Notes in Electrical Engineering 320, DOI: 10.1007/978-3-319-09900-2_7

The paper is organized as follows. In Section 2, recursive variable fractional order derivative definition is recalled, and its basic properties are given. Section 3 gives main result of this paper – a numerical method, based on matrix approach, for simulation of variable order systems. In Section 4, experimental results of comparison between analog model and proposed numerical implementation is presented.

2 Fractional Variable Order Grunwald-Letnikov Type Derivatives

As a base of generalization onto variable order derivative the following definition is taken into consideration:

Definition 1. *Fractional constant order derivative is defined as follows:*

$$ {}_0\mathrm{D}_t^{\alpha} f(t) = \lim_{h\to 0} \frac{1}{h^{\alpha}} \sum_{r=0}^{n} (-1)^r \binom{\alpha}{r} f(t-rh), $$

where $h>0$ is a step time, and $n=\lfloor t/h \rfloor$.

For the case of order changing with time (variable order case), different types of derivative definitions can be found in the literature [10], [11].

Definition 2. *The $\mathcal{B}$-type fractional variable order derivative is defined as follows*

$$ {}_0^{\mathcal{B}}\mathrm{D}_t^{\alpha(t)} f(t) = \lim_{h\to 0} \sum_{r=0}^{n} \frac{(-1)^r}{h^{\alpha(t-rh)}} \binom{\alpha(t-rh)}{r} f(t-rh). $$

The discrete realization of $\mathcal{B}$-type fractional variable order derivative is the following

$$ {}^{\mathcal{B}}\Delta^{\alpha_l} f_l = \sum_{j=0}^{l} \frac{(-1)^j}{h^{\alpha_{l-j}}} \binom{\alpha_{l-j}}{j} f_{l-j}. $$

The $\mathcal{B}$-type derivative assumes that coefficients for past samples are obtained for order that was present for these samples.

The alternative definition, that has different properties than $\mathcal{B}$-type of derivative, possess recursive nature, and has the following form:

Definition 3 ([12]). *The $\mathcal{E}$-type fractional variable order derivative is defined as follows:*

$$ {}_0^{\mathcal{E}}\mathrm{D}_t^{\alpha(t)} f(t) = \lim_{h\to 0} \left(\frac{f(t)}{h^{\alpha(t)}} - \sum_{j=1}^{n} (-1)^j \binom{-\alpha(t-jh)}{j} \frac{h^{\alpha(t-jh)}}{h^{\alpha(t)}} {}_0^{\mathcal{E}}\mathrm{D}_{t-jh}^{\alpha(t)} f(t) \right). $$

The discrete realization of $\mathcal{E}$-type fractional variable order derivative has the following form

$$^{\mathcal{E}}\!\Delta^{\alpha_l} f_l = \frac{x_l}{h^{\alpha_l}} - \sum_{j=1}^{l} (-1)^j \binom{-\alpha_{l-j}}{j} \frac{h^{\alpha_{l-j}}}{h^{\alpha_l}} \Delta^{\alpha_{l-j}} x_{l-j} \tag{1}$$

for $l = 0, 1, 2, \ldots, k$.

Remark 1. For a constant order $\alpha(t)$ = const we get the same results as for constant order derivative and difference definitions, that is

$$^{\mathcal{B}}_{0}\mathrm{D}_t^{\alpha} f(t) = {}^{\mathcal{E}}_{0}\mathrm{D}_t^{\alpha} f(t) = {}_0\mathrm{D}_t^{\alpha} f(t).$$

2.1 Properties of $\mathcal{E}$-Type Derivative Definition

As it was shown in [12], the $\mathcal{E}$-type derivative is equivalent to the switching scheme given in Fig. 1.

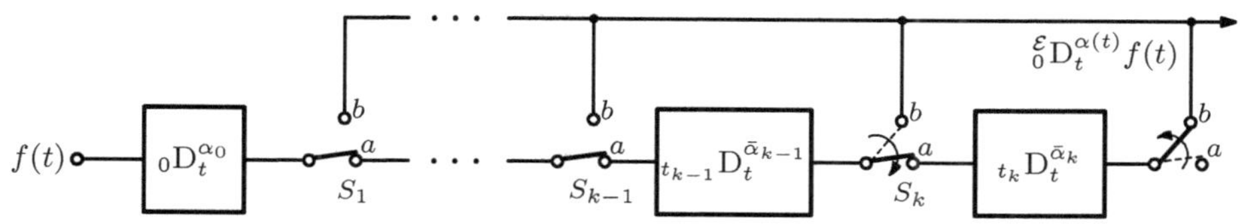

Fig. 1. Realization of $\mathcal{E}$-type derivative in the form of switching scheme, where $\bar{\alpha}_j = \alpha_j - \alpha_{j-1}$, $j = 1, \ldots, k$, (configuration at time $t = t_k$)

For the purpose of numerical calculations, the $\mathcal{E}$-type derivative definition can be rewritten in an alternative form:

Theorem 1 ([12]). *The $\mathcal{E}$-type fractional difference given by* (1) *can be expressed in the following matrix form:*

$$\begin{pmatrix} ^{\mathcal{E}}\!\Delta^{\alpha_0} x_0 \\ ^{\mathcal{E}}\!\Delta^{\alpha_1} x_1 \\ \vdots \\ ^{\mathcal{E}}\!\Delta^{\alpha_k} x_k \end{pmatrix} = \mathfrak{Q}_0^k \begin{pmatrix} x_0 \\ x_1 \\ \vdots \\ x_k \end{pmatrix}, \tag{2}$$

where

$$\mathfrak{Q}_0^k = \left(\begin{array}{c|c|c|c|c|c} h^{-\alpha_0} & 0 & 0 & \cdots & 0 & 0 \\ \hline \mathfrak{q}_{2,1} & h^{-\alpha_1} & 0 & \cdots & 0 & 0 \\ \hline \vdots & \vdots & \vdots & \ddots & \ddots & \vdots \\ \hline \mathfrak{q}_{k,1} & \mathfrak{q}_{k,2} & \mathfrak{q}_{k+1,3} & \cdots & h^{-\alpha_{k-1}} & 0 \\ \hline \mathfrak{q}_{k+1,1} & \mathfrak{q}_{k+1,2} & \mathfrak{q}_{k+1,3} & \cdots & \mathfrak{q}_{k+1,k} & h^{-\alpha_k} \end{array}\right), \tag{3}$$

where, for $i, j = 1, \ldots, k+1$,

$$\mathfrak{q}_{i,j} = \begin{cases} q_{i-1}(\mathfrak{q}_{1,j}, \ldots, \mathfrak{q}_{i-1,j})^T & \textit{for } i > j, \\ h^{-\alpha_i} & \textit{for } i = j, \\ 0 & \textit{for } i < j, \end{cases} \tag{4}$$

and for $r = 1, \ldots, k$

$$q_r = (-v_{-\alpha_0,r}, \ldots, -v_{-\alpha_{r-1},1}) \in \mathbb{R}^{1\times r}, \tag{5}$$

$$v_{-\alpha_{r-p},p} = (-1)^p \binom{-\alpha_{r-p}}{p} \frac{h^{\alpha_{r-p}}}{h^{-\alpha_r}}, \quad p = 1, \ldots, r, \tag{6}$$

that is, the m*-th element of* q_r, $m = 1, \ldots, r$, *is*

$$(q_r)_m = -v_{-\alpha_{m-1},r-m+1} = (-1)^{r-m+1}\binom{-\alpha_{m-1}}{r-m+1}. \tag{7}$$

Example 1. For $k = 2$, by Theorem 1, we get

$$q_1 = -h^{\alpha_0-\alpha_1}\alpha_0, \quad q_2 = \left(-h^{\alpha_0-\alpha_2}\binom{-\alpha_0}{2}, -h^{\alpha_1}h^{-\alpha_2}\alpha_1\right),$$

and

$$\mathfrak{Q}_0^2 = \begin{pmatrix} h^{-\alpha_0} & 0 & 0 \\ -h^{-\alpha_1}\alpha_0 & h^{-\alpha_1} & 0 \\ -h^{-\alpha_2}\left(\binom{-\alpha_0}{2} - \alpha_1\alpha_0\right) & -h^{-\alpha_2}\alpha_1 & h^{-\alpha_2} \end{pmatrix}. \tag{8}$$

Lemma 1 ([12]). *The following holds*

$$\mathfrak{Q}_0^k = \mathfrak{Q}(\alpha_k, k) \cdots \mathfrak{Q}(\alpha_1, 1)\mathfrak{Q}(\alpha_0, 0), \tag{9}$$

where, for $r = 0, \ldots, k$,

$$\mathfrak{Q}(\alpha_r, r) = \left(\begin{array}{c|c|c} I_{r,r} & 0_{r,1} & 0_{r,k-r} \\ \hline q_r & h^{-\alpha_r} & 0_{1,k-r} \\ \hline 0_{k-r,r} & 0_{k-r,1} & I_{k-r,k-r} \end{array}\right) \in \mathbb{R}^{(k+1)\times(k+1)},$$

where q_r *is given by* (5), *and* $0_{m,n} \in \mathbb{R}^{m\times n}$, $I_{m,n} \in \mathbb{R}^{m\times n}$ *stand for zero and identity matrix, respectively.*

Example 2. For $k = 2$, by Lemma 1, we get

$$\mathfrak{Q}(\alpha_0, 0) = \begin{pmatrix} h^{-\alpha_0} & 0 & 0 \\ 0 & 1 & 0 \\ 0 & 0 & 1 \end{pmatrix}, \quad \mathfrak{Q}(\alpha_1, 1) = \begin{pmatrix} 1 & 0 & 0 \\ -h^{\alpha_0-\alpha_1}\alpha_0 & h^{-\alpha_1} & 0 \\ 0 & 0 & 1 \end{pmatrix},$$

$$\mathfrak{Q}(\alpha_2, 2) = \begin{pmatrix} 1 & 0 & 0 \\ 0 & 1 & 0 \\ -h^{\alpha_0-\alpha_2}\binom{-\alpha_0}{2} & -h^{\alpha_1-\alpha_2}\alpha_1 & h^{-\alpha_2} \end{pmatrix},$$

which, after multiplication according to (9), equals to (8) obtained in Ex. 1.

Remark 2. Taking the limit $h \to 0$ we get the following form of the $\mathcal{E}$-type variable order derivative definition:

$$\begin{pmatrix} {}^{\mathcal{E}}_{0}\mathrm{D}^{\alpha(t)}_{0}x(t) \\ {}^{\mathcal{E}}_{0}\mathrm{D}^{\alpha(t)}_{h}x(t) \\ \vdots \\ {}^{\mathcal{E}}_{0}\mathrm{D}^{\alpha(t)}_{kh}x(t) \end{pmatrix} = \lim_{h\to 0} \mathfrak{Q}^{k}_{0} \begin{pmatrix} x(0) \\ x(h) \\ \vdots \\ x(kh) \end{pmatrix}, \tag{10}$$

where $\mathfrak{Q}^{k}_{0}$ is given by (3) or (9).

Remark 3. In general, the following relations occur

$${}^{\mathcal{B}}\Delta^{\pm\bar{\alpha}}\left({}^{\mathcal{B}}\Delta^{\mp\bar{\alpha}}x_k\right) \neq x_k, \quad {}^{\mathcal{B}}_{0}\mathrm{D}^{\pm\alpha(t)}_{t}\left({}^{\mathcal{B}}_{0}\mathrm{D}^{\mp\alpha(t)}_{t}f(t)\right) \neq f(t),$$

and

$${}^{\mathcal{E}}\Delta^{\pm\bar{\alpha}}\left({}^{\mathcal{E}}\Delta^{\mp\bar{\alpha}}x_k\right) \neq x_k, \quad {}^{\mathcal{E}}_{0}\mathrm{D}^{\pm\alpha(t)}_{t}\left({}^{\mathcal{E}}_{0}\mathrm{D}^{\mp\alpha(t)}_{t}f(t)\right) \neq f(t),$$

where $\bar{\alpha} = \{\alpha_0, \dots, \alpha_k\}$. It means that for a variable order difference (derivative) the semigroup property does not hold. However, for a constant-order, i.e., $\alpha_0 = \dots = \alpha_k$, and respectively $\alpha(t) = \text{const}$, this property holds.

Otherwise, between different types of definitions there exists a very special property, called duality [5, 13], which is given as follows:

$${}^{\mathcal{B}}\Delta^{\pm\bar{\alpha}}\left({}^{\mathcal{E}}\Delta^{\mp\bar{\alpha}}x_k\right) = x_k, \quad {}^{\mathcal{B}}_{0}\mathrm{D}^{\pm\alpha(t)}_{t}\left({}^{\mathcal{E}}_{0}\mathrm{D}^{\mp\alpha(t)}_{t}f(t)\right) = f(t),$$

$${}^{\mathcal{E}}\Delta^{\pm\bar{\alpha}}\left({}^{\mathcal{B}}\Delta^{\mp\bar{\alpha}}x_k\right) = x_k, \quad {}^{\mathcal{E}}_{0}\mathrm{D}^{\pm\alpha(t)}_{t}\left({}^{\mathcal{B}}_{0}\mathrm{D}^{\mp\alpha(t)}_{t}f(t)\right) = f(t).$$

3 Numerical Simulation Methodology

3.1 Solving Fractional Variable Order Differential Equation

The matrix form of the $\mathcal{E}$-type fractional variable order difference or derivative given by (2) and (10), respectively, can be used to solve numerically fractional variable order differential equations of $\mathcal{B}$-type definition.

Let us consider the following scalar fractional variable order differential equation

$${}^{\mathcal{B}}\mathrm{D}^{\alpha(t)}x = cx + u, \tag{11}$$

where $x = x(t)$ is the real valued unknown function, $u = u(t)$ is a real valued known function,

$$\alpha(t) = \alpha_i, \quad t_i \le t < t_{i+1}, \quad i = 0, \dots, N, \quad t_0 = 0, \tag{12}$$

and $c \in \mathbb{R}$ is a constant.

The approximation of differential equation (11) is given by the fractional variable order difference equation

$${}^{\mathcal{B}}\Delta^{\bar{\alpha}}x = cx + u, \tag{13}$$

where, abusing the notation, $x = (x_0, \ldots, x_k)^T \in \mathbb{R}^{k+1}$ is a values vector of unknown variable, $u = (u_1, \ldots, u_k)^T \in \mathbb{R}^{k+1}$ is a vector of known function values, $k = (t_{N+1}/h) - 1$, and $\bar{\alpha} = \{\bar{\alpha}_0, \ldots, \bar{\alpha}_k\}$, then ${}^{\mathcal{B}}\Delta^{\bar{\alpha}}x = ({}^{\mathcal{B}}\Delta^{\bar{\alpha}_0}x_0, \ldots, {}^{\mathcal{B}}\Delta^{\bar{\alpha}_k}x_k)^T$, where

$$\bar{\alpha}_j = \alpha_i, \quad t_i/h \leq j < t_{i+1}/h, \quad i = 0, \ldots, N, \tag{14}$$

assuming for simplicity that t_i's are integer-multiplicities of h, and $h \leq \min(t_i)$, i.e.,

$$\bar{\alpha} = \{\underbrace{\alpha_0, \ldots, \alpha_0}_{\rho_0\text{-times}}, \underbrace{\alpha_1, \ldots, \alpha_1}_{\rho_1\text{-times}}, \ldots, \underbrace{\alpha_N, \ldots, \alpha_N}_{\rho_N\text{-times}}\}, \tag{15}$$

where $\rho_i = (t_{i+1} - t_i)/h$, $i = 0, \ldots, N$, and $\sum \rho_i = k + 1$. Applying to both sides of (13) the difference operator ${}^{\mathcal{E}}\Delta^{-\bar{\alpha}}$, and using the duality property (see Remark 3) between the $\mathcal{B}$-type and $\mathcal{E}$-type definition, i.e., ${}^{\mathcal{E}}\Delta^{-\bar{\alpha}}\left({}^{\mathcal{B}}\Delta^{\bar{\alpha}}x\right) = x$, we get

$$x = {}^{\mathcal{E}}\Delta^{-\bar{\alpha}}cx + {}^{\mathcal{E}}\Delta^{-\bar{\alpha}}u. \tag{16}$$

By Theorem 1, difference equation (16) can be rewritten as

$$x = c\mathfrak{Q}_0^k x + \mathfrak{Q}_0^k u, \tag{17}$$

where $\mathfrak{Q}_0^k$ is given by (3), or equivalently by (9), calculated for $-\bar{\alpha}$, where $\bar{\alpha}$ is given by (14), i.e.,

$$\mathfrak{Q}_0^k = \prod_{j=0}^{k} \mathfrak{Q}(-\bar{\alpha}_j, j), \quad k = \frac{t_{N+1}}{h} - 1. \tag{18}$$

Thus, the solution of (13), and thereby approximated solution of (11), is the following (see also Fig. 2)

$$x = \left(I_{k+1,k+1} - c\mathfrak{Q}_0^k\right)^{-1} \mathfrak{Q}_0^k u, \tag{19}$$

which exists always provided that $h^{\alpha_j} \neq c$, $j = 0, \ldots, k+1$. Obviously, for $c \leq 0$, solution (19) exists always.

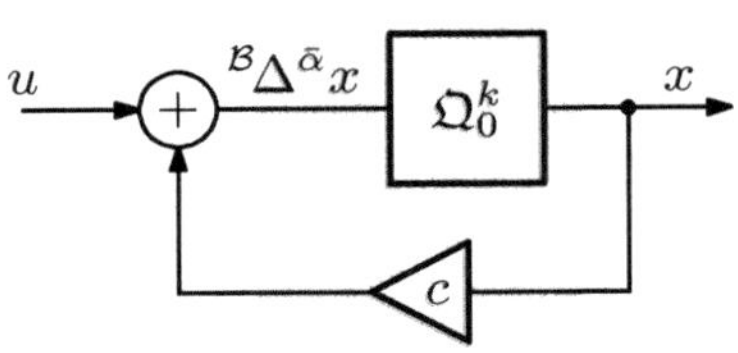

Fig. 2. Realization of difference equation (13)

In the case, when gain of integration action varies in time, e.g., takes the value $\lambda_i \in \mathbb{R}$ during $t_i \leq t < t_{i+1}$, $i = 0, \ldots, N$, which occurs, for example, in the analog realization of fractional variable order integral of the $\mathcal{E}$-type, matrix (3) has to be replaced by

$$\hat{\mathfrak{Q}}_0^k = \mathfrak{Q}_0^k \cdot \Lambda, \tag{20}$$

where

$$\Lambda = \text{diag}\{\underbrace{\lambda_0, \ldots, \lambda_0}_{\rho_0\text{-times}}, \underbrace{\lambda_1, \ldots, \lambda_1}_{\rho_1\text{-times}}, \ldots, \underbrace{\lambda_N, \ldots, \lambda_N}_{\rho_N\text{-times}}\}, \tag{21}$$

and $\rho_i = (t_{i+1} - t_i)/h$, $i = 0, \ldots, N$, $\sum \rho_i = k + 1$.

Example 3. Consider a time-variant differential equation

$$^{\mathcal{B}}\mathrm{D}^{\alpha(t)}x = \lambda(t)(cx + u), \tag{22}$$

where

$$\alpha(t) = \begin{cases} \alpha_0 & \text{for } t \in [0, T_1), \\ \alpha_1 & \text{for } t \in [T_1, T_2) \end{cases} \quad \text{and} \quad \lambda(t) = \begin{cases} \lambda_0 & \text{for } t \in [0, T_1), \\ \lambda_1 & \text{for } t \in [T_1, T_2). \end{cases} \tag{23}$$

Thus, the switch of the variable order (and gain coefficient) occurs at time-instant $t = T_1$, then $N = 1$. The approximated numerical solution of (22) is

$$x = \left(I_{k+1,k+1} - c\mathfrak{Q}_0^k \Lambda\right)^{-1} \mathfrak{Q}_0^k \Lambda u, \tag{24}$$

where, using (18),

$$\mathfrak{Q}_0^k = \prod_{j=0}^{k} \mathfrak{Q}(-\bar{\alpha}_j, j), \quad k = \frac{T_2}{h} - 1,$$

and, according to (14),

$$\bar{\alpha}_j = \alpha_i, \quad \frac{T_i}{h} \le j < \frac{T_{i+1}}{h}, \quad i = 0, 1, \quad T_0 = 0. \tag{25}$$

The matrix of gain coefficients is

$$\Lambda = \text{diag}\{\bar{\lambda}_0, \bar{\lambda}_1\}, \quad \bar{\lambda}_j = \{\underbrace{\lambda_j, \ldots, \lambda_j}_{\rho_j\text{-times}}\}, \quad j = 0, 1, \tag{26}$$

where $\rho_0 = T_1/h$, $\rho_1 = (T_2 - T_1)/h$, and then $\sum \rho_i = T_2/h = k + 1$. Solution plots of (22) are depicted in Fig. 4.

4 Experimental Results

An analog realization of the $\mathcal{E}$-type fractional variable order integral, based on switching scheme given in Fig. 1, is presented in Fig. 3. Detailed description of used experimental setup is given in [12]. Depending on switches position marked as S_i, $i = 1, 2$, in Fig. 3, the circuit can be described by fractional order

($\alpha = -0.5$) or integer order ($\alpha = -1$). In the first case, when S_1 and S_2 switches are connected to terminals marked as a, and in the second case as b, the following fractional order derivative functions have been obtained, respectively:

$$y_1(t) = \frac{1}{\tau_1} {}_0\mathrm{D}_t^{-0.5} u(t) \quad \text{and} \quad y_2(t) = \frac{1}{\tau_2} {}_0\mathrm{D}_t^{-1} u(t),$$

where τ_1 and τ_2 are constant coefficients.

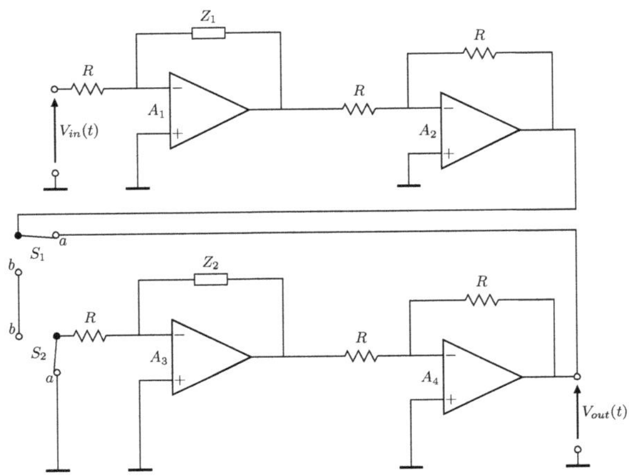

Fig. 3. Analog realization of the $\mathcal{E}$-type fractional variable order integral

4.1 Experimental Results of the $\mathcal{E}$-Type Integral System

The identification results for $\mathcal{E}$-type integral system were obtained by numerical minimization of time responses square error with sampling time $T_s = 0.01$ sec., and input signal $u(t) = 0.1H(t)$, where $H(t)$ is a Heaviside step function. After identification process, the following models for orders -0.5 and -1, respectively, in time domain, were obtained:

$$y_1(t) = {}_0\mathrm{D}_t^{-0.5} \lambda_0 u(t) = 1.35 {}_0\mathrm{D}_t^{-0.5} u(t),$$
$$y_2(t) = {}_0\mathrm{D}_t^{-1} \lambda_1 u(t) = 1.88 {}_0\mathrm{D}_t^{-1} u(t),$$

which gives rise to the following variable order integrator:

$$y(t) = {}_0^{\mathcal{E}}\mathrm{D}_t^{-\alpha(t)} \left(\lambda(t) u(t)\right).$$

In this case, the system identified by function $y_1(t)$ is switching to the system described by function $y_2(t)$, in switching time $T_1 = 0.1$ sec.

The experimental results compared to numerical implementation of the variable order integral system based on $\mathcal{E}$-type of definition are presented in Fig. 4a.

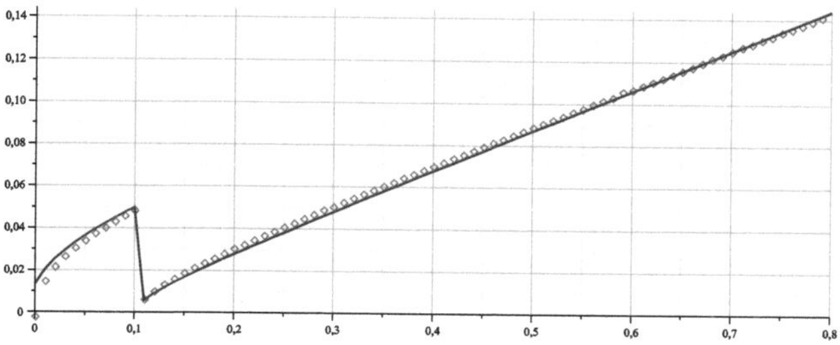

(a) For $c = 0$, i.e., ${}^{\mathcal{B}}\mathrm{D}^{\alpha(t)}x = \lambda(t)u$, where $u = 0.1$, $\lambda_0 = 1.35$, $\lambda_1 = 1.88$, $T_1 = 0.1$, $T_2 = 0.8$

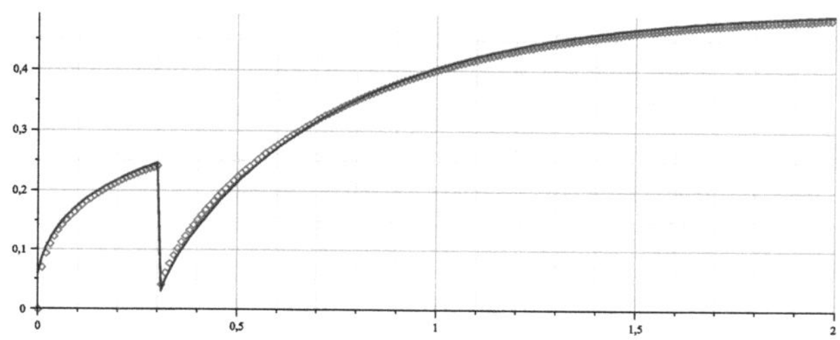

(b) For $c = -1$, i.e., ${}^{\mathcal{B}}\mathrm{D}^{\alpha(t)}x = \lambda(t)(-x + u)$, where $u = 0.5$, $\lambda_0 = 1.35$, $\lambda_1 = 1.95$, $T_1 = 0.3$, $T_2 = 2$

Fig. 4. Plots of experimental results (*diamonds*) and numerical solution (*solid line*) of (22), for $\alpha_0 = 0.5$, $\alpha_1 = 1$, $h = T_s = 0.01$

4.2 Experimental Results of the Fractional Variable Order Inertial System

The identification results for variable order inertial system

$$ {}^{\mathcal{B}}\mathrm{D}^{\alpha(t)}x = \lambda(t)(-x + u), \tag{27}$$

were obtained by numerical minimization of time responses square error with sampling time $T_s = 0.01$ sec., and input signal $u(t) = 0.5H(t)$. The order $\alpha(t)$ and parameter $\lambda(t)$ are changing their values according to (23), where for $\alpha_0 = 0.5$ and $\alpha_1 = 1$ we get $\lambda_0 = 1.35$ and $\lambda_1 = 1.95$, respectively. The experimental results compared to numerical implementation of the variable order inertial system are shown in Fig. 4b.

5 Conclusions

In the paper, an original numerical method for solving and simulating of linear fractional variable order systems, was presented. The proposed method is based on matrix approach and the duality property between two kinds of variable order derivatives. In order to confirm proposed methodology, results of numerical simulations were compared with experimental results obtained from analog model. Presented results fully confirm efficiency and correctness of proposed methods.

References

1. Ramirez, L.E.S., Coimbra, C.F.M.: On the Selection and Meaning of Variable Order Operators for Dynamic Modeling. International Journal of Differential Equations (2010)
2. Sierociuk, D., Malesza, W., Macias, M.: Derivation, interpretation and analog validation of the fractional variable order derivative definition. In: Proceedings of European Control Conference ECC 2013, Zurich, Switzerland, pp. 3464–3469 (2013)
3. Sierociuk, D., Malesza, W., Macias, M.: On a new definition of fractional variable-order derivative. In: Proceedings of the 14th International Carpathian Control Conference, ICCC 2013, Rytro, Poland, pp. 340–345 (2013)
4. Sierociuk, D., Malesza, W., Macias, M.: Switching scheme, equivalence, and analog validation of the alternative fractional variable-order derivative definition. In: Proceedings of the 52nd IEEE Conference on Decision and Control 2013, Florence, Italy, December 10-13 (2013)
5. Sierociuk, D., Malesza, W., Macias, M.: On the recursive fractional variable-order derivative – equivalent switching strategy, duality, and analog modeling. Circuits, Systems, and Signal Processing (accepted to, 2014)
6. Valerio, D., Vinagre, B., Domingues, J., da Costa, J.S.: Variable-order fractional derivatives and their numerical approximations I-real orders. In: Proceedings of the Symposium on Fractional Signals and Systems 2009, Caparica (2009)
7. Valerio, D., da Costa, J.S.: Variable-order fractional derivatives and their numerical approximations II-complex orders. In: Proceedings of the Symposium on Fractional Signals and Systems 2009, Caparica (2009)
8. Podlubny, I.: Matrix Approach to Discrete Fractional Calculus. Fractional Calculus and Applied Analysis 3, 359–386 (2000)
9. Podlubny, I., Chechkin, A., Skovranek, T., Chen, Y., Vinagre Jara, B.M.: Matrix approach to discrete fractional calculus II: Partial fractional differential equations. Journal of Computational Physics 228(8), 3137–3153 (2009)
10. Lorenzo, C., Hartley, T.: Variable order and distributed order fractional operators. Nonlinear Dynamics 29(1-4), 57–98 (2002)
11. Valerio, D., da Costa, J.S.: Variable-order fractional derivatives and their numerical approximations. Signal Processing 91(3, SI), 470–483 (2011)
12. Macias, M., Sierociuk, D.: An alternative recursive fractional variable-order derivative definition and its analog validation. In: Proceedings of International Conference on Fractional Differentiation and its Applications, Catania, Itally (2014)
13. Sierociuk, D., Twardy, M.: Duality of variable fractional order difference operators and its application to identification. Bulletin of the Polish Academy of Sciences: Technical Sciences (to appear, 2014)

Bi-Fractional Filters, Part 1: Left Half-Plane Case

Paweł Piątek, Jerzy Baranowski, Marta Zagórowska,
Waldemar Bauer, and Tomasz Dziwiński

AGH University of Science and Technology,
Department of Automatics and Biomedical Engineering,
Al. Mickiewicza 30, 30-059 Kraków
{ppi,jb,zagor,bauer,tdz}@agh.edu.pl

Abstract. In this paper new non-integer order filter is proposed. Considered filter is given by a transfer function $\frac{c}{s^{2\alpha}+bs^{\alpha}+c}$, with parameters b and c chosen in a way, that locates the eigenvalues of the system in left open complex half plane. Dependence of frequency characteristic of the system on parameters α, b and c is investigated. Also method for realisation in the form of non-integer order differential equations is analysed.

1 Introduction

Bi-fractional filters (BFF) are a class of non-integer filters fully characterised by three parameters:

- base order α
- damping coefficient b
- free coefficient c

and are given by the following transfer function

$$G(s) = \frac{c}{s^{2\alpha}+bs^{\alpha}+c} \tag{1}$$

In this paper filters with bounded impulse responses are considered, that is systems of base order $\alpha \in [1/2, 1)$. Another examples can be found in [3] or in [2].

Equivalent representation of (1) is the realisation in the form of a system of differential equations of order α. This system can take form (see [5])

$$\begin{aligned} {}_0^C\mathrm{D}_t^{\alpha}\mathbf{x}(t) &= \mathbf{A}\mathbf{x}(t) + \mathbf{B}u(t) \\ y(t) &= \mathbf{C}\mathbf{x}(t) \end{aligned} \tag{2}$$

with the following matrices

$$\mathbf{A} = \begin{bmatrix} 0 & 1 \\ -c & -b \end{bmatrix} \qquad \mathbf{B} = \begin{bmatrix} 0 \\ 1 \end{bmatrix} \qquad \mathbf{C} = \begin{bmatrix} c \; 0 \end{bmatrix} \tag{3}$$

© Springer International Publishing Switzerland 2015

K.J. Latawiec et al. (eds.), *Advances in Modeling and Control of Non-integer Order Systems*,
Lecture Notes in Electrical Engineering 320, DOI: 10.1007/978-3-319-09900-2_8

2 Stability

System (1) or (2) is asymptotically stable if and only if eigenvalues of matrix **A** are located in the sector presented in figure 1.

Stability of filter (1) can be determined with the use of the following theorem.

Theorem 1. *System in the form* (2) *or* (1) *for* $\alpha \in (0,1)$ *is asymptotically stable if and only if one of the following conditions holds*

1. $b \geq 0$ *and* $c > 0$
2. $b < 0$ *and*

$$c > \frac{b^2}{4}\left(\frac{\tau^2+1}{\tau^2-1}\right)^2$$

where

$$\tau = \tan\frac{\alpha\pi}{4}$$

Proof. Matrix **A** is in Frobenius form. Because of that its characteristic polynomial takes form

$$W(\lambda) = \lambda^2 + b\lambda + c$$

Using that fact one can verify the conditions.

1. From Hurwitz criterion [6] one can see that, if $b > 0$ and $c > 0$ then roots of the polynomial are in left open complex half plane, which is contained in stability region. If $b = 0$ and $c > 0$ then the roots of polynomial are $\lambda_{1,2} = \pm j\sqrt{c}$ and are located on the imaginary axis, which is also the part of characteristic polynomial.
2. Second condition regards the eigenvalues of **A** that are located in grayed sector of right complex half plane illustrated in the figure 1. Roots are located in that sector if the following conditions occur
   - They are both complex, which occurs iff

$$\Delta = b^2 - 4c < 0$$

   - The angle of roots is greater than $\frac{\alpha\pi}{2}$ so

$$\frac{\sqrt{|\Delta|}}{-b} > \tan\left(\frac{\alpha\pi}{2}\right)$$

   Because of condition for Δ and considering that b is negative then

$$\sqrt{4c-b^2} > -b\tan\left(\frac{\alpha\pi}{2}\right) = -b\frac{2\tan\left(\frac{\alpha\pi}{4}\right)}{1-\tan^2\left(\frac{\alpha\pi}{4}\right)}$$

$$4c - b^2 > b^2\frac{4\tan^2\left(\frac{\alpha\pi}{4}\right)}{\left(1-\tan^2\left(\frac{\alpha\pi}{4}\right)\right)^2}$$

$$c > \frac{b^2}{4}\frac{4\tan^2\left(\frac{\alpha\pi}{4}\right)+\left(1-\tan^2\left(\frac{\alpha\pi}{4}\right)\right)^2}{\left(1-\tan^2\left(\frac{\alpha\pi}{4}\right)\right)^2}$$

   Substitution of $\tau = \tan\frac{\alpha\pi}{4}$ and using the square of difference formula finishes the proof.

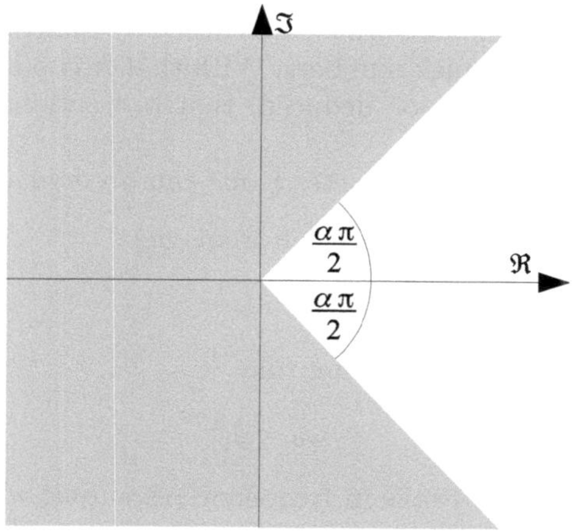

Fig. 1. Stability region on the complex plane

In certain situations, it is more convenient to use the parameters of BFF in the form $b = 2\xi\omega_0$ and $c = \omega_0^2$. Stability of such form is much easier to verify, using the following lemma.

Lemma 1. *Bi-fractional filter in the form*

$$G(s) = \frac{\omega_0^2}{s^{2\alpha} + 2\xi\omega_0 s^\alpha + \omega_0^2} \tag{4}$$

where $\omega_0 > 0$ *and* $\alpha \in (0, 1)$ *is asymptotically stable iff*

$$\xi > \frac{\tau^2 - 1}{\tau^2 + 1}$$

where

$$\tau = \tan\frac{\alpha\pi}{4}$$

Proof. Proof is based on substitution of $b = 2\xi\omega_0$ and $c = \omega_0^2$ in theorem 1 and observing the fact that $\tau^2 - 1$ is always less than zero.

In this paper systems covered by the first case of theorem 1 are considered (or with $\xi > 0$).

3 Frequency Response

Frequency responses of non-integer order filters can be computed with use of principal values of complex number:

$$\mathrm{Pv}(j\omega)^\alpha = |\omega|^\alpha \left(\cos\left(\frac{\pi\alpha}{2}\right) + j\sin\left(\frac{\pi\alpha}{2}\right)\right) \tag{5}$$

this allows avoiding the problems which are caused by multi-value nature of non-integer order powers of complex numbers. With that it is possible to analyse the behaviour of the filter. It can be deduced, that bi-fractional filters (4) have a generally low pass character.

Depending on the value of parameter ξ one can observe analyse two cases.

1. $\xi < 1$ In this case BFF has one cutoff frequency
$$f_0 = \omega^{\frac{1}{\alpha}} \tag{6}$$
which at the same time is a tuning rule
$$\omega_0 = f_0^{\alpha} \tag{7}$$
Such filters can exhibit peaks in frequency responses, which will be further analysed .
2. $\xi > 1$
This case is similar to the behaviour of second order filter with two real poles. In this case amplitude characteristic has two bending points at two frequencies f_1 and f_2. Here only the tuning rule will be presented. For given order α introduce two auxiliary variables
$$q_1 = f_1^{\alpha} \tag{8}$$
$$q_2 = f_2^{\alpha} \tag{9}$$
using which filter parameters are determined i.e.
$$\omega_0 = \sqrt{q_1 q_2} \tag{10}$$
$$\xi = \frac{q_1 + q_2}{2\omega_0} \tag{11}$$
In this case ξ is always greater than 1.

In figures 2 and 3 one can observe the evolution of frequency responses with changing parameters for both cases. In the figure 2 cutoff frequency was set constant at $f_0 = 1$ rad/s and parameters were varying. As one can see frequency responses for the same cutoff frequency can exhibit resonance peaks. Thorough discussion of existence of peaks is presented in the next section.

In the figure 3 tha case of $\xi > 1$ is covered. In that case for chosen frequencies of $f_1 = 0.01$ rad/s and $f_2 = 100$ were kept and the filter order was varying. It should be especially noted that phase characteristics becomes flatter when order decreases.

3.1 Peak in the Amplitude Response

As it can be observed in earlier figures some amplitude responses for $\xi < 1$ exhibit a peak near cutoff frequency. Determination of actual conditions for presence of such peaks is difficult, but authors present a necessary condition and numerical results below.

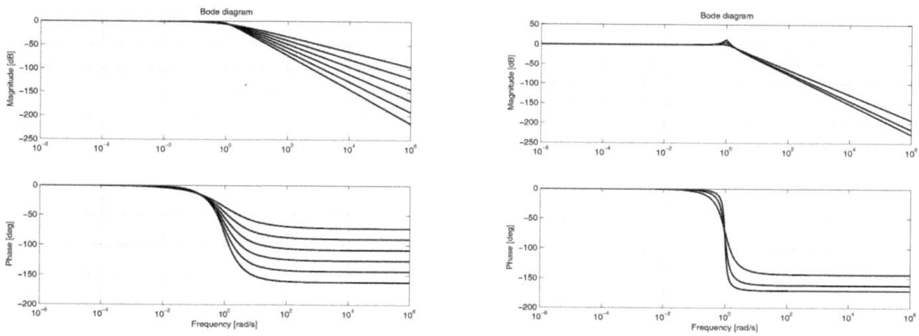

(a) Frequency responses without peaks (b) Frequency responses with peaks

Fig. 2. Frequency characteristics for $\xi < 1$

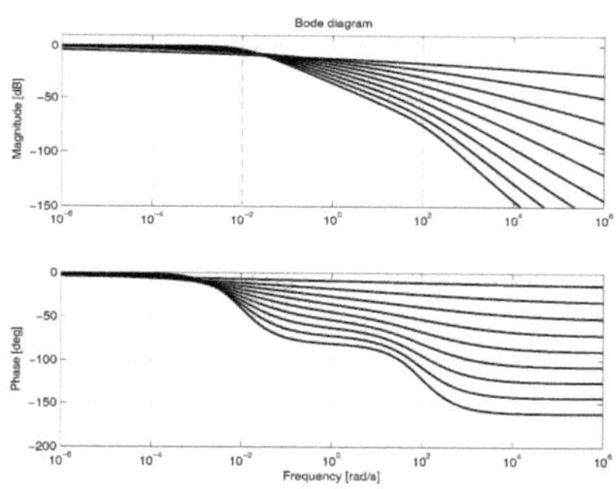

Fig. 3. Bode characteristic of BFF with $f_1 = 0.01$ rad/s, $f_2 = 100$ rad/s and varying α from 0.1 to 0.9

Theorem 2 (Peaks in frequency response). *Necessary condition for frequency response of bi-fractional filter* (4) *with* $\xi > 0$ *to have a maximal value, for certain* $\omega^* > 0$ *is*

$$\xi < \sqrt{\frac{1}{6} - \frac{\cos(\pi\alpha)}{2}} \quad \text{and} \quad \cos(\pi\alpha) < \frac{1}{3} \tag{12}$$

Proof. The modulus of frequency characteristic of BFF takes the following form

$$|G(j\omega)| = \frac{\omega_0^2}{\sqrt{\sum_{i=0}^{4} a_i(\alpha, \xi, \omega_0) q^i}} = \frac{\omega_0^2}{\sqrt{P(q)}}$$

where $q = |\omega|^\alpha$ and $a_i(\alpha, \xi, \omega_0)$ are functions of coefficients. It can be observed that because $|G(j\omega)|$ is decreasing for sufficiently large ω then if it has any extrema at least one of them has to be a maximum. It should be noted, that

$|G(j\omega)|$ has extremum if its derivative becomes zero for any positive ω. Elementary calculations show, that zeros of derivative of $|G(j\omega)|$ coincide with positive zeros of the derivative of $P(q)$ i.e.

$$|G(j\omega)|' = \frac{-P'(q)}{2P(q)}, \quad q = |\omega|^{\alpha}$$

$P'(q) = \sum_{i=0}^{3} p_i q^i$ is a polynomial of third order with coefficients p_i given by:

$$\begin{bmatrix} p_3 \\ p_2 \\ p_1 \\ p_0 \end{bmatrix} = \begin{bmatrix} 2 \\ 6\omega_0 \xi \cos\left(\frac{\pi\alpha}{2}\right) \\ 4{\omega_0}^2 \left(2\xi^2 + \cos\left(\pi\alpha\right)\right) \\ 2{\omega_0}^3 \xi \cos\left(\frac{\pi\alpha}{2}\right) \end{bmatrix}$$

In order to determine existence of positive zeros one can use Hurwitz criterion. From analysis, of the coefficients, two inequalities can show when $P'(q)$ has zeros in right half plane. They are

$$\begin{aligned} p_2 &= 2\xi^2 + \cos\left(\pi\alpha\right) < 0 \\ p_1 p_2 - p_0 p_3 &= 8\omega_0^3 \xi \cos\left(\frac{\pi\alpha}{2}\right)\left(6\xi^2 + 3\cos\left(\pi\alpha\right) - 1\right) < 0 \end{aligned}$$

second inequality, can be reduced to

$$6\xi^2 + 3\cos\left(\pi\alpha\right) - 1 < 0$$

Both these inequalities are solvable with solutions:

$$\xi < \sqrt{-\frac{\cos(\pi\alpha)}{2}} \quad \text{and} \quad \alpha > \frac{1}{2} \tag{13}$$

and

$$\xi < \sqrt{\frac{1}{6} - \frac{\cos(\pi\alpha)}{2}} \quad \text{and} \quad \cos(\pi\alpha) < \frac{1}{3} \tag{14}$$

One can observe these solutions graphically in the figure 4. As it can be seen if the inequality (14) is fulfilled so is inequality (13). It concludes the proof of necessary condition for maximal value.

Remark 1. Sufficient condition for the maximal value is much more difficult to prove, and probably will not lead to any useful formula. It is a known property of polynomials (see [9]) that the free term of polynomial of any order is the product of its roots with sign depending on order i.e.

$$p_0 = (-1)^3 q_1 q_2 q_3$$

On the other hand, one can see that because for all considered systems ξ is positive, ω_0 is positive and $\cos\left(\frac{\pi\alpha}{2}\right)$ is positive, then p_0 is negative.

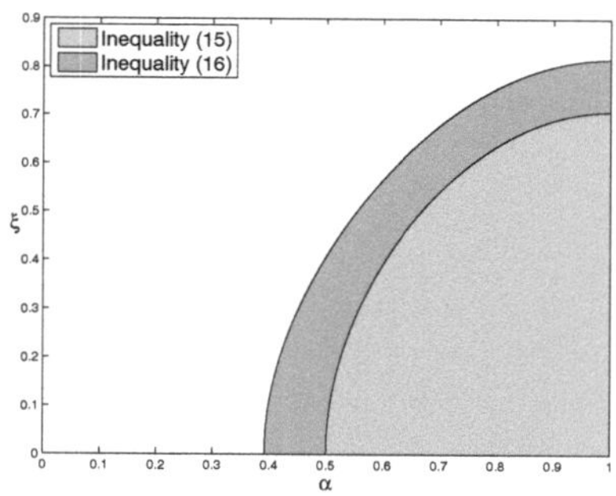

Fig. 4. Inequalities (13) and (14)

Now one needs to consider two cases.

1. $q_1 \in \mathbb{R}$ and $q_2 = q_3^* \in \mathbb{C}$. In that case sign of p_0 is the opposite to sign of the real root. In that case if p_0 is positive then real root is negative and there is no maximum.
2. $q_1,\ q_2,\ q_3 \in \mathbb{R}$. In this case, if p_0 is positive, then there have to be either two positive roots and one negative or all three are negative. If necessary condition is fulfilled then only pair of positive real roots and one negative root are possible and then maximum exists.

In order to obtain a sufficient condition one has to guarantee, that $P'(q)$ has only real roots, which requires analysis of discriminant i.e. if

$$\Delta = 18p_0p_1p_2p_3 - 4p_2^3p_0 + p_2^2p_1^2 - 4p_3p_1^3 - 27p_3^2p_0^2 > 0$$

then roots are real. This formula is not very practical, and solution requires analysis of polynomial inequalities of high order. However, knowing the area of parameters from the necessary conditions authors have obtained a visualisation of area of sufficient condition through use of Montecarlo methods. Such area is presented in the figure 5.

It should be noted that in the figure 2(b) parameters were chosen from a line that approaches the interior of area marked in the figure 5 and in the figure 2(a) they were taken from an orthogonal line.

3.2 Asymptotic Behaviour of Bode Plots

Similar to the integer order case one can create approximate amplitude characteristics of BFF. This is based on the asymptotic behaviour for low and high frequencies. Two cases have to be distinguished.

1. $\xi < 1$
   For filters where $\xi < 1$ these approximations consist of two straight lines in the logarithmic scale. These lines are:

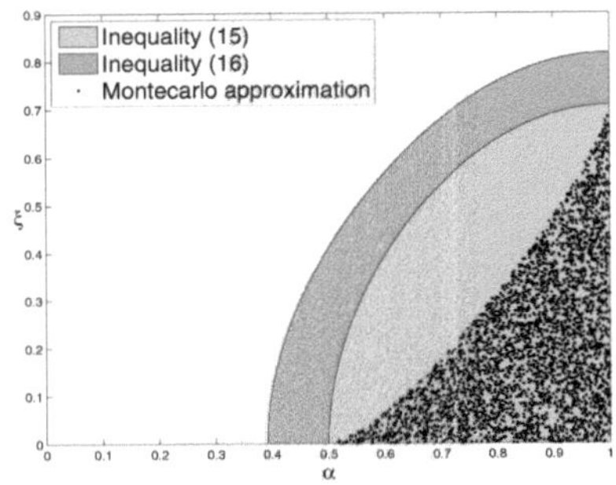

Fig. 5. Montecarlo approximation of sufficient condition area with inequalities (13) and (14)

(a) Flat line (slope 0) from 0 to the cutoff frequency $f_0 = \omega_0^{\frac{1}{\alpha}}$
(b) Line with a slope of 40α dB/decade starting from f_0 at 0.
This approximation does not include the resonance peak, as its exact position and amplitude is very difficult to determine analytically. This approximation is very efficient for a wide area of parameters. Example of approximate Bode plot is presented in the figure

2. $\xi \geq 1$
For filters where $\xi > 1$ these approximations consist of three straight lines in the logarithmic scale. These lines are:
(a) Flat line (slope 0) from 0 to the first cutoff frequency f_1
(b) Line with a slope of 20α dB/decade spanning from f_1 at 0 to f_2 at $-20\alpha \log_{10}(f_2/f_1)$.
(c) Line with a slope of 40α dB/decade starting from f_2 at $-20\alpha \log_{10}(f_2/f_1)$.

Both types of approximation are presented in the figure 6.

4 Impulse Response

Impulse response of BFF can be expressed analytically with use of the (2) form of system. Using the formula given by [5] one can express it as:

$$g(t) = t^{\alpha-1}\mathbf{C}\mathbf{E}_{\alpha,\alpha}(\mathbf{A}t^{\alpha})\mathbf{B} \tag{15}$$

where $\mathbf{E}_{\alpha,\beta}(z)$ is the two parameter Mittag-Leffler function given by

$$\mathbf{E}_{\alpha,\beta}(z) = \sum_{k=0}^{\infty} \frac{z^k}{\Gamma(\alpha k + \beta)} \tag{16}$$

Values of Mittag-Leffler functions are obtainable through available programs, one of the popular version given by Igor Podlubny [8]. In this paper an original

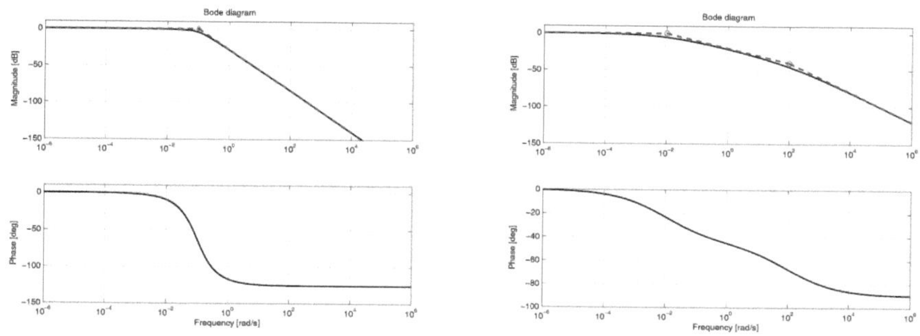

(a) Frequency responses without peaks (b) Approximation for $f_1 = 0.01$ rad/s, $f_2 = 100$ rad/s, $\alpha = 0.5$

Fig. 6. Asymptotic approximations of Bode plots

program is used, based on the definition of the function. The reason for new code is the difference in purposes of both programs. Code by Podlubny is used for computation of function values at one given value of z (even complex) with arbitrary precision. Authors' code needs to compute a time series of values of only real arguments. That is why code using series expansion of order 100 was deemed satisfactory. It allows computation of impulse responses with reasonable precision on short time intervals.

Behaviour of impulse response at $t = 0$ requires special consideration. One can easily observe with Laplace's initial value theorem that for $\alpha < 1/2$ impulse response is unbounded at 0, for $\alpha = 1/2$ it takes a bounded nonzero value and for $\alpha > 1/2$ its value is zero. On the other hand, formula (15) cannot be computed for $t = 0$ for all $\alpha < 1$ because of division by zero. Its value at zero can be however derived separately, and substituted keeping the continuity. In

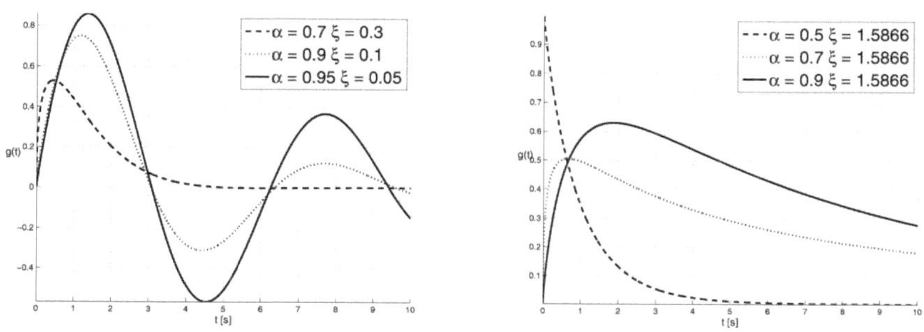

(a) Parameters approaching the interior of resonance peak area (b) $f_1 = 0.1$ rad/s and $f_2 = 1$ rad/s and increasing order of filter

Fig. 7. Impulse responses

this paper only bounded impulse responses are considered. In the figure 7(a) one can observe impulse responses of BFFs, for which parameters are approaching interior of the resonance peak area (see figure 5) on a straight line. As one can see after entering the area impulse response becomes oscillatory. In the figure 7(b) impulse responses of BFFs with $\xi > 1$ are considered. In particular filters designed for cutoff frequencies $f_1 = 0.1$ rad/s and $f_2 = 1$ rad/s are presented for increasing order.

5 Conclusion

This paper illustrates a work in progress on a new type of filters, which can prove to be very beneficial for use in signal processing. There is much potential in non-integer order filters, both for offline and online processing of signals. For online processing filter has to be approximated with integer order one. Among the methods of such approximation one should verify methods of Oustaloup [7], Charef [4] or a method developed by authors [1,3].

Work realised in the scope of project titled "Design and application of non-integer order subsystems in control systems". Project was financed by National Science Centre on the base of decision no. DEC-2013/09/D/ST7/03960.

References

1. Bania, P., Baranowski, J.: Laguerre polynomial approximation of fractional order linear systems. In: Mitkowski, W., Kacprzyk, J., Baranowski, J. (eds.) Theory & Appl. of Non-integer Order Syst. Lecture Notes in Electrical Engineering, vol. 257, pp. 171–182. Springer, Heidelberg (2013)
2. Baranowski, J., Piatek, P., Bauer, W., Dziwinski, T., Zagorowska, M.: Bi-fractional filters, part 2: Right half-plane case. In: 2014 19th International Conference on Methods and Models in Automation and Robotics, MMAR (accepted, 2014)
3. Baranowski, J., Zagorowska, M., Bania, P., Bauer, W., Dziwinski, T., Piatek, P.: Impulse response approximation method for bi-fractional filter. In: 2014 19th International Conference on Methods and Models in Automation and Robotics, MMAR (accepted 2014)
4. Djouambi, A., Charef, A., Besancon, A.V.: Approximation and synthesis of non integer order systems. In: 2nd IFAC Workshop on Fractional Differentiation and its Applications, FDA 2006. Portugal, Porto (July 2006)
5. Kaczorek, T.: Selected Problems of Fractional Systems Theory. LNCIS, vol. 411. Springer, Berlin (2011)
6. Mitkowski, W.: Stabilizacja systemów dynamicznych. Wydawnictwa Naukowo-Techniczne, Warszawa (1991)
7. Monje, C.A., Chen, Y., Vinagre, B.M., Xue, D., Feliu, V.: Fractional-order systems and controls. Fundamentals and applications. Advances in Industrial Control. Springer, London (2010)
8. Podlubny, I.: Mittag-leffler function, `http://www.mathworks.com/matlabcentral/fileexchange/8738-mittag-leffler-function` (acccessed: April 21, 2014)
9. Turowicz, A.: Geometria zer wielomianów. PWN, Warszawa (1967)

A Comparative Analysis of Selected Integer-Order and Noninteger-Order Linear Models of Complex Dynamical Systems

Marek Rydel, Rafał Stanisławski,
Włodzimierz Stanisławski, and Krzysztof J. Latawiec

Department of Electrical, Control and Computer Engineering
Opole University of Technology
ul. Prószkowska 76, 45-758 Opole, Poland
{m.rydel,r.stanislawski,w.stanislawski,k.latawiec}@po.opole.pl

Abstract. This paper presents a comparison of four model order reduction algorithms applied to modeling of evaporating tubes system in the BP–1150 steam boiler. The following model reduction techniques are compared: Frequency Weighted, Rational Krylov, Frequency Weighted with time delay and non-integer order transfer function with and without time delay. Optimal reduction parameters and values of f-zeros and f-poles of non-integer order transfer function are obtained using evolutionary algorithm.

Keywords: model order reduction, fractional order model, time delay, evolutionary algorithm.

1 Introduction

A variety of applications use mathematical models of complex physical and technical processes. A mathematical model of a system can be obtained analytically, if the governing physical laws are well known. Often accurate description of dynamic behavior of complex dynamical systems causes higher complexity of the mathematical models, especially for nonlinear systems (in a linearized form, around a specific operating point) and obtained by using the Finite Element Method. Moreover many real physical systems contain pure time delays. When delay is present in the underlying physical system, this often leads to a very high order for the rational model.

Despite increasing computational speed of computers, simulation, optimization or controller designing for large–scale systems is difficult because of system requirements, long time simulation and numerical errors. For this reason, an ability to properly reduce the model complexity without the loss of its dominant dynamic behavior becomes highly significant [1].

There are several techniques for complex model reduction [1, 4, 11]. Most of contemporarily used methods of linear models reduction do not change the class

© Springer International Publishing Switzerland 2015
K.J. Latawiec et al. (eds.), *Advances in Modeling and Control of Non-integer Order Systems*,
Lecture Notes in Electrical Engineering 320, DOI: 10.1007/978-3-319-09900-2_9

of the model, i.e. reduction of the high-order rational model of order n, Eqn. (1), gives a low-order rational model of order $k < n$, Eqn. (2).

$$\begin{aligned} \dot{x}(t) &= \mathbf{A}\, x(t) + \mathbf{B}\, u(t) \\ y(t) &= \mathbf{C}\, x(t) \end{aligned} \tag{1}$$

$$\begin{aligned} \dot{x_r}(t) &= \mathbf{A}_r\, x_r(t) + \mathbf{B}_r\, u(t) \\ y_r(t) &= \mathbf{C}_r\, x(t) \end{aligned} \tag{2}$$

Among reduction methods, a great attention has been given to the SVD-based methods, which use the balanced model realization theory (especially *Frequency Weighted* methods, which introduce frequency weight functions) and the Krylov-based approximation methods, based on moment matching of the impulse response [1].
Proper selection of weight functions for the *Frequency Weighted* methods as well as a value of the expansion point and the number of moments for every expansion point for the *Rational Krylov* methods enables a significant improvement of model approximation results for a given frequency scope [14, 18]. The optimization of reduction parameters is not a trivial task, as it is a non-convex problem. Therefore, it is necessary to apply the algorithms of global optimization, e.g. evolutionary algorithms [13, 14].

In many cases, a simpler model with a pure time delay describes more accurately the physical reality than a rational model of a very high order. Of course, in continuous time, the introduction of an irrational quantity e^{-sT} into a transfer function may cause substantial analytical problems in design.

A reduced model of very low order, especially First Order Plus Time Delay (FOPTD) or Second Order Plus Time Delay (SOPTD) [12], which contains only a low, integer order transfer function, in many cases gives severely poor description of the system, due to large modeling errors. However, it is well known that the fractional order dynamics can model very high order transfer functions [8]. Logically therefore, if a reduced integer order model cannot properly describe a high order system, a fractional order technique can be used to achieve a better accuracy in the model reduction. During the reduction, the original high order model is approximated by much lower fractional order transfer function, which ensures a required accuracy of approximation.

2 Model Order Reduction Methods

2.1 SVD-Based Methods

The SVD-based model reduction methods were introduced in the Moore's works [10]. The concept of balanced model realization is an easy way to determine the dominating part of the model and reduction by 'cutting' the matrices, describing dynamics of the model in state space (*Balanced Truncation Approximation*).

In this case, the reduction of a high order model is based on the controllability and observability Gramians and the linear state transformation $x \to \mathbf{T}x$ to obtain a balanced realization (3) of the model. As a result, the controllability $\overline{\mathbf{P}}$ and observability $\overline{\mathbf{Q}}$ Gramians are identical diagonal matrices, with decreasing Hankel singular values at the main diagonal. The states with very low singular values have a negligible influence on properties of the model and can be removed. The dominating part of the balanced model $\mathbf{A}_{11}$, $\mathbf{B}_1$, $\mathbf{C}_1$ creates the reduced model (of order $k < n$).

$$\begin{aligned} \dot{x}(t) &= \mathbf{TAT}^{-1}\, x(t) + \mathbf{TB}\, u(t) = \begin{bmatrix} \bar{\mathbf{A}}_{11} & \bar{\mathbf{A}}_{12} \\ \bar{\mathbf{A}}_{21} & \bar{\mathbf{A}}_{22} \end{bmatrix} \begin{bmatrix} x_1 \\ x_2 \end{bmatrix} + \begin{bmatrix} \bar{\mathbf{B}}_1 \\ \bar{\mathbf{B}}_2 \end{bmatrix} u \\ y(t) &= \mathbf{CT}^{-1}\, x(t) = \begin{bmatrix} \bar{\mathbf{C}}_1 & \bar{\mathbf{C}}_2 \end{bmatrix} \begin{bmatrix} x_1 \\ x_2 \end{bmatrix} \end{aligned} \tag{3}$$

There are many algorithms that can be applied to determine the matrix $\mathbf{T}$ [1,11] The *Frequency Weighted* methods, which introduce frequency weight functions, are more general than the BTA method. The weighting functions can be applied on the input or on the output of the model or, simultaneously on the input and output. The first type of such a method has been proposed by Enns. The Enns algorithm guarantees stability of the reduced model for one-sided weighting functions only [1].

A proper selection of weight functions enables significant improvement of the model approximation results for a given frequency scope [14,18]. A disadvantage of the SVD methods is a high degree of calculation complexity. In spite of their unquestionable advantages, these methods are not usually used for models whose number of state variables is higher than 10^4 [1].

2.2 Krylov-Based Methods

For systems with more than 10^4 state variables the Krylov-based approximation methods are proposed [1]. They are characterized by lower calculation complexity. The main disadvantage of these reduction methods is a lack of guarantee of preserving the stability of the reduced model, as well as less accurate approximation of the frequency response as compared to the BTA methods, in particular those implementing frequency weighs [6]. Moment matching is a key ingredient of the Krylov-based methods. The idea is to match moments of the original higher-order model (1) with the moments of a lower-order model (2). This is achieved by iteratively constructing matrices that span the Krylov subspaces.

Moment matching model reduction of a system is an approximation of its transfer function $\mathbf{G}(s)$ by a rational function of a lower degree. This can be done by matching some (k) terms of the Laurent series expansion of $\mathbf{G}(s)$ at various points of the complex plane.

$$\mathbf{G}(s) = \sum_{i=0}^{\infty} \eta_i (s - s_0)^i, \quad \mathbf{G_r}(s) = \sum_{i=0}^{k} \eta_i (s - s_0)^i \tag{4}$$

where: η_i are moment of an impulse response of the system $g(t)$ around the arbitrary point $s = s_0$:

$$\eta_i(s_0) = \int_0^{\infty} t^i g(t) e^{-s_0 t} dt$$

or the ith derivative of the transfer function around the point $s = s_0$:

$$\eta_i(s_0) = (-1)^i \frac{d^i}{ds^i} G(s)\bigg|_{s=s_0} = \mathbf{C}\left(s_0\mathbf{I} - \mathbf{A}\right)^{-(i+1)} \mathbf{B} \quad \text{for} \quad i = 0, 1, 2, \ldots$$

In many cases the direct computation of the moments is numerically problematic because of numerical reasons. Algorithms based on determination of a Krylov subspace are much better [1]. The methods require determining of the orthonormal basis. Two algorithms are most important: the Arnoldi algorithm and the asymmetrical Lanczos algorithm.

The *Rational Krylov* method is a generalized version of the standard Krylov-based approximation methods. Instead of choosing one expansion point, multiple expansion points are chosen [1,5,7]. The choice of a set of points $\{s_1, s_2, \ldots, s_j\}$ makes it possible for the reduced order model to match the frequency response of the original system in a wide range, from the steady-state response to high frequencies.

The reduced system is not guaranteed to be stable and no global error bounds exist. Moreover, the selection of expansion points, which determines the reduced model, is not an automated process and has to be figured out by the user by trial and error. However, this algorithm can be applied to the system of a very high order.

2.3 Approximation of High-Order Model by Low-Order Model with Time Delays

If it is known from other sources that the system described by a high–order transfer function $\mathbf{G}(s)$ is a system with a certain time delay at the input/output, it is possible to approximate $\mathbf{G}(s)$ by a kth ($k < n$) order model with a pure time delay τ [9].

The problem of reduction can be changed and instead of approximation of $\mathbf{G}(s)$, we can reduce a stable rational approximation of $e^{sT}\mathbf{G}(s)$.

$$\left\|\mathbf{G}(s) - e^{-sT}\mathbf{G_r}(s)\right\|_{\infty} = \left\|e^{sT}\mathbf{G}(s) - \mathbf{G_r}(s)\right\|_{\infty} \tag{5}$$

At first we must determine an approximation of $e^{sT}\mathbf{G}(s)$. It could be done by expanding e^{sT}, as a Taylor series or Padé approximations [9]. However, Padé approximations of e^{sT} using all-pass functions will have all unstable poles, just as Padé approximations of e^{-sT} have all stable poles. Therefore, it is necessary to form an approximation of $e^{sT}\mathbf{G}(s)$ using a Padé approximation and then select the strictly proper and stable part. Now the approximation of $e^{sT}\mathbf{G}(s)$ can be reduced with using the SVD-based or Krylov-based methods.

2.4 Approximation of High-Order Systems by Non-integer Low-Order Transfer Function with and without Time Delay

In many conventional process control applications high-order process models are approximated using simple FOPTD and SOPTD structures given by:

$$\mathbf{G_r}(s) = \frac{K}{Ts+1}e^{-\tau s}, \quad \mathbf{G_r}(s) = \frac{K}{s^2+2\zeta\omega s+\omega^2}e^{-\tau s}$$

For complex models these structures give large approximation errors and this proves the inadequacy of model reduction. Hence, to obtain better accuracy of the reduced order models, two structures involving fractional order (FO) elements, were proposed [12]:

$$\mathbf{G_r}(s) = \frac{K}{Ts^\alpha+1}e^{-\tau s}, \quad \mathbf{G_r}(s) = \frac{K}{s^\beta+2\zeta\omega s^\alpha+\omega^2}e^{-\tau s}$$

As a generalization of these forms, a new structure of the non-integer order models is proposed here:

$$\mathbf{G_r}(s) = \frac{K\sum_{i=1}^{Z_1}\left(T_i^N s^\alpha+1\right)\sum_{i=1}^{Z_2}\left(s^{2\alpha}+2\zeta_i^N\omega_i^N s^\alpha+\left(\omega_i^N\right)^2\right)}{\sum_{i=1}^{P_1}\left(T_i^D s^\alpha+1\right)\sum_{i=1}^{P_2}\left(s^{2\alpha}+2\zeta_i^D\omega_i^D s^\alpha+\left(\omega_i^D\right)^2\right)}e^{-\tau s} \tag{6}$$

The numbers of f-zeros and f-poles [15,16], (Z_1, Z_2) and (P_1, P_2) respectively, can be chosen arbitrary, with the f-order of the reduced model (related to f-poles) being $k = P_1+2P_2$. The most suitable structure of the reduced model can be decided for the minimum value of approximation error of the reduced model which can be accepted by the designer with regard to its future use for control or simulation. A suitable set of values of reduced model parameters $\{T_i^N, \zeta_i^N, \omega_i^N\}$ for the numerator, $\{T_i^D, \zeta_i^D, \omega_i^D\}$ for the denominator and $\{K, \alpha, \tau\}$ can be obtained by using evolutionary algorithms or the Nelder–Mead Simplex algorithm implemented in the Matlab's Optimization Toolbox.

3 Evolutionary Algorithm for Determination of Method Reduction Parameters

Based on presented the characteristics, it can be concluded that the optimization of the reduction parameters is not a trivial task. Determination of optimal parameters of the reduction is a non-convex problem and the objective function (norm of an approximation error of the reduced model) contains many local minima. It is therefore necessary to apply algorithms of global optimization. To solve this problem, the evolutionary algorithm based on evolution strategies has been adopted [2,3,13]. Evolutionary algorithms carry out the adaptation process in order to find a better solution than the one produced so far. However, it is not possible to guarantee that the algorithm will find the best possible solution. By increasing the number of the iterations, we increase only a probability of receiving a global solution.

Evolutionary algorithms have arosen as a result of inspiration from genetics and evolution, which supplied a special terminology that links the languages of biology and computer science. Each of the solutions is called individual (or phenotype). The algorithm processes the population of individuals. The individual is equipped with information constituting its genotype (a set of features). Every feature is called a chromosome.

For a reduction method, the genotype consists of chromosomes encoding independent variables:

- *Frequency Weighted* methods [13]:
  $\{n_f, \omega_f\}$ – orders and cutoff frequencies of low-pass filters, which creates frequency weighted functions applied on the input or/and the output of the primary model
- *Rational Krylov* methods [14]:
  $\{s_1^V, s_2^V, \ldots, s_J^V\}$, $\{s_1^W, s_2^W, \ldots, s_J^W\}$ – the value of expansion points and $\{I_1^V, I_2^V, \ldots, I_J^V\}$, $\{I_1^W, I_2^W, \ldots, I_J^W\}$ – the number of moments for every expansion point, which are used to create input ($\mathbf{V}$) and output ($\mathbf{W}$) Krylov subspaces.
- Approximation of high-order systems by integer and non-integer low-order transfer function with or without time delays:
  $\{K, T_i^N, \zeta_i^N, \omega_i^N\}$ – for numerator parameters of reduced model,
  $\{T_i^D, \zeta_i^D, \omega_i^D\}$ – for denominator parameters of reduced model,
  α – only for non-integer reduced models,
  τ – only for reduced models with time delays.

4 Analysis of Model Reduction Results

All model reduction results are shown on the example of the evaporating tubes of the BP–1150 steam boiler. The model of subsystems of evaporating tubes is the distributed parameter systems along with the length and circumference of the tubes and the connecting blades creating the furnace walls. The model of the evaporating tubes consists four subsystems which are elaborated as models with lumped parameters, using the following numbers of finite elements:

- One-phase zone - 150 finite elements,
- Two-phase zone I – 50 finite elements,
- Two-phase zone II – 75 finite elements,
- Two-phase zone III – 200 finite elements.

Each finite element is described by 15 state variables. The mathematical models of the boiler subsystems are described in detail in Refs. [17, 18]. The evaporating tube model has a limited adequacy scope due to the following simplifying assumptions:

– steam-water mixture is in a state of the thermodynamic balance,
– the simplification of heat energy transport along the pipe wall model,
– applied density of space variable digitizing.

For this reasons, we can assume, that adequacy scope of the evaporating tube models is around 1 [rad/s]. Within this frequency scope, there are dynamic processes related to the accumulation of heat in the pipe walls and in the working medium, heat exchange between the pipe walls and the working medium, as well as transport of working medium along the evaporating tubes, and changes in working medium enthalpy [17,18].

MIMO models are often characterized by significant differences of transfer functions for individual input/output pairs. Moreover, the magnitudes of frequency responses for individual input/output pairs may differ significantly with changes in frequency. Therefore, applying the Hankel norm exclusively does not always allow for the appropriate assessment of the reduced model [18]. The most advantageous choice for the model of the BP–1150 steam boiler is the average square relative approximation error of frequency responses, for the frequency scope of the model adequacy:

$$\Delta = \sqrt{\frac{1}{N}\sum_{j=1}^{N}\left(\frac{|G(i\omega_j) - G_r(i\omega_j)|}{|G(i\omega_j)|}\right)^2} \tag{7}$$

where N - number of approximation points in the frequency domain.

The modeling performance criterion Δ is valid for a SISO system, that is a selected input-output channel of the MIMO plant. Here, we will not discuss the MIMO extension issue for Δ. Rather, we will illustrate the usefulness of Δ in case of modeling of a selected SISO channel of the considered MIMO plant, in particular the channel $h_{in} \rightarrow M_{out}$, that is input enthalpy of the working medium $\rightarrow$ mass flow of the working medium on the outlet of the evaporating tube.

The resulting reduced models of orders $k = 3$ and $k = 5$ are shown in Fig. 1. We note that the obtained results are close to each other and the models quite well approximate the primary model in the adequacy scope, in particular the first peaks are well approximated. The best approximation at low frequencies is obtained for models derived with the FW and RK methods. This is because of the fact that the values of interpolation points for RK are very low and the weight function is in the shape of low-pass filters for the FW methods.

The primary model of the evaporating tubes contains a time delay of some $15s$ distributed along the length of the tube. As a result of modeling, this time delay has initially been covered by the high-order model. However, during the reduction it is possible to replace this distributed time delay by a pure time delay.

As we can see, better reduction results can be obtained with use of the methods which introduce pure time delay into a reduced model (FWPTD–*Frequency Weighted plus time delay*, IOPD–*integer order plus time delay*, NIOPTD–*non-integer order plus time delay*). The best overall results are obtained using the fractional order technique. The optimal parameters used to obtain reduced order models are presented in Table 1.

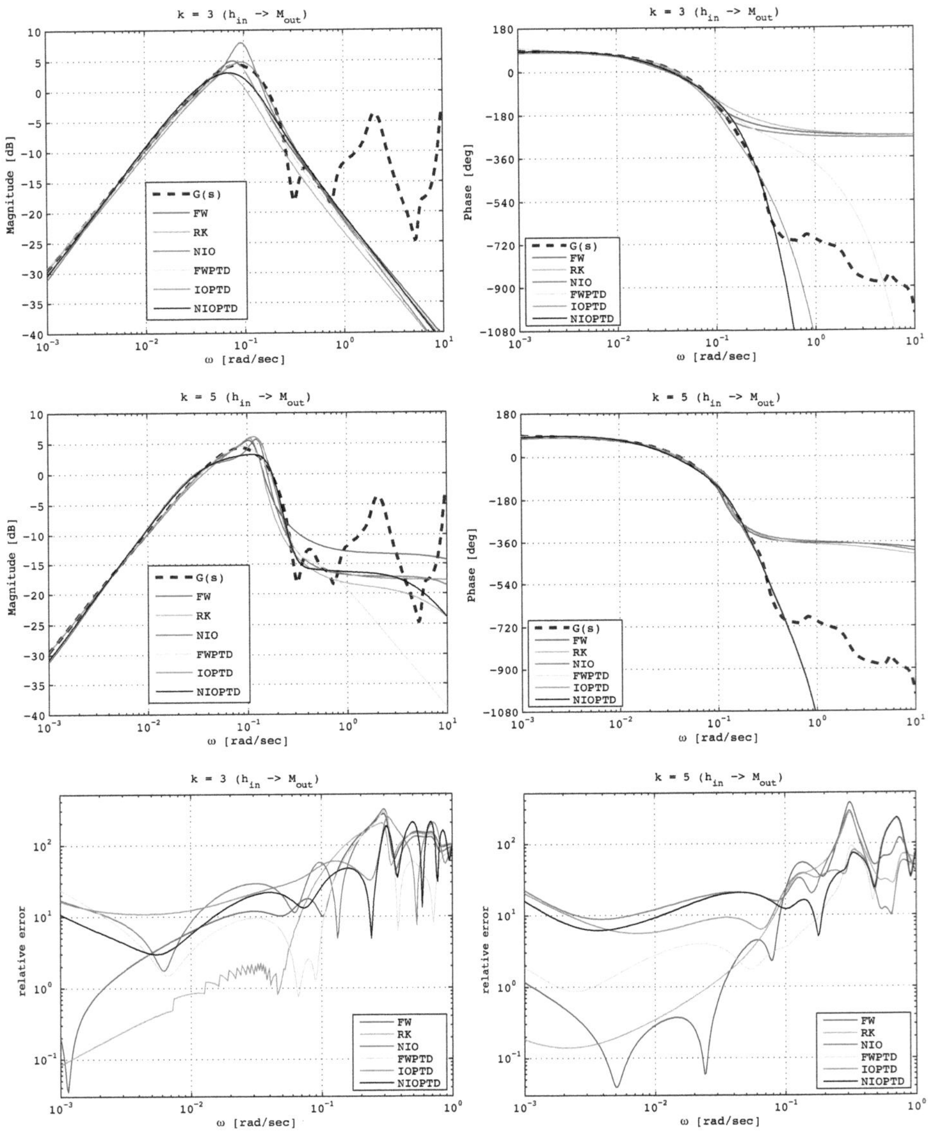

Fig. 1. Frequency responses and relative approximation errors for reduced evaporating tube models ($k = 3$ and $k = 5$) for $h_{in} \rightarrow M_{out}$

Table 1. Model reduction parameters and reduction results of order $k = 3$ and $k = 5$

$k = 3$	Δ	parameters
FW	0.9255	$n_{fin} = 24$, $\omega_{fin} = 0.05$, $n_{fout} = 16$, $\omega_{fout} = 0.185$
RK	1.1471	$\{s_1^V\} = \{0\}$, $\{I_1^V\} = \{3\}$, $\{s_1^W\} = \{0\}$, $\{I_1^W\} = \{3\}$
NIO	0.8788	$\alpha = 1.0925$, $K = 2.959 \times 10^{-5}$, $T_1^N = 10779$, $T_2^N = -16.34$, $T_1^D = 63.29$, $\zeta_1^D = 0.4367$, $\omega_1^D = 0.07785$
FWPTD	0,7945	$n_{fin} = 18$, $\omega_{fin} = 0.128$, $n_{fout} = 16$, $\omega_{fout} = 0.097$, $\tau = 2.23$s
IOPTD	0.7349	$K = 5.528 \times 10^{-5}$, $T_1^N = 6931$, $T_2^N = -3.904$, $T_1^D = 17.73$, $\zeta_1^D = 0.5038$, $\omega_1^D = 0.1125$, $\tau = 15.63$s
NIOPTD	0.7125	$\alpha = 1.0662$, $K = 2.936 \times 10^{-5}$, $T_1^N = 11468$, $T_2^N = -7.971$, $T_1^D = 28.97$, $\zeta_1^D = 0.6635$, $\omega_1^D = -0.08425$, $\tau = 33.95$s
$k = 5$	Δ	parameters
FW	0.7245	$n_{fin} = 25$, $\omega_{fin} = 0.041$, $n_{fout} = 8$, $\omega_{fout} = 4.52$
RK	0.6436	$\{s_1^V, s_2^V\} = \{0, 0.103\}$, $\{I_1^V, I_2^V\} = \{2, 3\}$, $\{s_1^W, s_2^W\} = \{0.104, 1.211\}$, $\{I_1^W, I_2^W\} = \{4, 1\}$
NIO	0.5982	$\alpha = 1.1102$, $K = 3.886 \times 10^{-3}$, $T_1^N = 6443$, $T_2^N = 4.188$, $\zeta_1^N = -0.6710$, $\omega_1^N = 0.1440$, $T_1^D = 46.53$, $\zeta_1^D = 0.3823$, $\omega_1^D = 0.1043$, $\zeta_2^D = 9.281$, $\omega_2^D = 0.9036$
FWPTD	0,5166	$n_{fin} = 4$, $\omega_{fin} = 0.188$, $\tau = 12.12$s
IOPTD	0.5288	$K = 2.553 \times 10^{-2}$, $\tau = 13.19$s, $T_1^N = 4912$, $T_2^N = 1.461$, $\zeta_1^N = -0.3656$, $\omega_1^N = 0.3016$, $T_1^D = 19.72$, $\zeta_1^D = 0.3308$, $\omega_1^D = 0.1366$, $\zeta_2^D = 8.311$, $\omega_2^D = 4.309$
NIOPTD	0.5155	$\alpha = 1.0898$, $K = 1.593 \times 10^{-3}$, $\tau = 13.04$s, $T_1^N = 7972$, $T_2^N = 2.609$,, $\zeta_1^N = -0.2337$, $\omega_1^N = 0.2684$, $T_1^D = 40.45$, $\zeta_1^D = 0.5230$, $\omega_1^D = 0.1485$, $\zeta_2^D = 3.249$, $\omega_2^D = 0.8789$

5 Conclusions

The task of accurate modeling of complex dynamical systems may involve prohibitive computational burdens. For this reason, an ability to properly reduce model complexity becomes highly significant.

Good reduction results for LTI models of control plants with a limited adequacy scope can be achieved by proper selection of parameters of model reduction (e.g. *FW*, *RK* methods) as well as by selection of model parameters (e.g. zeros, poles). Estimation of those parameters is not a trivial task, as it is a non–convex optimization problem. Therefore, it is necessary to apply the algorithms for global optimization, e.g. evolutionary algorithms.

The main result of our comparative frequency domain analysis of low-order modeling of an exemplary, complex dynamical plant is that non-integer (fractional) order models outperform some other popular model reduction techniques, including the Frequency Weighted and Rational Krylov methods.

References

1. Antoulas, A.: Approximation of Large-Scale Dynamical System. Society for Industrial and Applied Mathematics, Philadelphia (2005)
2. Arabas, J.: Lectures on evolutionary algorithms. WNT, Warsaw (2001) (in Polish)
3. Coello, C.A., Lamont, G.B., Van Veldhuizen, D.A.: Evolutionary algorithms for solving multi-objective problems. Springer, New York (2007)
4. Fortuna, I., Nunnari, G., Gallo, A.: Model order reduction techniques with Applications in electrical engineering. Springer, London (1992)
5. Grimme, E.: Krylov projection methods for model reduction. PhD thesis, University of Illinois at Urbana (1997)
6. Gugercin, S., Antoulas, A.: A comparative study of 7 algorithms for model reduction. In: 39th IEEE Conference on Decision and Control, vol. 3, pp. 2367–2372 (2000)
7. Heres, P.J.: Robust and efficient Krylov subspace methods for Model Order Reduction. PhD thesis, Eindhoven University of Technology (2005)
8. Krishna, B.T.: Studies on fractional order differentiators and integrators: A survey. Signal Processing 91(3), 386–426 (2011)
9. Liu, Y., Anderson, B.: Model reduction with time delay. IEE Proceedings of Control Theory and Applications 134(6), 349–367 (1987)
10. Moore, B.: Principal component analysis in linear systems: Controllability, observability and model reduction. IEEE Transactions on Automatic Control AC–26(1), 17–32 (1981)
11. Obinata, G., Anderson, B.: Model Reduction for Control System Design. Springer, London (2001)
12. Pan, I., Das, S.: Intelligent fractional order systems and control: An Introduction. Springer, Berlin (2013)
13. Rydel, M., Stanisławski, W.: Optimization of reduction parameters of svd based reduction methods for complex plant mimo lti models. Methods of Applied Computer Science 24(3), 197–220 (2010) (in Polish)
14. Rydel, M., Stanisławski, W.: Selection of reduction parameters of rational krylov methods for complex mimo lti models using the evolutionary algorithm. In: 19th International Conference on Methods and Models in Automation and Robotics (2014)
15. Stanisławski, R.: Advances in Modeling of Fractional Difference Systems - New Accuracy, Stability and Computational Results. Opole University of Technology Press, Opole (2013)
16. Stanisławski, R., Latawiec, K.J.: Stability analysis for discrete-time fractional-order LTI state-space systems. Part II: New stability criterion for FD-based systems. Bulletin of the Polish Academy of Sciences, Technical Sciences 61(2), 362–370 (2013)
17. Stanisławski, W.: Modelling and Computer simulation of flow Power Boilers. Opole University of Technology Press, Opole (2001) (in Polish)
18. Stanisławski, W., Rydel, M.: Hierarchical models of complex plants on basis of power boiler example. Archives of Control Sciences 20(LVI)(4), 381–416 (2010)

Modeling and Identification of Fractional-Order Discrete-Time Laguerre-Based Feedback-Nonlinear Systems

Rafał Stanisławski, Krzysztof J. Latawiec,
Marcin Gałek, and Marian Łukaniszyn

Department of Electrical,
Control and Computer Engineering, Opole University of Technology,
ul. Sosnkowskiego 31, 45-272 Opole, Poland
{r.stanislawski,k.latawiec,m.lukaniszyn}@po.opole.pl

Abstract. This paper presents a new implementable strategy for modeling and identification of a fractional-order discrete-time block-oriented feedback-nonlinear system. Two different concepts of orthonormal basis functions (OBF) are used to model a linear dynamic part, namely "regular" OBF and inverse IOBF. It is shown that the IOBF concept enables to separate linear and nonlinear submodels, which leads to a linear regression formulation of the parameter estimation problem, with the detrimental bilinearity effect totally eliminated. Finally, Laguerre filters are uniquely embedded in modeling of the fractional-order dynamics. Unlike for regular OBF, simulation experiments show a very good identification performance for an IOBF-structured, fractional-order Laguerre-based feedback-nonlinear model, both in terms of low prediction errors and accurate reconstruction of the actual system characteristics.

1 Introduction

Nonlinear block-oriented systems, including the Hammerstein, Wiener and feedback-nonlinear ones, have attracted considerable research interest both from the industrial and academic environments [1,2,3,4,5]. On the other hand, it is well known that orthonormal basis functions (OBF) have proved to be useful in identification and control of dynamical systems, including nonlinear block-oriented systems [6,7,8,9]. In particular, an inverse OBF (IOBF) modeling approach has been effective in identification of a linear dynamic part of the Hammerstein system [6]. The approach provides the so-called separability in estimation of linear and nonlinear submodels [7], thus eliminating the bilinearity issue detrimentally affecting e.g. the ARX-based modeling schemes. The IOBF modeling approach is continued to be efficiently used here to model a linear fractional-order dynamic part of the feedback-nonlinear system.

Recently, fractional-order dynamics have been given a huge research interest, mostly for linear systems [10,11,12,13,14,15,16,17,18].

Discrete-time fractional-order OBF-based modeling is a new research area and there is a few papers on the topic that has up to date been available

© Springer International Publishing Switzerland 2015
K.J. Latawiec et al. (eds.), *Advances in Modeling and Control of Non-integer Order Systems*,
Lecture Notes in Electrical Engineering 320, DOI: 10.1007/978-3-319-09900-2_10

[19,20,21,22,23]. Those papers illustrate that fractional-order discrete Laguerre filters can be very effective in modeling of dynamical systems.

This paper presents a new strategy for feedback-nonlinear system identification, which is a combination of the inverse-OBF modeling concept and fractional-order generalization of discrete Laguerre filters. The effective combination gives rise to the introduction of a powerful method for identification of the fractional-order feedback-nonlinear system.

2 Fractional-Order Discrete-Time Difference

A simple generalization of the familiar Grünwald-Letnikov difference [12] is the fractional difference (FD) in discrete time t, described by equation [10,14,15,24]

$$\Delta^{\alpha}x(t) = \sum_{j=0}^{t} P_j(\alpha)x(t)q^{-j} = x(t) + \sum_{j=1}^{t} P_j(\alpha)x(t)q^{-j} \quad t = 0,1,\dots \tag{1}$$

where $\alpha \in (0,2)$ is the fractional order, q^{-1} is the backward shift operator and

$$P_j(\alpha) = (-1)^j \gamma_j(\alpha) \tag{2}$$

with

$$\gamma_j(\alpha) = \binom{\alpha}{j} = \begin{cases} 1 & j = 0 \\ \frac{\alpha(\alpha-1)\dots(\alpha-j+1)}{j!} & j > 0 \end{cases} \tag{3}$$

Note that each element in Eqn. (1) from time t back to 0 is nonzero so that each incoming sample of the signal $x(t)$ increases the complication of the model equation. In the limit, with $t \to +\infty$, we end up with computational explosion. Therefore in [25], truncated or finite fractional difference (FFD) has been considered for practical, feasibility reasons. Finite fractional difference (FFD) is defined as

$$\Delta^{\alpha}x(t,J) = x(t) + \sum_{j=1}^{J} P_j(\alpha)x(t)q^{-j} \tag{4}$$

where $J = \min(t, \overline{J})$ and $\overline{J}$ is the upper bound for j when $t > \overline{J}$.

In this paper, we assume that α is known.

Remark 1. Possible accounting for the sampling period T when transferring from the Grünwald-Letnikov continuous-time derivative to the Grünwald-Letnikov discrete-time difference results in dividing the right-hand side of Eqns. (1) and (4) by T^{α}. Operating without T^{α} as in the sequel corresponds to putting $T = 1$ or to the substitution of $P_j(\alpha)$ for $\frac{P_j(\alpha)}{T^{\alpha}}$, $j = 0, \dots, t$.

3 Fractional-Order Discrete-Time Laguerre Filters

A classical (or integer-order, or "regular") OBF model of a dynamical system, or shortly, OBF system, can be presented in form

$$y(t) = \sum_{i=1}^{K} C_i L_i(q) u(t) + e(t) \tag{5}$$

where $u(t)$, $y(t)$ and $e(t)$ are the system input, output and disturbance, respectively, $L_i(z)$ and C_i, $i = 1, ..., K$, are orthonormal transfer functions and weighting parameters, respectively and $e(t)$ is the output error. In case of use of discrete Laguerre filters we have

$$L_i(z) = \frac{k}{z-P}\Big(\frac{-Pz+1}{z-P}\Big)^{i-1} \qquad i = 1, ..., K \tag{6}$$

where $k = \sqrt{1-P^2}$ and P is a dominant pole. In the sequel, we limit our interest to the practically justified case of $P > 0$. The unknown parameters C_i, $i = 1, \ldots, K$, can be easily estimated using e.g. Recursive Least Squares (RLS) or Least Mean Squares (LMS) algorithms formalized in a linear regression fashion [26]. In our examples, RLS estimation is used. Pursuing an optimal Laguerre pole P_{opt} has been well established [8,9,27,28,29].

The Laguerre filters presented in Eqn. (6), can be factorized to the form [25,21,22]

$$L_i(q^{-1}) = G_L(q^{-1})\big(G_R(q^{-1}) - P\big)^{i-1} \quad i = 1, ..., K \tag{7}$$

with

$$G_L(q^{-1}) = \frac{kq^{-1}}{1 - Pq^{-1}} \tag{8}$$

$$G_R(q^{-1}) = \frac{k^2 q^{-1}}{1 - Pq^{-1}} = kG_L(q^{-1}) \tag{9}$$

and the consecutive filter outputs being $y_L(t) = G_L(q^{-1})u(t)$ and $y_R^i(t) = G_R(q^{-1})U_i(t)$, $i = 1, ..., K-1$, with

$$U_i(t) = \begin{cases} y_L(t) & i = 1 \\ y_R^{i-1}(t) - PU_{i-1}(t) & i = 2, ..., K \end{cases} \tag{10}$$

The two filters can also be described as

$$\begin{aligned} &G_L^f : \\ &\Delta y_L(t) = (P-1) y_L(t) q^{-1} + k u(t) q^{-1} \end{aligned} \tag{11}$$

$$\begin{aligned} &G_R^f : \\ &\Delta y_R^i(t) = (P-1) y_R^i(t) q^{-1} + k^2 U_i(t) q^{-1} \end{aligned} \tag{12}$$

where $\Delta y_L(t) = y_L(t) - y_L(t-1)$ and similar is $\Delta y_R^i(t)$, $i = 1, ..., K$.

The outstanding value of the factorization (7) of the expression (6) is that $G_L(q^{-1})$ and $G_R(q^{-1})$ are the first-order filters that can be easily adopted to the fractional-order form. The fraction-formalized filters $G_L^f(q^{-1})$ and $G_R^f(q^{-1})$ can now be described as

$$\begin{aligned}&G_L^f:\\&\Delta^{\alpha} y_L(t) = (P-1)y_L(t)q^{-1} + ku(t)q^{-1}\end{aligned} \tag{13}$$

$$\begin{aligned}&G_R^f:\\&\Delta^{\alpha} y_R^i(t) = (P-1)y_R^i(t)q^{-1} + k^2U_i(t)q^{-1}\end{aligned} \tag{14}$$

where $U_i(t)$ is as in Eqn. (10). Finally, the outputs from the FD versions of the $G_L^f(q^{-1})$ and $G_R^f(q^{-1})$ filters can be obtained as

$$\begin{aligned}&G_L^f:\\&y_L(t) = (P-1)y_L(t)q^{-1} + ku(t)q^{-1} - \textstyle\sum_{j=1}^{t} P_j(\alpha)y_L(t)q^{-j}\end{aligned} \tag{15}$$

$$\begin{aligned}&G_R^f:\\&y_R^i(t) = (P-1)y_R^i(t)q^{-1} + k^2U_i(t)q^{-1} - \textstyle\sum_{j=1}^{t} P_j(\alpha)y_R^i(t)q^{-j}\end{aligned} \tag{16}$$

The outputs for FFD versions of the $G_L^f(q^{-1})$ and $G_R^f(q^{-1})$ filters can be calculated as

$$\begin{aligned}&G_L^f:\\&y_L(t) = (P-1)y_L(t)q^{-1} + ku(t)q^{-1} - \textstyle\sum_{j=1}^{J} P_j(\alpha)y_L(t)q^{-j}\end{aligned} \tag{17}$$

$$\begin{aligned}&G_R^f:\\&y_R^i(t) = (P-1)y_R^i(t)q^{-1} + k^2U_i(t)q^{-1} - \textstyle\sum_{j=1}^{J} P_j(\alpha)y_R^i(t)q^{-j}\end{aligned} \tag{18}$$

Remark 2. Possible accounting for the sampling period T when transferring from the Grünwald-Letnikov continuous-time derivative to the Grünwald-Letnikov discrete-time difference results in multiplication of the two first components at the right-hand sides of Eqns. (17) and (16) by T^{α}.

Finally, the output from the fractional-order Laguerre model is computed as

$$\hat{y}(t) = \sum_{i=1}^{K} C_i U_i(t) \tag{19}$$

where $U_i(t)$ calculated in Eqn. (10), with $y_L(t) = G_L^f(q^{-1})u(t)$ and $y_R^i(t) = G_R^f(q^{-1})U_i(t)$, $i = 1, ..., K-1$, respectively (see Eqns. (15), (16), (17) and (18)).

4 System Description

4.1 Non-fractional Case [7,30,31]

In the block-oriented feedback-nonlinear system (Fig. 1), the output of the linear dynamic part is fed (negatively) back to the input through the static nonlinearity, so that the whole system can be described by the equation

$$\begin{aligned} y(t) &= G(q)\left[u(t) - f(y(t)) + e_F(t)\right] \\ &= G(q)\left[u(t) - x(t) + e_F(t)\right] \end{aligned} \tag{20}$$

where $G(q)$ models a linear dynamic part, $f(.)$ describes a nonlinear static function and $e_F(t)$ is the error/disturbance term.

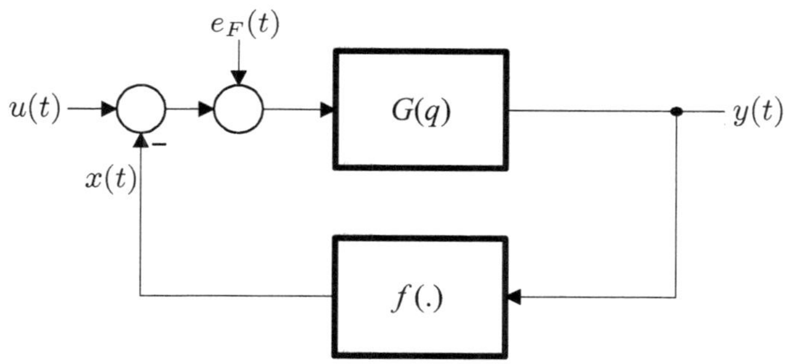

Fig. 1. Feedback-nonlinear system

Two different concepts for modeling of a linear dynamic part by means of OBF are examined here. The first one is the "regular" OBF modeling approach and the second one is the IOBF concept in which the inverse of a dynamic element is OBF-modeled.

Regular OBF Modeling of Feedback-Nonlinear System. It is presented in Section 3 that an open-loop stable linear discrete-time system described by the transfer function $G(q)$ can be represented with an arbitrary accuracy by the model $\hat{G}(q) = \sum_{i=1}^{M} C_i L_i(q)$, including a series of orthonormal transfer functions $L_i(q)$ and the weighting parameters C_i, $i = 1, ..., M$, characterizing the model dynamics. Thus, the model of the system (20) can be written as [7,30,31]

$$\hat{y}(t) = \sum_{i=1}^{M} C_i L_i(q)\left(u(t) - f[y(t)]\right) \tag{21}$$

Various OBF can be used in (21). Two commonly used sets of OBF are simple Laguerre and Kautz functions. These functions are characterized by the 'dominant' dynamics of a system, which is given by a single real pole (p) or a pair of complex ones (p, $p*$), respectively.

The nonlinear part of the feedback-nonlinear system $f(.)$ can be approximated e.g. with the polynomial expansion

$$f\big(y(t)\big) = a_1 y(t) + a_2 y^2(t) + ... + a_m y^m(t) \tag{22}$$

with a_i, $i = 1, ..., m$, being the unknown model parameters.

Combining equations (21) and (22) we arrive at the equation describing the model output of the whole feedback-nonlinear system

$$\hat{y}(t) = \sum_{i=1}^{M} C_i L_i(q) u(t) - \sum_{i=1}^{M} C_i L_i(q) \sum_{j=1}^{m} a_j y^j(t) \tag{23}$$

Inverse OBF Modeling for Feedback-Nonlinear System. In case of use of the IOBF concept to model a linear dynamic part, the feedback-nonlinear model equation can be presented in form

$$R(q)\hat{y}(t) = u(t) \; - x(t) \tag{24}$$

where the FIR model $R(q) = r_0 q^d + r_1 q^{d-1} + ... + r_d + r_{d+1} q^{-1} ... + r_{L-1} q^{-L+d+1}$ is the (approximate) inverse of the system model $\hat{G}(q)$, with d being the time delay. In the IOBF concept, the inverse $R(q)$ of the system is modeled using OBF. An OBF modeling approach can now be applied to equation (4) and finally we can present equation (20) in the following form [30]

$$y\,(t) + \sum_{i=1}^{M} C_i L_i(q)\, y\,(t) = \beta_0[u(t-d) - x(t-d)] + e(t) \tag{25}$$

where $\beta_0 = r_0^{-1}$ and $e(t) = \beta_0 e_F(t)$ is the equation error.

Now, accounting for the polynomial expansion of the nonlinear element (Eqn. (22)) we arrive at the equation describing the whole IOBF-related feedback-nonlinear system

$$y\,(t) \; + \; \sum_{i=1}^{M} C_i L_i(q)\, y\,(t) \;\; = \;\; \beta_0 \left[u(t-d) - \sum_{j=1}^{m} a_j y^j(t-d) \right] \; + \; e(t) \tag{26}$$

Putting $\underline{a}_j = \beta_0 a_j$, $j = 1, ..., m$, the output from the feedback-nonlinear system can be finally given as

$$y\,(t) \; = \beta_0 u(t-d) - \sum_{i=1}^{M} C_i L_i(q)\, y\,(t) - \sum_{j=1}^{m} \underline{a}_j y^j(t-d) + e(t) \tag{27}$$

Remark 3. It is essential that the Laguerre filters are, in the IOBF framework, driven by $y(t)$. This means that in order to calculate the fractional-order output equation in the IOBF fashion we have substituted $y(t)$ for $u(t)$ in Eqns. (25), (26) and (27).

Discriminating between the Gains. The challenge of modeling and parameter estimation of the feedback-nonlinear system is that without certain, sometimes unrealistic assumptions it is not possible to accurately reconstruct properties of the two internal blocks from input-output measurements only [30,31]. For example, an assumption on knowledge of a gain of the linear (or nonlinear) subsystem enables to estimate parameters of the models of the two blocks. The problem is that such a knowledge is rather seldom available. The whole issue arises from the fact that estimators based on input-output measurements only cannot discriminate between the gains of linear and nonlinear contributors to the system. Here we present a method for the recovery of a gain of the linear subsystem in case of a specific form of a nonlinear static characteristic. The method constitutes a preliminary stage for our new identification strategy for feedback-nonlinear systems. We firstly assume that a nonlinear static characteristic can be linearized around u=0. We also assume that an identification experiment with a low-range input signal is feasible, in which case only the linear mode of a system is excited, so that all nonlinear components can be neglected. Now, equation (26) can be written as

$$y\,(t) + \sum_{i=1}^{M} C_i L_i(q)\, y\,(t) = \beta_0 \big[u(t-d) - a y(t-d)\big] + e(t) \tag{28}$$

where $a = a_1$ is the gain of the nonlinear submodel around $u = 0$. Rewrite the last equation as

$$\hat{y}(t) + \underline{a}_1 y(t-d) = \underline{\varphi}^T(t)\underline{\Theta} \tag{29}$$

where $\underline{a}_1 = \beta_0 a$, $\underline{\Theta}^T = [C_1 \cdots C_M\, \beta_0]$, $\underline{\varphi}^T(t) = [-v_1(t) \cdots - v_m(t)\; u(t-d)]$ and $v_i(t) = L_i(q)y(t)$. We can now estimate the parameter vector $\underline{\Theta}$ provided that $\underline{a}_1$ is "guessed" or fixed, or just tuned (compare [30,31]). (Once again, it is not possible to simultaneously estimate all the unknown parameters $\underline{a}_1$ and $\underline{\Theta}$ from input-output measurements only.) Tuning of the gain parameter $\underline{a}_1$ is quite easy and we have even recommended to use the tuning method in case of time-varying parameters [30,31]. Now, we can precisely recover the gains of linear and nonlinear parts of the system model. (Note: We refer here to the IOBF approach as the regular OBF one produces much poorer results.)

4.2 Fractional-Order Case

Regular OBF Modeling for Fractional-Order Feedback-Nonlinear System. Equation (23) can now be rewritten in form

$$\hat{y}(t) = \sum_{i=1}^{M} C_i U_i(t) - \sum_{i=1}^{M} C_i \sum_{j=1}^{m} a_j \underline{U}_{i,j}(t) \tag{30}$$

where $U_i(t)$ calculated in Eqn. (10), with $y_L(t) = G_L^f(q^{-1})u(t)$ and $y_R^i(t) = G_R^f(q^{-1})U_i(t)$, $i = 1, ..., K-1$, respectively, and

$$\underline{U}_{i,j}(t) = \begin{cases} G_L^f(q)y^j(t) & i = 1 \\ y_R^{i-1}(t) - PU_{i-1}(t) & i = 2, ..., K \end{cases} \tag{31}$$

Model equation as in (31) can be presented in a linear regression form

$$\hat{y}(t) = \varphi^T(t)\Theta \tag{32}$$

where $\varphi^T(t) = [U_1(t) ... U_M(t) \; -\underline{U}_{1,1}(t) ... -\underline{U}_{M,1}(t) ... -\underline{U}_{1,m}(t) ... -\underline{U}_{M,m}(t)]$. The unknown parameter vector is $\theta^T = [C_1 \; ... \; C_M \; w_{11} \; ... \; w_{1M} \; w_{21} \; ... \; w_{2M} \; ... \; w_{m1} \; ... \; w_{mM}]$, where the $w_{ji} = C_i a_j$, $i = 1, ..., M$ and $j = 1, ..., m$. Unfortunately, the bilinearity issue is present here in that the products of the parameters of linear and nonlinear submodels unnecessarily appear in the unknown parameter vector to be estimated. This copies the drawback of the regular OBF-based approaches to Hammerstein system identification as in [7,30,31].

Fractional-Order Inverse OBF Modeling for Feedback-Nonlinear System. We assume now that a linear dynamics is of fractional order. Referring to Eqns. (27) and (31), the model ouput of the fractional-order feedback-nonlinear system can be presented as

$$\hat{y}(t) + \underline{a}_1 y(t-d) = \beta_0 u(t-d) - \sum_{i=1}^{M} C_i \underline{U}_{i,1} - \sum_{j=2}^{m} \underline{a}_j y^j(t-d) \tag{33}$$

where $\underline{U}_{i,1}$, $i = 1, ..., M$ are as in Eqn. (31). The model of Eqn. (33) can be easily presented in a linear regresion form

$$\hat{y}(t) + \underline{a}_1 y(t-d) = \varphi^T(t)\Theta \tag{34}$$

with $\Theta^T = [\beta_0 \; C_1 \; ... \; C_M \; \underline{a}_2 \; ... \; \underline{a}_m]$ and $\varphi^T(t) = [u(t-d) \; -\underline{U}_1(t) \; ... \; -\underline{U}_M(t) \; y^2(t) \; ... \; y^m(t)]$ with $\underline{a}_i = \beta_0 a_i$ and $\underline{U}_{i,1}(t)$, $i = 1, ..., M$, driven by $y(t)$ as in Eqn. (31) and $\underline{a}_1$ fixed/tuned in a similar way as for Eqn. (29). Now, the parameters Θ can be easily estimated using e.g. the RLS algorithm (or its adaptive version ALS).

Remark 4. Note that, the IOBF fractional-order feedback-nonlinear model is much simpler than the regular OBF one, in terms of a number of estimated parameters. In the fractional-order feedback-nonlinear regular Laguerre-based model we have to estimate $(m+1)M$ parameters, whereas in the fractional-order feedback-nonlinear inverse Laguerre-based model (we estimate) $m+M+1$ parameters only.

5 Simulation Experiments

Example 1. Consider a discrete-time fractional-order feedback-nonlinear system, with a static nonlinearity $f(u(t)) = u^3(t)$ and a fractional-order dynamic part described in state-space

$$\Delta^\alpha x(t+1) = A_f x(t) + Bu(t), \tag{35}$$

$$y(t) = Cx(t) + Du(t) \tag{36}$$

with

$$A_f = \begin{bmatrix} -0.4 & -0.03 \\ 1 & 0 \end{bmatrix}, \; B = \begin{bmatrix} 1 \\ 0 \end{bmatrix},$$
$$C = [0 \quad 0.23], D = [0],$$
$$\alpha = 0.5$$

The dynamic part is described by the FFD-based fractional-order IOBF-structured Laguerre model, with $P = 0.49$, $M = 8$, $m = 4$, $\alpha = 0.7$ and various implementation lengths of the FFD approximation ($\overline{J}$). MSPE is used to evaluate the accuracy of modeling. Selected results are presented in Table 1. (We refrain from showing quite similar results for the regular OBF-structured model whose computational inefficiency has earlier been indicated.)

Table 1. MSPE for feedback-nonlinear system with FFD-based Laguerre model

$\overline{J}$	50	100	200	1000
MSPE	8.36e-3	3.96e-3	2.24e-3	2.00e-3

Fig. 2 presents the results of modeling in terms of (indistinguishable) time plots of the actual and modeled outputs of the fractional-order feedback-nonlinear system for some random input signal.

It can be concluded from Fig. 2 and Table 1 that the introduced fractional-order IOBF-based feedback-nonlinear model can be very effective in modeling of the class of fractional-order block-oriented nonlinear systems.

This is also illustrated by the reconstructed nonlinear static characteristic

$$\hat{f}(y) = 5.05\; 10^{-4}\; y^2 - 1.00\; y^3 - 3.58\; 10^{-5}\; y^4 \tag{37}$$

confirming a very good identification performance.

However, to obtain high modeling accuracies we have to use high implementation lengths of the FFD approximation. This inconvenience can be essentially reduced by making use of our computationally more efficient approximations to FD, that is AFFD, PFFD [32], FLD and, in particular, FFLD [19], the task being a subject of our current and future research works.

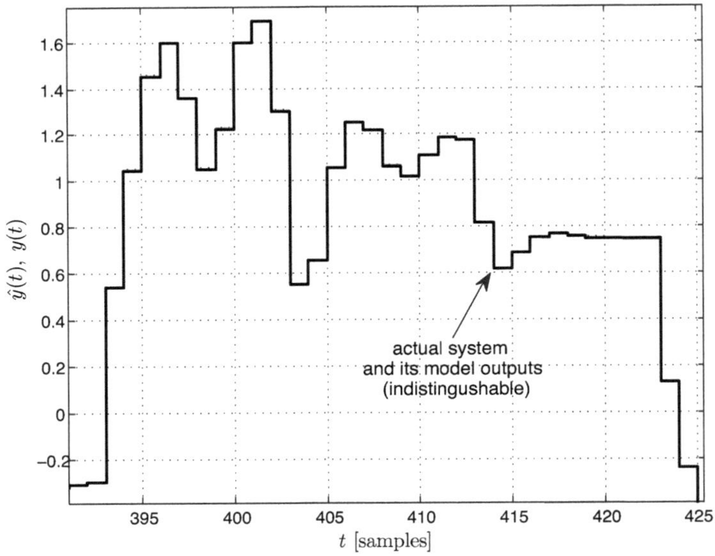

Fig. 2. Time plots of actual and modeled outputs of the fractional-order feedback-nonlinear system

6 Conclusion

The paper has presented a new, simple, analytical solution to the nonlinear identification problem for the fractional-order block-oriented feedback-nonlinear system using fractional-order Laguerre-based models. We have demonstrated that a combination of the inverse OBF modeling concept and fractional-order Laguerre filters can provide high-performance identification of the class of fractional-order nonlinear systems. Simulation examples show that low prediction errors and accurate reconstructions of both nonlinear and linear blocks of the system have been achieved for the introduced models.

References

1. Astrom, K.J., Bell, R.D.: A nonlinear model for steam generation process. In: Preprints IFAC 12th World Congress, Sydney, Australia (1993)
2. Hasiewicz, Z.: Hammerstein system identification by the haar multiresolution approximation. Int J. Adaptive Control and Signal Processing 74, 191–217 (1999)
3. Greblicki, W.: Stochastic approximation in nonparametric identification of Hammerstein system. IEEE Transactions on Automatic Control 47, 1800–1810 (2002)
4. Ninness, B., Gibson, S., Weller, S.: Practical aspects of using orthonormal system parameterisations in estimation problems. In: Proc. 12th IFAC Symp. on System Identification (SYSID 2000), Santa Barbara, CA, pp. 463–468 (2000)

5. Hanczewski, S., Stasiak, M., Weissenberg, J.: The queueing model of a multiservice system with dynamic resource sharing for each class of calls. In: Kwiecień, A., Gaj, P., Stera, P. (eds.) CN 2013. CCIS, vol. 370, pp. 436–445. Springer, Heidelberg (2013)
6. Latawiec, K., Marciak, C., Hunek, W., Stanisławski, R.: A new analytical design methodology for adaptive control of nonlinear block-oriented systems. In: Proceedings the 7th World Multi-Conference on Systemics, Cybernetics and Informatics, Orlando, Florida, vol. 11, pp. 215–220 (2003)
7. Latawiec, K.J., Marciak, C., Stanislawski, R., Oliveira, G.H.C.: The mode separability principle in modeling of linear and nonlinear block-oriented systems. In: Proc. 10th IEEE MMAR Conference (MMAR 2004), Miedzyzdroje, Poland, pp. 479–484 (2004)
8. Heuberger, P.S.C., Van den Hof, P.M.J., Wahlberg, B.: Modeling and identification with rational orthogonal basis functions. Springer, London (2005)
9. Stanisławski, R.: Hammerstein system identification by means of orthonormal basis functions and radial basis functions. In: Pennacchio, S. (ed.) Emerging Technologies, Robotics and Control Systems, Internationalsar, Italy, pp. 69–73 (2007)
10. Kaczorek, T.: Selected Problems of Fractional Systems Theory. Springer, Berlin (2011)
11. Matignon, D.: Stability results for fractional differential equations with applications to control processing. In: Computational Engineering in Systems and Applications Multiconference, Lille, France, vol. 2, pp. 963–968 (1996)
12. Miller, K., Ross, B.: An Introduction to the fractional calculus and fractional differential equations. Willey, New York (1993)
13. Monje, C., Chen, Y., Vinagre, B., Xue, D., Feliu, V.: Fractional-order Systems and Controls. Springer, London (2010)
14. Ostalczyk, P.: Equivalent descriptions of a discrete-time fractional-order linear system and its stability domains. International Journal of Applied Mathematics and Computer Science 22(3), 533–538 (2012)
15. Oldham, K., Spanier, J.: The fractional calculus. Academic Press, Orlando (1974)
16. Podlubny, I.: Fractional differential equations. Academic Press, Orlando (1999)
17. Stanisławski, R., Latawiec, K.J.: Stability analysis for discrete-time fractional-order LTI state-space systems. Part I: New necessary and sufficient conditions for asymptotic stability. Bulletin of the Polish Academy of Sciences, Technical Sciences 61(2), 353–361 (2013)
18. Stanisławski, R., Latawiec, K.J.: Stability analysis for discrete-time fractional-order LTI state-space systems. Part II: New stability criterion for FD-based systems. Bulletin of the Polish Academy of Sciences, Technical Sciences 61(2), 362–370 (2013)
19. Stanisławski, R.: New Laguerre filter approximators to the Grünwald-Letnikov fractional difference. Mathematical Problems in Engineering 2012 (2012), Paper ID: 732917
20. Stanisławski, R.: Advances in Modeling of Fractional Difference Systems - New Accuracy, Stability and Computational Results. Opole University of Technology Press, Opole (2013)
21. Stanisławski, R., Hunek, W.P., Latawiec, K.J.: Normalized finite fractional discrete-time derivative a new concept and its application to obf modeling. Measurements, Automation and Monitoring 57(3), 241–243 (2011)

22. Stanisławski, R., Latawiec, K.J.: Modeling of open-loop stable linear systems using a combination of a finite fractional derivative and orthonormal basis functions. In: Proceedings of the 15th International Conference on Methods and Models in Automation and Robotics, Miedzyzdroje, Poland, pp. 411–414 (2010)
23. Latawiec, K.J., Stanisławski, R., Hunek, W.P., Łukaniszyn, M.: Laguerre-based modeling of fractional-order LTI SISO systems. In: Proceedings of the 18th International Conference on Methods and Models in Automation and Robotics, Miedzyzdroje, Poland, pp. 64–69 (2013)
24. Sierociuk, D., Dzieliński, A.: Fractional Kalman filter algorithm for states, parameters and order of fractional system estimation. International Journal of Applied Matchematics and Computer Science 16(1), 101–112 (2006)
25. Stanisławski, R.: Identification of open-loop stable linear systems using fractional orthonormal basis functions. In: Proceedings of the 14th International Conference on Methods and Models in Automation and Robotics, Miedzyzdroje, Poland, pp. 935–985 (2009)
26. Ljung, L.: System Identification. Prentice-Hall, Englewood Cliffs (1987)
27. Boukis, C., Mandic, D.P., Constantinides, A.G., Polymenakos, L.C.: A novel algorithm for the adaptation of the pole of Laguerre filters. IEEE Signal Processing Letters 13(7), 429–432 (2006)
28. Oliveira, S.T.: Optimal pole conditions for Laguerre and two-parameter Kautz models: A survey of known results. In: 12th IFAC Symposium on System Identification (SYSID 2000), Santa Barbara, CA, pp. 457–462 (2000)
29. Stanisławski, R., Hunek, W.P., Latawiec, K.J.: Modeling of non-linear block-oriented systems using orthonormal basis and radial basis functions. Systems Science 35(2), 11–18 (2009)
30. Stanisławski, R., Latawiec, K.J.: Regular vs. inverse orthonormal basis functions concepts in identification of feedback-nonlinear systens by means of radial basis functions. International Journal of Factory Automation, Robotics and Soft Computing 2007(4), 64–68 (2007)
31. Latawiec, K.J.: The Power of Inverse Systems in Linear and Nonlinear Modeling and Control. Opole University of Technology Press, Opole (2004)
32. Stanisławski, R., Latawiec, K.J.: Normalized finite fractional differences: the computational and accuracy breakthroughs. International Journal of Applied Mathematics and Computer Science 22(4), 907–919 (2012)

Impulse Response Approximation Method for "Fractional Order Lag"

Marta Zagórowska, Jerzy Baranowski, Piotr Bania, Paweł Piątek, Waldemar Bauer, and Tomasz Dziwiński

AGH University of Science and Technology,
Department of Automatics and Biomedical Engineering,
Al. Mickiewicza 30, 30-059 Kraków
{zagor,jb,pba,ppi,bauer,tdz}@agh.edu.pl

Abstract. "Fractional order lag" is a system that is popular in multiple applications. In this paper, authors consider a new method for approximation of this system based on its impulse response. Certain assumptions of the approximation method are verified and algorithm is presented. Also certain problems with this system analysis are discussed, especially its realisation in the form of non-integer order differential equations.

1 Introduction

Non-integer order systems, often called "fractional", gain more and more interest from various scientists, engineers and researchers. These systems might be useful for modelling processes difficult or even impossible to analyse with classical methods. Some basic notions for fractional calculus can be found in [5,9,10]. One of the problems with these systeme is that they are quite complicated in numerical implementation. Therefore, there are papers considering various methods of approximation both in time and frequency domain, e.g. [1–4, 6, 8].

In this paper, the analysed system cannot be formulated in terms of differential equations. It can be written as a following formula

$$x(t) = u * g = \int_0^t u(t-\theta)g(\theta)\mathrm{d}\theta. \tag{1}$$

The transfer function has the form

$$\hat{g}(s) = \frac{\hat{h}(s)}{\hat{f}(s)}, \tag{2}$$

where $\hat{h}(s)$ and $\hat{f}(s)$ denote respectively the numerator and denominator.

© Springer International Publishing Switzerland 2015
K.J. Latawiec et al. (eds.), *Advances in Modeling and Control of Non-integer Order Systems*,
Lecture Notes in Electrical Engineering 320, DOI: 10.1007/978-3-319-09900-2_11

2 Finite Dimensional Approximation of Fractional Differential Equations

It can be shown that the solution of equation (2) can be approximated with a solution of a system of n linear ordinary differential equation [1]. The approximation uses an orthonormal basis in $\mathcal{L}_2(0, \infty)$

$$e_k(\theta, \mu) = \sqrt{2\mu}\mathrm{e}^{-\mu\theta} L_k(2\mu\theta), \quad k = 0, 1, 2, \dots, \tag{3}$$

where μ is an arbitrary positive constant and L_k is k-th Laguerre polynomial of form

$$L_k(z) = \frac{\mathrm{e}^z}{k!}\frac{\mathrm{d}^k}{\mathrm{d}z^k}(\mathrm{e}^{-z}z^k). \tag{4}$$

Theorem 1 gives the conditions that must be fulfiled in order to find the approximation with minimal error.

Theorem 1. *If $g \in \mathcal{L}_1(0, \infty) \cup \mathcal{L}_2(0, \infty)$ and $|u(t)| \leq u_{max}$ then:*

1. *The solution of* (2) *can be approximated with*

$$x_n(t) = \sum_{k=0}^{n} \beta_k \xi_k(t), \tag{5}$$

where functions $\xi_k(t) : [0, \infty) \rightarrow$ are solution of a system

$$\dot{\xi}_k = -\mu\xi_k - 2\mu\sum_{i=0}^{k-1}\xi_i + \sqrt{2\mu}u, \quad \xi_k(0) = 0,\ k = 0, 1, 2, \dots, n \tag{6}$$

and

$$\beta_k = \int_0^\infty g(\theta)e_k(\theta, \mu)\mathrm{d}\theta. \tag{7}$$

2. *For every $\varepsilon > 0$ there exists a number n_0 dependant on g, ε and u_{max} that approximation error $d_n(t) = x(t) - x_n(t)$ fulfils the inequality*

$$|d_n(t)| < \varepsilon \tag{8}$$

for all $n \geq n_0$ and $t \geq 0$

Proof. For proof of theorem 1 see [1].

The formula (7) for calculating the coefficients is not convenient for numerical implementation. In [1] the authors presented the following recurrence formula

$$\beta_k = \frac{\sqrt{2\mu}}{k!}\sum_{j=0}^{k}\binom{k}{j}c_j^k(\mu)\hat{g}^{(k-j)}(\mu), \tag{9}$$

where

$$c_j^k = \frac{k-j+1}{2\mu} c_{j-1}^k,\ c_0^k(\mu) = (2\mu)^k,\ j = 1, 2, \ldots, k \tag{10}$$

and $\hat{g}^{(j)}(s) = \frac{\mathrm{d}^j \hat{g}(s)}{\mathrm{d}s^j}$. The derivatives of transfer function can be computed using the following recurrence relation.

Let $\hat{g}(s) = \frac{\hat{h}(s)}{\hat{f}(s)}$, then $\hat{g}(s)\hat{f}(s) = \hat{h}(s)$. Differentiation with respect to s and using Leibnitz formula for function product gives

$$\hat{g}^{(k)}(s) = \frac{\hat{h}^{(k)}(s) - \sum_{j=1}^{k} \binom{k}{j} \hat{g}^{(k-j)}(s) \hat{f}^{(j)}(s)}{\hat{f}(s)}. \tag{11}$$

2.1 Choice of Parameter μ

The performace of approximation method depends on parameter μ. One of the methods for computing the parameter is presented below. The main goal of this method is to minimize approximation error d_n, that can be computed as follows (argument t will be dropped in order to simplify the notation)

$$\begin{aligned}
\|d_n\| &= \|g - g_n\|_2^2 \\
&= \int_0^\infty (g - g_n)^2 \mathrm{d}t = \\
&= \int_0^\infty g^2 \mathrm{d}t - 2\int_0^\infty g g_n \mathrm{d}t + \int_0^\infty g_n^2 \mathrm{d}t = \\
&= \|g\|_2^2 + \sum_{k=0}^{n} \beta_k^2(\mu) - 2\int_0^\infty g \sum_{k=0}^{n} \beta_k^2 e_i \mathrm{d}t = \\
&= \|g\|_2^2 + \sum_{k=0}^{n} \beta_k^2(\mu) - 2\sum_{k=0}^{n} \beta_k^2(\mu) = \\
&= \|g\|_2^2 - \sum_{k=0}^{n} \beta_k^2(\mu),
\end{aligned} \tag{12}$$

where

$$g_n(\theta) = \sum_{k=0}^{n} \beta_k e_k(\theta, \mu) \tag{13}$$

denotes the approximation of impulse response. Therefore, μ should be chosen to maximize the function

$$J(\mu) = \sum_{k=0}^{n} \beta_k^2(\mu). \tag{14}$$

3 "Fractional Order Lag"

This class of systems is very popular in applications, usually without serious analysis. These systems are classified as non-integer order, however their realisation in the form of non-integer differential equations (at least not their finite number). The only representation that can be used in the time domain is equation (1). Their transfer function take form

$$\hat{g}(s) = \frac{1}{(Ts+1)^{\alpha}}, \tag{15}$$

where T is a constant. Impulse response for (15) can be found using inverse Laplace transform

$$g(t) = \mathcal{L}^{-1}\left\{\frac{1}{(Ts+1)^{\alpha}}\right\} \tag{16}$$

$$= \mathcal{L}^{-1}\left\{\frac{1}{T^{\alpha}(s+\frac{1}{T})^{\alpha}}\right\} \tag{17}$$

$$= \frac{1}{T^{\alpha}} \cdot \mathcal{L}^{-1}\left\{\frac{1}{(s+\frac{1}{T})^{\alpha}}\right\} \tag{18}$$

$$= \frac{1}{T^{\alpha}} \cdot \mathrm{e}^{-\frac{1}{T}}\mathcal{L}^{-1}\left\{\frac{1}{s^{\alpha}}\right\} \tag{19}$$

$$= \frac{1}{T^{\alpha}} \cdot \frac{t^{\alpha-1}\mathrm{e}^{-\frac{t}{T}}}{\Gamma(\alpha)}, \tag{20}$$

where $\Gamma(x)$ denotes gamma function [9].

3.1 Verification of Assumptions

In order to use the decsribed method, it is necessary to check if all the assumptions are fulfiled. First of all, the norm of impulse response in $\mathcal{L}_1(0,\infty)$ must be bounded. The impulse response is positive for all $t \geq 0$, therefore, the absolute value in norm can be omitted.

$$\int_0^t |g(t)|\mathrm{d}t = \int_0^t g(t)\mathrm{d}t = h(t)$$

Therefore

$$\int_0^{\infty} |g(t)|dt = \lim_{t\to\infty} h(t) = \lim_{s\to 0} H(s)$$

and it is sufficient to use initial and final value theorems:

$$\lim_{t\to\infty} x(t) = \lim_{s\to 0}\left(s \cdot \frac{1}{s(Ts+1)^{\alpha}}\right) = 1 \tag{21}$$

and

$$\lim_{t\to 0} x(t) = \lim_{s\to\infty}\left(s\cdot\frac{1}{s(Ts+1)^{\alpha}}\right) = 0 \tag{22}$$

Therefore, $\|x(t)\|_1 = 1$ is bounded. Second assumption is that impulse response is bounded in $\mathcal{L}_2(0,\infty)$. The following calculations are valid

$$\begin{aligned}\int_0^{infty} g^2(t)\mathrm{d}t &= \frac{1}{T^{2\alpha}\cdot\Gamma^2(\alpha)}\int_0^{infty} t^{2\alpha-2}\mathrm{e}^{-\frac{2t}{T}}\mathrm{d}t\\ &= -\frac{2^{1-2\alpha}T^{2\alpha-1}}{T^{2\alpha}\cdot\Gamma^2(\alpha)}\cdot\Gamma\left(2\alpha-1,\frac{2t}{T}\right)\Big|_0^{\infty},\end{aligned} \tag{23}$$

where

$$\Gamma(a,x) = \int_x^{\infty} t^{a-1}\mathrm{e}^{-t}\mathrm{d}t \tag{24}$$

is the "upper" incomplete gamma function and $\mathrm{Re}\, a > 0$.

The limit of (23) for $t\to 0$ can be written as

$$\begin{aligned}\lim_{t\to 0}\left(-\frac{2^{1-2\alpha}T^{\alpha-1}}{\Gamma(\alpha)}\cdot\Gamma\left(2\alpha-1,\frac{2t}{T}\right)\right) &= -\frac{2^{1-2\alpha}T^{\alpha-1}}{\Gamma(\alpha)}\lim_{t\to 0}\Gamma\left(2\alpha-1,\frac{2t}{T}\right)\\ &= -\frac{2^{1-2\alpha}T^{\alpha-1}}{\Gamma(\alpha)}\Gamma(2\alpha-1,0)\\ &= -\frac{2^{1-2\alpha}T^{\alpha-1}}{\Gamma(\alpha)}\Gamma(2\alpha-1)\end{aligned} \tag{25}$$

and is bounded for $2\alpha-1>0$, that is $\alpha>\frac{1}{2}$. Same analysis was conducted for the second limit $t\to\infty$

$$\begin{aligned}&\lim_{t\to\infty}\left(-\frac{2^{1-2\alpha}T^{\alpha-1}}{\Gamma(\alpha)}\cdot\Gamma\left(2\alpha-1,\frac{2t}{T}\right)\right) = -\frac{2^{1-2\alpha}T^{\alpha-1}}{\Gamma(\alpha)}\lim_{t\to\infty}\Gamma\left(2\alpha-1,\frac{2t}{T}\right) =\\ &= -\frac{2^{1-2\alpha}T^{\alpha-1}}{\Gamma(\alpha)}\lim_{t\to\infty}\left(\int_0^{\infty} x^{2\alpha-2}\mathrm{e}^{-x}\mathrm{d}x - \int_0^{\frac{2t}{T}} x^{2\alpha-2}\mathrm{e}^{-x}\mathrm{d}x\right)\\ &= 0.\end{aligned} \tag{26}$$

From formulas (25) and (26), one obtains that the norm in $\mathcal{L}_2(0,\infty)$ is bounded for $\alpha\in\left(\frac{1}{2},\infty\right)$.

The third assumption refers to limitations of impulse response. It can be observed that (20) can be unbounded only for $t=0$ and $\alpha<1$. For $\alpha>1$ formula (20) is bounded for all t.

From these three considerations, it is obtained that the assumptions of 1 are fulfiled for $\alpha>1$.

The derivative of (15) can be calculated using (11) with the following assumptions

$$\hat{h}(s) = 1 \Longrightarrow \hat{h}^{(k)}(s) = 0$$
$$\hat{f}(s) = T^{\alpha} \cdot \left(s + \frac{1}{T}\right)^{\alpha} \Longrightarrow$$
$$\hat{f}^{(k)}(s) = T^{\alpha}\alpha(\alpha-1)\ldots(\alpha-k+1)\left(s + \frac{1}{T}\right)^{\alpha-k}$$
$$= T^{\alpha}k!\binom{\alpha}{k}\left(s + \frac{1}{T}\right)^{\alpha-k}, \; k \in \mathbb{N}.$$

It can be easily observed in figure 1 that the approximation gives the correct answer for most cases. When the approximation order is high $n = 30$, parameter μ is of little importance. However, when the approximation order is small $n = 5$, the choice of parameter cannot be neglected. Using (14), it is obtained that the optimal value for μ is 1.134 (for $\alpha = 1.8$, $n = 30$, $T = 2$). The figure 2 shows the plot of performance index as function of μ.

Error of approximation in both $\mathcal{L}_1(0,\infty)$ and $\mathcal{L}_2(0,\infty)$ spaces, for different orders can be seen in figure 3. As it can be seen method reduces the error with increasing order. It can be however observed that rate of improvement falls. It is due to certain numerical considerations.

3.2 Numerical Considerations

Approximation error d_n can be written in two ways. First, it is the difference between impulse response and its approximation with system of equations. In the same time, it can be written as

$$\|d_n\| = \|g\|^2 - \sum_{k=0}^{n} \beta_k^2. \tag{27}$$

Comparing these two formulas ($\alpha = 1.8$, approximation order $n = 30$, $T = 2$), one obtains

$$10^{-7} = \|g - g_n\|^2 = \|d_n\| = \|g\|^2 - \sum_{k=0}^{n} \beta_k^2 = -1.4293 \times 10^{-7}, \tag{28}$$

where g is the impulse response for system (15), g_n is the impulse response of approximating equations and β_k are given by (7). It can be easily observed that the difference between the values comes from numerical errors.

It can be observed that for order greater than 29 parameters cease to decrease. According to approximation properties, parameters β_k should tend to 0. However, due to numerical errors, probably in numerical computation of factorial, this is not the case.

Figure 4(a) shows how the difference between two sides of equation (28) increases with approximation order.

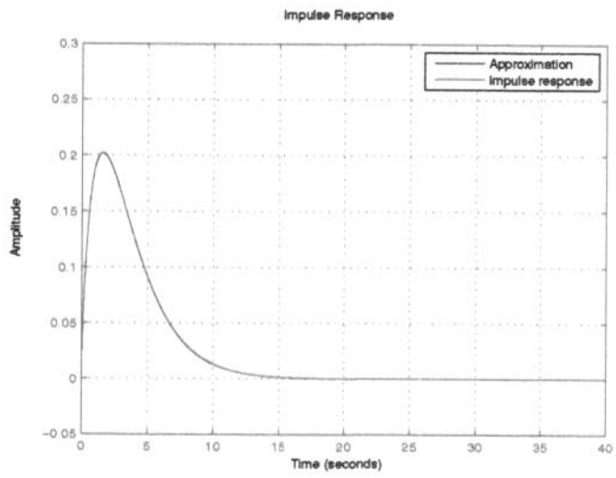

(a) $n = 30$, $\mu = 1.134$

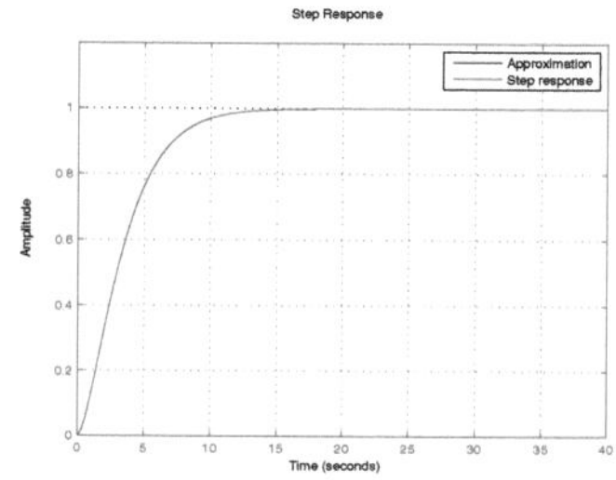

(b) $n = 30$, $\mu = 1.134$

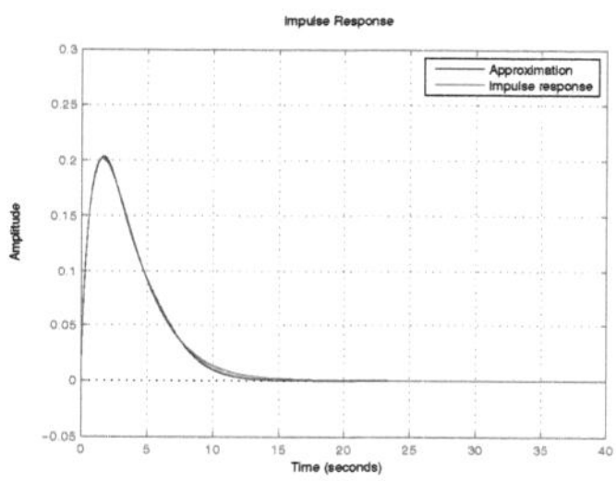

(c) $n = 5$, $\mu = 1.134$

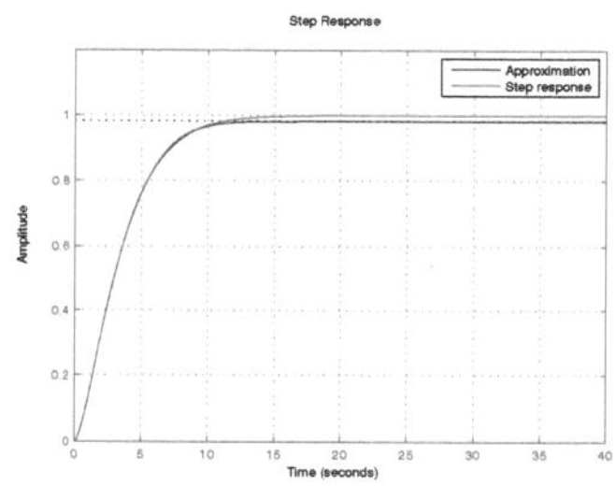

(d) $n = 5$, $\mu = 1.134$

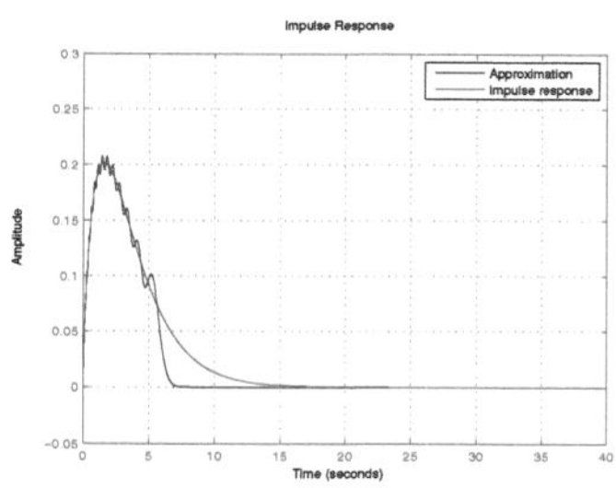

(e) $n = 30$, $\mu = 10$

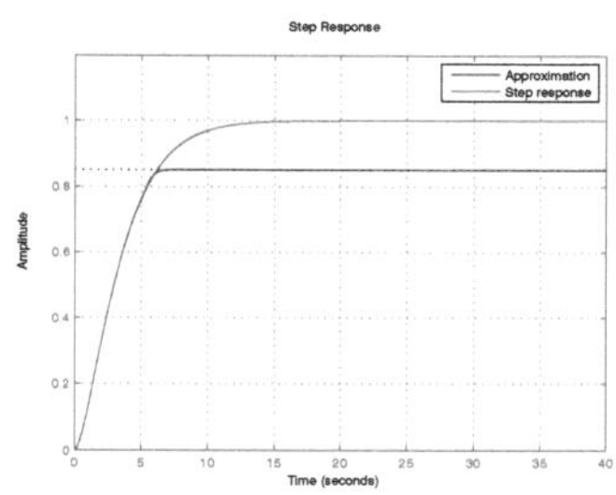

(f) $n = 30$, $\mu = 10$

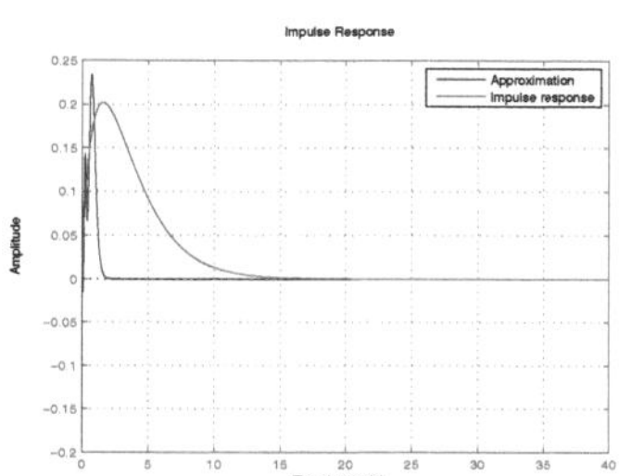

(g) $n = 5$, $\mu = 10$

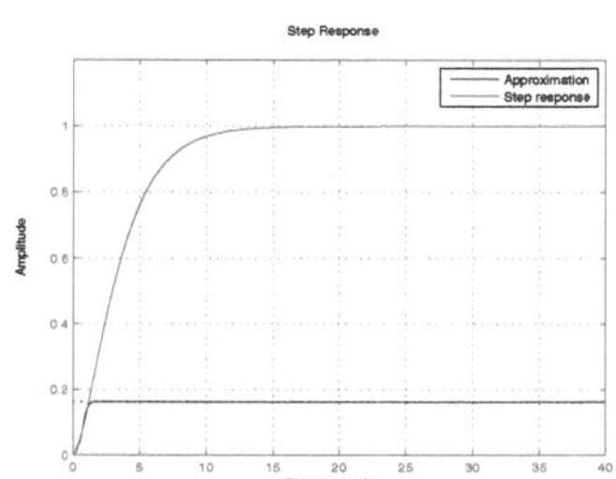

(h) $n = 5$, $\mu = 10$

Fig. 1. Impulse (left) and step response (right) for different approximation order n and parameter μ. Parameters $\alpha = 1.8$ and $T = 2$.

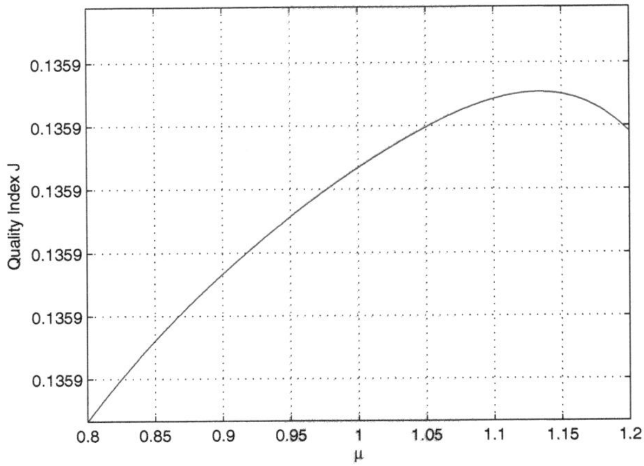

Fig. 2. Quality index $J(\mu)$

The lefthand side of equation (28) can be analysed as function of approximation order n. As shown in figure 4(b) it is a decreasing function (although not strictly decreasing) for both $\mathcal{L}_1(0,\infty)$ and $\mathcal{L}_2(0,\infty)$ norm. The analysis was conducted for different values from 1 to 25. For higher order approximation the numerical error was too large for further consideration. Numerical errors imply also that for higher order approximation, it decreases slower than for orders lower than 10.

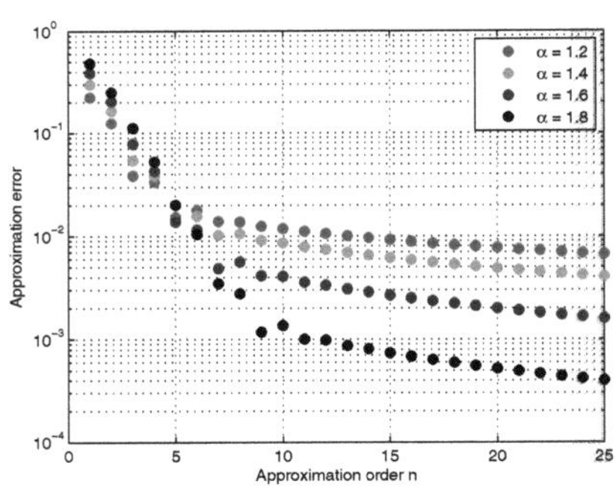

(a) Error for different α in $\mathcal{L}_1(0,\infty)$

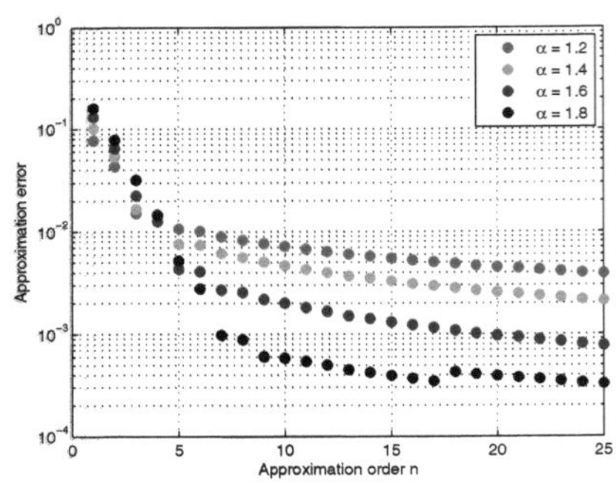

(b) Error for different α in $\mathcal{L}_2(0,\infty)$

Fig. 3. Norms of approximation errors depending on order

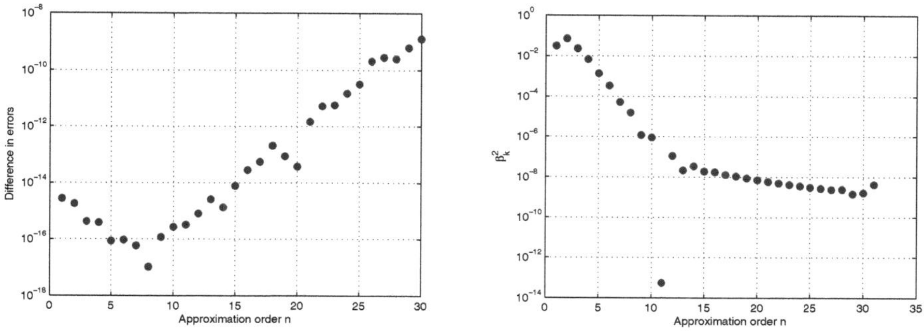

(a) Error difference (b) Different values of β_k^2

Fig. 4. Numerical considerations of error

3.3 Example with $\alpha \in (\frac{1}{2}, 1)$

It can be observed that the method presented in theorem 1 works also for functions that do not fulfil all assumptions. When $\alpha \in (\frac{1}{2}, 1)$, the impulse response $x(t)$ is unbounded at 0. However, for $\alpha = 0.85$, $T = 2$, $\mu = 1.134$ and $n = 10$ one obtains a very good approximation. Impulse and step response of the system and its approximation are in the figure 5.

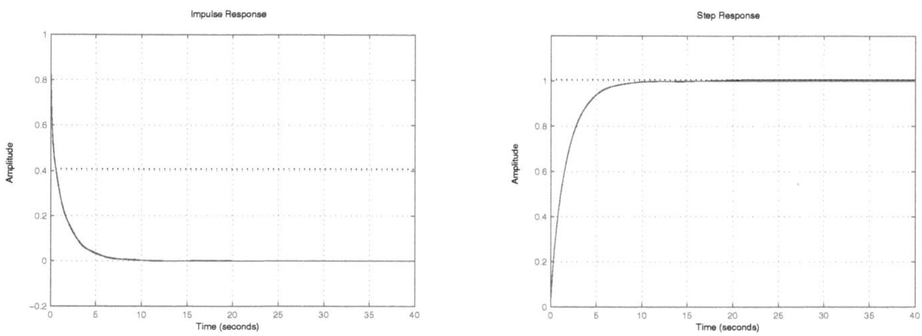

(a) Impulse response (b) Step response

Fig. 5. System with $\alpha = 0.85$, $T = 2$ and its approximation with $\mu = 1.134$ and $n = 10$

4 Conclusion

In this paper, the authors considered a novel approach to non-integer order system approximation. The novelty consists of the application of the method from [1] for a system that cannot be described with fractional order differential equations only with a transfer function. The theorem was presented, that allows to approximate fractional order systems in time domain. Numerical analysis was also conducted for this method.

The approximation method proved very useful for considered system. Regardles, that this system is non-integer "in-name-only" method can also be used for much wider class of systems. However, it must be noticed that for higher order the numerical error is of significant importance and cannot be neglected. Further research will include analysis of method operation in the frequency domain and detailed comparison with other methods of approximation. In particular methods by Oustaloup [7] and Charef [3] will be investigated.

Acknowledgment. Work realised in the scope of project titled "Design and application of noninteger order subsystems in control systems". Project was financed by National Science Centre on the base of decision no. DEC-2013/09/D/ST7/03960.

References

1. Bania, P., Baranowski, J.: Laguerre polynomial approximation of fractional order linear systems. In: Mitkowski, W., Kacprzyk, J., Baranowski, J. (eds.) Theory & Appl. of Non-integer Order Syst. LNEE, vol. 257, pp. 171–182. Springer, Heidelberg (2013)
2. Baranowski, J.: Legendre polynomial approximations of time delay systems. In: Materiały XII Międzynarodowych Warsztatów Doktoranckich OWD, Wisła 23-26, pp. 15–20 (2010)
3. Djouambi, A., Charef, A., Besancon, A.V.: Approximation and synthesis of non integer order systems. In: 2nd IFAC Workshop on Fractional Differentiation and its Applications, FDA 2006, Portugal, Porto (July 2006)
4. Mitkowski, W.: Approximation of fractional diffusion-wave equation. Acta Mechanica et Automatica 5, 65–68 (2011)
5. Oldham, K., Spanier, J.: The fractional calculus. Academic Press, New York (1974)
6. Oustaloup, A.: La commande CRONE: commande robuste d'ordre non entier. Hermes (1991)
7. Oustaloup, A., Levron, F., Mathieu, B., Nanot, F.M.: Frequency-band complex noninteger differentiator: Characterization and synthesis. IEEE Transactions on Circuits and Systems I: Fundamental Theory and Applications 47(1), 25–39 (2000)
8. Piątek, P., Zagórowska, M., Baranowski, J., Bauer, W., Dziwiński, T.: Discretisation of different non-integer order approximations. In: 2014 19th International Conference on Methods and Models in Automation and Robotics, MMAR (accepted 2014)
9. Podlubny, I.: Fractional Differential Equations: An Introduction to Fractional Derivatives, Fractional Differential Equations, to Methods of Their Solution and Some of Their Applications. Mathematics in Science and Engineering. Elsevier Science (1998)
10. Podlubny, I., Chechkin, A., Skovranek, T., Chen, Y., Jara, B.M.V.: Matrix approach to discrete fractional calculus ii: Partial fractional differential equations. Journal of Computational Physics 228(8), 3137–3153 (2009)

Part IV
Controllability and Control

Comparison of Performance Indices for Tuning of $PI^{\lambda}D^{\mu}$ Controller for Magnetic Levitation System

Waldemar Bauer, Tomasz Dziwiński, Jerzy Baranowski, Paweł Piątek, and Marta Zagórowska

AGH University of Science and Technology,
Department of Automatics and Biomedical Engineering,
Al. Mickiewicza 30, 30-059 Kraków
{bauer,tdz,jb,ppi,zagor}@agh.edu.pl

Abstract. Control of active magnetic bearings is an important area of research. The laboratory magnetic levitation system can be interpreted as a model of a single axis of bearings and is a useful testbed for control algorithms. The mathematical model of this system is highly nonlinear and requires careful analysis and identification. In this paper authors compare performance indices for tuning of $PI^{\lambda}D^{\mu}$ controller for this system. It is a part of an ongoing research on non integer controller tuning rules.

1 Introduction

Magnetic levitation systems have many varied uses such as in frictionless bearings, high-speed maglev passenger trains, levitation of wind tunnel models, vibration isolation of sensitive machinery, levitation of molten metal in induction furnaces and the levitation of metal slabs during manufacture, see [5]. Much interest is recently focused on active magnetic bearings. These bearings are considered to be superior over conventional bearings because the friction losses are significantly reduced due to contactless operation. The bearings can also give high speed and are also able to eliminate lubrication and moreover, operation will be quiet, see [1]. Magnetic bearings are increasingly used in industrial machines such as compressors, turbines, pumps, motors and generators. Very interesting are also their applications in artificial hearts. Also important, especially in current popularity of "green" energy solutions, is the flywheel energy storage system.

Flywheel energy systems are now considered as enabling technology for many applications including space satellite low earth orbits, hybrid electric vehicles (see [6]), and many stationary applications. Such mechanical batteries normally consist of a high speed inertial composite rotor, a magnetic bearings support and a control system, an integral drive motor/generator, power electronics for electrical conversion, and so on. One of the advantages over chemical batteries is that the design life has no degradation during its entire cycle life, and current testing indicates that flywheels are not damaged by repetitively deep discharge. Also,

© Springer International Publishing Switzerland 2015
K.J. Latawiec et al. (eds.), *Advances in Modeling and Control of Non-integer Order Systems*, Lecture Notes in Electrical Engineering 320, DOI: 10.1007/978-3-319-09900-2_12

the contactless nature of magnetic bearings brings up higher energy efficiency, lower wear, longer life span, absence of lubrication and mechanical maintenance, and wider range of work temperature. Moreover, the closed-loop control of magnetic bearings enables active vibration suppression and on-line control of bearing stiffness (see [18]).

Control of magnetic levitation system was analysed by many researchers focusing on different approaches. A linearising feedback control was considered among the others by Barie and Chiasson (see [5]), Joo and Seo (see [10] and [14]). Different approach to feedback linearisation of mag-lev (see [2]). The comparison of this approach with Takagi-Sugeno fuzzy control (see [9]). The cascade variant of the linearising feedback was also discussed by Baranowski and Piątek (see [3]). Real time neural feedforward control was considered by Bloch (see [11]). Practically efficient results were also obtained by Piątek (see [13]) with very fast linear control based on FPGA circuits. Piłat in [15] considered a non-integer order PD controller.

In this paper we discuss an application of tuning non-integer $\mathrm{PI}^{\lambda}\mathrm{D}^{\mu}$ controller, when control signal is not disturbed and disturbed. This is a continuation of authors earlier works (see [3,8,16]).

1.1 The Mathematical Model of the System

We consider the magnetic levitation system consisting of the electromagnet, the ferromagnetic sphere (which is later referred to as the "ball"), the current driver and the position measurement system.

To construct the mathematical model of the plant we will rely on a basic relation of Newton's second law, in this case:

$$m\ddot{x}_1(t) = F_l(x_1(t), x_3(t)) + mg \tag{1}$$

where $x_1(t)$ is the gap between the ball and the electromagnet, $x_3(t)$ is the electromagnet coil current, $F_l(x_1(t), x_3(t))$ is the force generated by the electromagnet, m is the mass of the ball and g is the gravitational acceleration. It is widely known that the force generated by the electromagnet is given by the following relation

$$F_l(x_1(t), x_3(t)) = \frac{1}{2} \cdot \frac{\mathrm{d}l(x_1(t))}{\mathrm{d}x_1(t)} x_3^2(t) \tag{2}$$

where $l(x_1(t))$ is the electromagnet inductance. Commonly, the inductance is considered for cuboidally shaped gaps as a hyperbolical function, as for an example (time argument was omitted)

$$l(x_1) = l_1 + \frac{\mu l_0}{\mu + x_1} \tag{3}$$

where l_0, l_1 and μ are positive constants. Expressions of this type were considered among the others by Barie and Chiasson (see [5]). What should be noted is that levitation systems such as considered have gaps of a different shape because a levitating object is round. That is why we consider the approximation developed by Piłat (in [14]) in a form of the following exponential function

$$l(x_1) \approx a \exp\left(-\frac{x_1(t)}{b}\right) \tag{4}$$

where a and b are positive constants. This approximation was obtained and verified experimentally and leads to very good results. Parameters a and b were determined by analysis of series of steady state points of the system with a closed stabilising feedback loop. Exponential function was fitted into these points through a least squares minimisation. For details see [14].

The coil current in the system usually is influenced by many factors like changes in inductance, velocity and others. However, our system includes a current driver, which has its own feedback loop. This solution is very popular (see [7]) because it leads to either lower order or simpler model structure. In optimal situation the driver should allow full current control, however in real situations it introduces its own dynamics. For considered system, this dynamics can be sufficiently modelled by a first order dynamical system given by the following equation

$$\dot{x}_3(t) = \frac{1}{T_s}(k_s u(t) - i_s - x_3(t)) \tag{5}$$

where $u(t)$ is the control voltage, k_s is the gain of current controller, T_s is the time constant of the current driver and is i_s the zero error of current driver.

Velocity of the ball x_2 is the first derivative of position, so we can construct the state space model. Let us introduce state space vector $\mathbf{x}$ given by

$$\mathbf{x} = [x_1 \ x_2 \ x_3]^T \tag{6}$$

which can be used to formulate the model of the system as the following system of first order differential equations

$$\dot{\mathbf{x}}(t) = \mathbf{f}(\mathbf{x}(t), u(t)) \tag{7}$$

where

$$\mathbf{f}(\mathbf{x}(t), u(t)) = \begin{bmatrix} x_2(t) \\ -\frac{a}{2mb}\exp\left(-\frac{x_1(t)}{b}\right)x_3^2 + g \\ \frac{1}{T_s}(k_s u(t) - i_s - x_3(t)) \end{bmatrix} \tag{8}$$

1.2 Nonlinear Feedforward

It is a known fact that the linear controller can operate properly in the neighbourhood of a chosen steady state. Performance of classical PID can be strongly improved, if the appropriate reference control value corresponding to a reference value is added to the generated control signal. Authors tested this solution with non-integer $PI^{\lambda}D^{\mu}$ controller.

Let us consider control structures presented in figures 1. Let us assume, that set point signal is piecewise constant. This goal can be satisfied then function $\Psi(w_r)$ have form:

$$\mathbf{f}(\mathbf{x}_r, \Psi(w_r)) = 0 \tag{9}$$

where $x_r = [w_r \; 0 \; x_{3r}]^{\mathrm{T}}$, w_r is constant value of $w(t)$, $\mathbf{f}$ is given by (8) and x_{3r} is the value of current corresponding to $w(t)$. Such function (along with x_{3r}) can be obtained by solving (9) and is given by the following formula:

$$\Psi(w_r) = \frac{1}{k_s}\left(i_s + \sqrt{\frac{2mbg}{a} \cdot \mathrm{e}^{\frac{w_r}{b}}}\right) \tag{10}$$

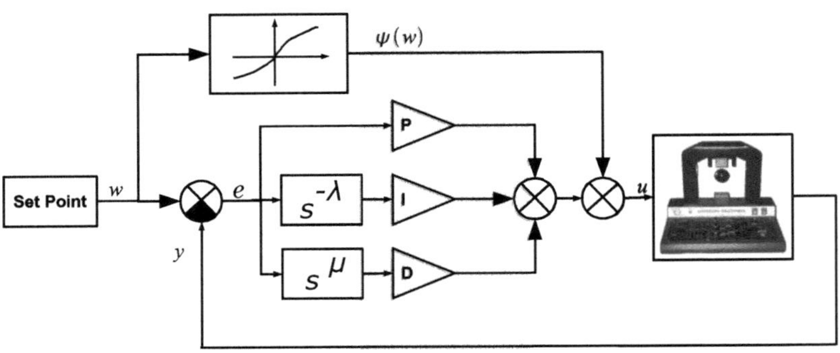

Fig. 1. Magnetic levitation with $PI^{\lambda}D^{\mu}$

2 Non-integer $PI^{\lambda}D^{\mu}$

This section describes a more generalized structure for the classical PID controller. Podlubny proposed a generalization of the PID, namely the $PI^{\lambda}D^{\mu}$ controller, involving an integrator of order λ and a differentiator of order μ. In time domain the equation for the $PI^{\lambda}D^{\mu}$ controller's output has the form (see [17]):

$$u(t) = K_p e(t) + K_i \, {}_0^C\mathrm{D}_t^{-\lambda} e(t) + K_d \, {}_0^C\mathrm{D}_t^{\mu} e(t) \tag{11}$$

Where:

- K_p is proportional gain
- K_i is integral gain
- K_d is derivative gain
- $e(t)$ is control deviation in time t
- λ, $\mu > 0$

And the transfer function formula is given by the equation:

$$G(s) = K_p + K_i s^{-\lambda} + K_d s^{\mu} \tag{12}$$

As can be observed, when $\lambda = 1$ and $\mu = 1$ we obtain a classical PID controller, similar when $\lambda = 0$ and $\mu = 1$ give PD, $\lambda = 0$ and $\mu = 0$ give P, $\lambda = 1$ and $\mu = 0$ give PI.

All these classical types of PID are the particular cases of the fractional PI$^\lambda$D$^\mu$. However, the PI$^\lambda$D$^\mu$ is more flexible.

For all numerical experiments the Simulated Annealing optimization method has been chosen for tuning PI$^\lambda$D$^\mu$ controller parameters. In this case we can define the decision variables as: K_p, K_i, K_d, λ and μ. The tests will be conducted for the following quality index:

Table 1. Result of tuning system without disturbance

Quality index	K_p	K_i	λ	K_d	μ	Quality value
$\int_0^T te^2(t)dt$	517.017	116.408	0.917	20.6418	0.6796	$2.69 \cdot 10^{-3}$
$\int_0^T e^2(t)dt$	475.1759	63.0862	0.2555	4.7824	0.7788	$3.69 \cdot 10^{-2}$
$\int_0^T \lvert e(t)\rvert dt$	553.146	91.828	0.786	4.336	0.77	0.112
$\int_0^T \left(e^2(t) + x_2^2(t)\right) dt$	498.241	68.415	0.777	66.5197	0.997	$1.33 \cdot 10^{-2}$

Table 2. Result of tuning with load disturbance

Quality index	K_p	K_i	λ	K_d	μ	Quality value
$\int_0^T te^2(t)dt$	481.202	291.560	0.0104	38.773	0.6429	$7.95 \cdot 10^{-2}$
$\int_0^T e^2(t)dt$	582.31	118.476	0.087	47.023	0.585	$5.42 \cdot 10^{-2}$
$\int_0^T \lvert e(t)\rvert dt$	491.63	54.99	0.0237	57.378	0.59	0.313
$\int_0^T \left(e^2(t) + x_2^2(t)\right) dt$	553.333	−38.514	0.997	94.882	1	9.88

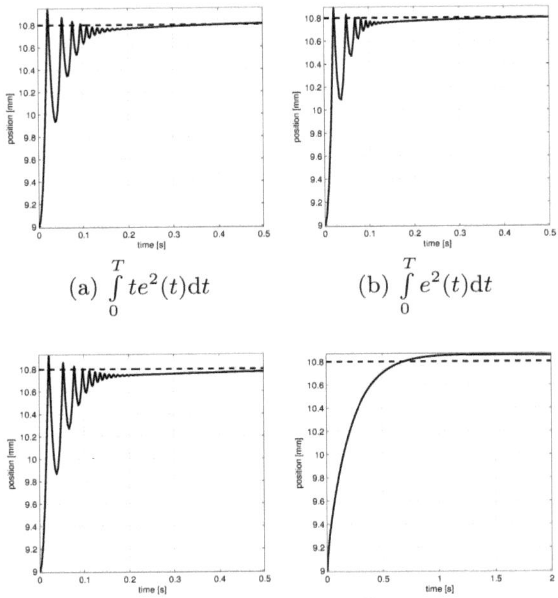

(a) $\int_0^T te^2(t)\mathrm{d}t$ (b) $\int_0^T e^2(t)\mathrm{d}t$

(c) $\int_0^T |e(t)|\mathrm{d}t$ (d) $\int_0^T e^2(t) + x_2^2(t)\mathrm{d}t$

Fig. 2. Result of tuning system without disturbance for differing quality index

- $\int_0^T te^2(t)\mathrm{d}t$
- $\int_0^T e^2(t)\mathrm{d}t$
- $\int_0^T |e(t)|\mathrm{d}t$
- $\int_0^T \left(e^2(t) + x_2^2(t)\right)\mathrm{d}t$

where $e(t) = w_r - x_1(t)$.

The controller was implemented with Oustalup method. For the fractional-order operator $G(s) = s^\alpha$, the continued fraction expansion can be written as (see [12]):

$$G_t(s) = K \prod_{i=1}^{N} \frac{s + \omega_i'}{s + \omega_i} \tag{13}$$

where:

$$\omega_i' = \omega_{\min}\omega_u^{(2i-1-\alpha)/N} \tag{14}$$

$$\omega_i = \omega_{\min}\omega_u^{(2i-1+\alpha)/N} \tag{15}$$

$$K = \omega_{\max}^{\alpha} \tag{16}$$

$$\omega_u = \sqrt{\frac{\omega_{\max}}{\omega_{\min}}} \tag{17}$$

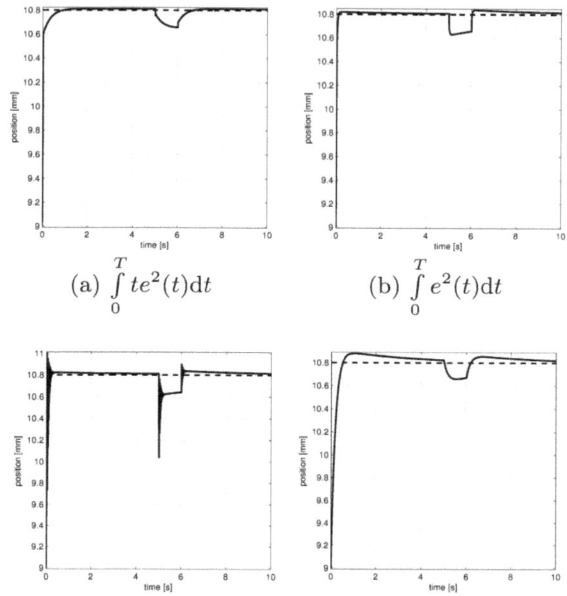

(a) $\int\limits_0^T te^2(t)\mathrm{d}t$ (b) $\int\limits_0^T e^2(t)\mathrm{d}t$

(c) $\int\limits_0^T |e(t)|\mathrm{d}t$ (d) $\int\limits_0^T e^2(t) + x_2^2(t)\mathrm{d}t$

Fig. 3. Result of tuning system with disturbance in control signal for differing quality index

2.1 Results

In all experiments, values of approximation parameters are:

- $N = 3$,
- $\omega_{\min} = 10^{-}6$,
- $\omega_{\min} = 10^6$,

and initial points have value:

- $K_p = 500$
- $K_i = 100$
- $K_d = 6$
- $\lambda, \mu = 0.5$

The optimal PI$^\lambda$D$^\mu$ settings for the system without disturbance are collected in table 1 and for the system with load disturbance settings are collected in table 2. Position states of the magnetic levitation were shown in figures 2 and 3.

How can see the best results have been achieved when quality indices of form $\int\limits_0^T e^2(t)\mathrm{d}t$ or $\int\limits_0^T te^2(t)\mathrm{d}t$ have been used (see figures 2(b) and 3(b)) (see figures 2(a) and 3(a)).

3 Conclusion and Further Research

It has been shown that fractional-order $PI^{\alpha}D^{\mu}$ controller is suitable for control of magnetic levitation systems. The paper has shown that simulated annealing optimisation method could be helpful in the tuning process. The authors tested also some quality indices for tuning the controller.

The further research is planned to implement $PI^{\alpha}D^{\mu}$ controller in digital real-time environment, based on RT-DAC board and MATLAB/RT-CON library, and to conduct experiments on physical plant.

Acknowledgement. Work realised in the scope of project titled "Design and application of non-integer order subsystems in control systems". Project was financed by National Science Centre on the base of decision no. DEC-2013/09/D/ ST7/03960.

References

1. Agarwal, P.K., Chand, S.: Fault-tolerant control of three-pole active magnetic bearing. Expert Systems with Applications (36), 12592–12604 (2009)
2. Bania, P.: Stabilizujące sprzężenie zwrotne dla systemu magnetycznej lewitacji. Automatyka 4(2), 117–139 (2000)
3. Baranowski, J., Piątek, P.: Nonlinear dynamical feedback for motion control of magnetic levitation system. In: Proceedings of 13th International Power Electronics and Motion Control Conference EPE-PEMC, Poznań, September 1-3, pp. 1469–1476 (2008)
4. Baranowski, J., Piátek, P.: Observer-based feedback for the magnetic levitation system. Transactions of the Institute of Measurement and Control 34(4), 422–435 (2012)
5. Barie, W., Chiasson, J.: Linear and nonlinear state-space controllers for magnetic levitation. International Journal of Systems Science 27(11), 1153–1163 (1996)
6. Beno, J., Thompson, R., Hebner, R.: Flywheel batteries for vehicles. In: Proceedings of the 2002 Workshop on Autonomous Underwater Vehicles, pp. 99–101 (2002)
7. Duan, G., Wu, Z., Howe, D.: Robust control of a magnetic-bearing flywheel using dynamical compensators. Transactions of the Institute of Measurement and Control 4(23), 249–278 (2001)
8. Dziwiński, T., Bauer, W., Baranowski, J., Piątek, P., Zagórowska, M.: Robust non-integer order controller for air heater. In: 2014 19th International Conference on Methods and Models in Automation and Robotics, MMAR (accepted 2014)
9. Grega, W., Piłat, A.: A comparison of nonlinear controllers for magnetic levitation system. In: 5th World Multiconference on Systemics, Cybernetics and Informatics, Orlando-Florida, July 22-25, pp. 188–193 (2001)
10. Joo, S., Seo, J.: Design and analysis of the nonlinear feedback linearizing control for an electromagnetic suspension system. IEEE Transactions on Control Systems Technology 5(1), 135–144 (1996)
11. Lairi, M., Bloch, G.: A neural network with minimal structure for maglev system modeling and control. In: Proceedings of the 1999 IEEE International Symposium on Intelligent Control/Intelligent Systems and Semiotics, 1999, pp. 40–45 (1999)

12. Oustaloup, A., Levron, F., Mathieu, B., Nanot, F.M.: Frequency-band complex noninteger differentiator: characterization and synthesis. IEEE Transactions on Circuits and Systems I: Fundamental Theory and Applications 47(1), 25–39 (2000)
13. Piątek, P.: Wykorzystanie specjalizowanych architektur sprzętowych do realizacji krytycznych czasowo zadań sterowania Application of Specialized Hardware Architectures for Realization of Time-critical Control Tasks. PhD thesis, Akademia Górniczo-Hutnicza im. Stanisława Staszica, Wydział Elektrotechniki, Automatyki, Informatyki i Elektroniki, Kraków, Promotor: W. Grega (2007)
14. Piłat, A.: Sterowanie układami magnetycznej lewitacji. PhD thesis, Akademia Górniczo-Hutnicza im. Stanisława Staszica, Wydział Elektrotechniki, Automatyki, Informatyki i Elektroniki, Kraków, Promotor: W. Grega (2002)
15. Pilat, A.: A comparative study of pi$^\lambda$d$^\mu$ controller approximations exemplified by active magnetic levitation system. In: Mitkowski, W., Kacprzyk, J., Baranowski, J. (eds.) Theory & Appl. of Non-integer Order Syst. LNEE, vol. 257, pp. 231–241. Springer, Heidelberg (2013)
16. Piątek, P., Zagórowska, M., Baranowski, J., Bauer, W., Dziwiński, T.: Discretisation of different non-integer order approximations. In: 2014 19th International Conference on Methods and Models in Automation and Robotics, MMAR (accepted 2014)
17. Podlubny, I.: Fractional Differential Equations: An Introduction to Fractional Derivatives, Fractional Differential Equations, to Methods of Their Solution and Some of Their Applications. Mathematics in Science and Engineering. Elsevier Science (1998)
18. Zhu, K., Xiao, Y., Optimal, R.K.Y.: control of the magnetic bearings for a flywheel energy storage system. Mechatronics (19), 1221–1235 (2009)

Fractional-Order Model Predictive Control with Small Set of Coincidence Points

Stefan Domek

Department of Control Engineering and Robotics
West Pomeranian University of Technology, Szczecin
ul. Sikorskiego 37, 70-313 Szczecin, Poland
stefan.domek@zut.edu.pl

Abstract. In the paper the possibility and conditions for employing the fractional-order differential calculus theory in the model predictive control are analyzed. First, the principle of the integer-order linear predictive control and theoretical foundations of the fractional-order differential calculus are reminded. Using the presented theoretical foundations attention is focused further on the possibility of employing the fractional-order calculus for model predictive control with a small set of coincidence points. The introduction of the fractional-order differential calculus at the stage of synthesizing the control algorithm offers an additional degree of freedom in tuning a control loop. The discussion is illustrated with results of simulation tests.

Keywords: non-integer order systems, model predictive control, fractional-order differential calculus.

1 Introduction

The idea of predictive control, which was put forward several dozen years ago and has been intensely developed since then, is considered to be, after many years of operating experience in industry, as one of the most universal and effective control methods. It can handle in natural way multivariable systems and moreover, it can take into account explicitly constraints on input and output signals [1–3]. In recent years, the concept of the non-integer order derivative/integral, having been known for a very long time, has found increasing use in automation, mainly as a result of intensive theoretical and practical research. Therefore, there are many works which applied the fractional-order differential calculus to the control theory, which should contribute to the development of new control algorithms significantly different from the well-known integer-order algorithms, and thus, by implication, provide potentially new opportunities for control performance and robustness [4–7]. Allowing integration/differentiation of arbitrary orders in classic control algorithms results in increasing the number of degrees of freedom in parameter tuning, and thus creates new potentialities as to control performance and robustness [8–10]. An excellent example here is the fractional-order digital $PI^{\lambda}D^{\mu}$ algorithm [6], already regarded as a standard one, but also the fractional-order dead-beat algorithms, sliding mode control, linear-quadratic control, model

© Springer International Publishing Switzerland 2015

K.J. Latawiec et al. (eds.), *Advances in Modeling and Control of Non-integer Order Systems*, Lecture Notes in Electrical Engineering 320, DOI: 10.1007/978-3-319-09900-2_13

reference adaptive control and iterative learning control, having been proposed in later years [4, 8–10]. Consequently, there are recently papers which join the predictive control with the fractional-order calculus [1, 11, 12].

2 Fractional Order Calculus

The generalized operator of integro-differentiation ${}_{t_0}D_t^\alpha$ or, alternatively, the integro-derivative operator of order $\alpha \in \mathbb{R}$ of the function $f(t)$ defined on the interval $[t_0, t]$ may be written as:

$$ {}_{t_0}D_t^\alpha f(t) = \begin{cases} \frac{d^\alpha f(t)}{dt^n} & \text{for } \alpha > 0 \\ f(t) & \text{for } \alpha = 0 \\ {}_{t_0}I_t^{-\alpha} f(t) & \text{for } \alpha < 0 \end{cases} \tag{1} $$

under the assumption that the real function $f(t)$ is defined almost everywhere for $t \geqslant 0$, and is multiple differentiable and integrable (depending on the order α) within every interval $[0, T], T > 0$ and exponentially-restricted, i.e., there exist such real numbers $\rho, M \in \mathbb{R}$ that the following inequality $|f(t)| \leqslant Me^{\rho t}$ holds for $t > 0$, and also that the integral $\int_{t_0}^{t} (t-\tau)^\mu f(\tau) d\tau$ exists for $\mu \in \mathbb{R}, t_0 < t$ [5]. ${}_{t_0}I_t^\alpha f(t)$ is the integration operator of order $\alpha > 0$ (summation operator of order $\alpha > 0$ in discrete case, respectively). There are known several definitions of the fractional-order derivative of function $f(t)$. The one most common adopted is that introduced by Grünwald and Letnikov. A derivative of fractional order α of function $f(t)$ is here defined as

$$ {}_{t_0}^{GL}D_t^\alpha f(t) = \lim_{h \to 0} h^{-\alpha} \sum_{j=0}^{\left[\frac{t-t_0}{h}\right]} (-1)^j \binom{\alpha}{j} f(t - jh) \tag{2} $$

where the so-called generalized Newton symbol is given by

$$ \binom{\alpha}{j} = \begin{cases} 1 & \text{for } j = 0 \\ \frac{\alpha(\alpha-1)\ldots(\alpha-j+1)}{j!} & \text{for } j = 1, 2, 3, \ldots \end{cases} \tag{3} $$

Equation (2) may be rewritten as

$$ {}_{t_0}^{GL}D_t^\alpha f(t) = \lim_{h \to 0} h^{-\alpha} \sum_{j=0}^{\left[\frac{t-t_0}{h}\right]} c_j^\alpha f(t - jh) \tag{4} $$

where

$$ c_j^\alpha = (-1)^j \binom{\alpha}{j}, \qquad j = 0, 1, 2, \ldots \tag{5} $$

The Grünwald-Letnikov definition of the fractional-order derivative is particularly popular for reasons of application, especially in digital control, where discretization of an $f(t)$ function taken with a sampling period h fits naturally into this art of control. If it is considered that the number of summands in the sum (4) is unavoidably finite in practice, then eq. (2) may be replaced by its approximation

$$ {}_{t_0}^{GL}D_t^{\alpha} f(t) \approx h^{-\alpha} \sum_{j=0}^{L} c_j^{\alpha} f(t - jh) \tag{6} $$

where the number of samples L of the function $f(t)$ (or length of memory where the samples are stored in practical realization) should be chosen so that the truncation error does not exceed a given value ε. Assuming the value of the function does not exceed the value M at any point, the number of samples may be estimated from the inequality [9]

$$ L \geqslant \left(\frac{M}{\varepsilon |\Gamma(1-\alpha)|} \right)^{\frac{1}{\alpha}} \tag{7} $$

3 Fractional-Order Model Predictive Control

In the predictive control algorithm of integer order the cost function depends on the sum of the weighted squared prediction errors over the prediction horizon from N_1 to N_2 and on the sum of the weighted squared control signal increments to be sought within the control horizon from 0 to $N_u - 1$

$$ \mathbf{J}(t) = \sum_{j=N_1}^{N_2} \mu(j)[y^p(t+j\,|t) - y^r(t+j\,|t)]^2 + \sum_{j=0}^{N_u-1} \lambda(j)[\Delta u(t+j\,|t)]^2 \tag{8} $$

where $y^p(t+j\,|t)$ denotes the predicted outputs within the prediction horizon, $y^r(t+j\,|t)$ denotes the reference values within the same horizon, $\Delta u(t+j\,|t)$ are the control signal increments within the control horizon and $\mu(j), \lambda(j) \geqslant 0$ are the weight coefficients.

The cost function (8) may be expressed alternatively by using discrete summation operators

$$ \mathbf{J}(t) = {}_{t+N_1}I_{t+N_2}^{1} \mu(t)[e^p(t)]^2 + {}_t I_{t+N_u-1}^{1} \lambda(t)[\Delta u(t)]^2 \tag{9} $$

where

$$ e^p(t) = y^p(t) - y^r(t) \tag{10} $$

is the prediction error.

In 2010 it was proposed in [11] to introduce a formal generalization of integer-order sums in the cost function (9) as those of fractional order

$$ \mathbf{J}(t) = {}_{t+N_1}I_{t+N_2}^{\alpha_1} [e^p(t)]^2 + {}_t I_{t+N_u-1}^{\alpha_2} [\Delta u(t)]^2 \tag{11} $$

where $\alpha_1, \alpha_2 \in \mathbb{R}_+$ are fractional orders of integration (summation) defined in (1), applied to squared prediction errors and squared control increments, respectively. By this means the principles of a fractional-order discrete predictive algorithm have been defined by adding fractional orders α_1, α_2 as two new tuning parameters in addition to those typical ones, namely the ranges of the prediction horizon N_1, N_2 and the length of the control horizon N_u [2, 3].

Considering a more general case where weighting coefficients $\mu(j)$ and $\lambda(j)$ known from integer-order control are left in the cost function (11)

$$\mathbf{J}(t) = {}_{t+N_1}I^{\alpha_1}_{t+N_2}\mu(t)[e^p(t)]^2 + {}_tI^{\alpha_2}_{t+N_u-1}\lambda(t)[\Delta u(t)]^2 \tag{12}$$

one may find the minimum of the cost function (12) with respect to the control increment $\Delta u(t)$ in the future (within the control horizon). On taking into consideration the rule governing concatenation of fractional-order integration/differentiation operators, we obtain

$${}_{t_0}I^{\alpha}_t f(t) = {}_{t_0}I^1_t[{}_{t_0}D^{1-\alpha}_t f(t)], \qquad \alpha > 0 \tag{13}$$

Assuming for simplicity that the step of discretization $h = 1$, the cost function (12) may be rewritten in vector-matrix form

$$\mathbf{J}(t) = [E^p_{\leftrightarrow}(t)]^T \mathbf{M}_\infty [E^p_{\leftrightarrow}(t)] + [\Delta U_{\leftrightarrow}(t)]^T \mathbf{\Lambda}_\infty [\Delta U_{\leftrightarrow}(t)] \tag{14}$$

where

$$\Delta U_{\leftrightarrow}(t) = \begin{bmatrix} \vdots \\ \Delta u(t-1\,|t) \\ \hline \Delta u(t\,|t) \\ \Delta u(t+1\,|t) \\ \vdots \\ \Delta u(t+N_u-1\,|t) \end{bmatrix} = \left[\frac{\Delta U_{\leftarrow}(t)}{\Delta U_{\rightarrow}(t)}\right] \in \mathbb{R}^\infty \tag{15}$$

$$E^p_{\leftrightarrow}(t) = \begin{bmatrix} \vdots \\ e^p(t-1\,|t) \\ e^p(t\,|t) \\ \hline e^p(t+1\,|t) \\ \vdots \\ e^p(t+N_1-1\,|t) \\ \hdashline e^p(t+N_1\,|t) \\ e^p(t+N_1+1\,|t) \\ \vdots \\ e^p(t+N_2\,|t) \end{bmatrix} = \left[\frac{E^p_{\leftarrow}(t)}{E^p_{\rightarrow}(t)}\right] \in \mathbb{R}^\infty \tag{16}$$

$$\mathbf{M}_\infty = diag[\cdots\ w_\mu(0)|w_\mu(1)\ \cdots\ w_\mu(N_1-1)\,\vdots\,w_\mu(N_1)\ \cdots\ w_\mu(N_2)] \tag{17}$$

$$\mathbf{\Lambda}_\infty = diag[\cdots\ w_\lambda(-1)|w_\lambda(0)\ w_\lambda(1)\ \cdots\ w_\lambda(N_u-1)] \tag{18}$$

Taking into account (12) and (13) we get [9]

$$w_\mu(j) = \begin{cases} c^{-\alpha_1}_{N_2-j} - c^{-\alpha_1}_{N_1-j-1} & \text{for} \quad j < N_1 \\ \mu(j)c^{-\alpha_1}_{N_2-j} & \text{for} \quad N_1 \leqslant j \leqslant N_2 \end{cases} \tag{19}$$

$$w_\lambda(j) = \begin{cases} c_{N_u-1-j}^{-\alpha_2} - c_{-1-j}^{-\alpha_2} & \text{for} \quad j < 0 \\ \lambda(j) c_{N_u-1-j}^{-\alpha_2} & \text{for} \quad 0 \leqslant j \leqslant N_u - 1 \end{cases} \tag{20}$$

The diagonal infinite-dimensional weight matrices (17) and (18) may be presented equivalently as

$$\mathbf{M}_\infty = diag[\mathbf{M}_\leftarrow \mid \mathbf{M}_\rightarrow] \tag{21}$$

$$\mathbf{\Lambda}_\infty = diag[\mathbf{\Lambda}_\leftarrow \mid \mathbf{\Lambda}_\rightarrow] \tag{22}$$

From (16) it follows that

$$Y_\leftrightarrow^p(t) = \begin{bmatrix} \vdots \\ y^p(t-1\,|t) \\ y^p(t\,|t) \\ \hline y^p(t+1\,|t) \\ \vdots \\ y^p(t+N_1-1\,|t) \\ \hdashline y^p(t+N_1\,|t) \\ y^p(t+N_1+1\,|t) \\ \vdots \\ y^p(t+N_2\,|t) \end{bmatrix} = \begin{bmatrix} Y_\leftarrow^p(t) \\ \hline Y_\rightarrow^p(t) \end{bmatrix} \in \mathbb{R}^\infty \tag{23}$$

and

$$Y_\leftrightarrow^r(t) = \begin{bmatrix} \vdots \\ y^r(t-1\,|t) \\ y^r(t\,|t) \\ \hline y^r(t+1\,|t) \\ \vdots \\ y^r(t+N_1-1\,|t) \\ \hdashline y^r(t+N_1\,|t) \\ y^r(t+N_1+1\,|t) \\ \vdots \\ y^r(t+N_2\,|t) \end{bmatrix} = \begin{bmatrix} Y_\leftarrow^r(t) \\ \hline Y_\rightarrow^r(t) \end{bmatrix} \in \mathbb{R}^\infty \tag{24}$$

With the expression for the prediction vector

$$Y_\rightarrow^p(t) = \mathbf{E}\Delta U_\rightarrow(t) + Y_\rightarrow^0(t) \tag{25}$$

where $\mathbf{E}$ is the process dynamics matrix [3], the cost function (14) may be written as

$$\mathbf{J}(t) = \begin{bmatrix} Y_\leftarrow^p(t) \\ \mathbf{E}\Delta U_\rightarrow(t) + Y_\rightarrow^0(t) - Y_\rightarrow^r(t) \end{bmatrix}^T \begin{bmatrix} \mathbf{M}_\leftarrow & \mathbf{0} \\ \mathbf{0} & \mathbf{M}_\rightarrow \end{bmatrix} \begin{bmatrix} Y_\leftarrow^p(t) \\ \mathbf{E}\Delta U_\rightarrow(t) + Y_\rightarrow^0(t) - Y_\rightarrow^r(t) \end{bmatrix} + \begin{bmatrix} \Delta U_\leftarrow(t) \\ \Delta U_\rightarrow(t) \end{bmatrix}^T \begin{bmatrix} \mathbf{\Lambda}_\leftarrow & \mathbf{0} \\ \mathbf{0} & \mathbf{\Lambda}_\rightarrow \end{bmatrix} \begin{bmatrix} \Delta U_\leftarrow(t) \\ \Delta U_\rightarrow(t) \end{bmatrix} \tag{26}$$

The above quadratic problem may be solved by setting the gradient

$$\frac{\partial \mathbf{J}(t)}{\partial \Delta U_{\rightarrow}(t)} = 2(\mathbf{E}^T\mathbf{M}_{\rightarrow}\mathbf{E} + \mathbf{\Lambda}_{\rightarrow})\Delta U_{\rightarrow}(t) + 2\mathbf{E}^T\mathbf{M}_{\rightarrow}[Y^0_{\rightarrow}(t) - Y^r_{\rightarrow}(t)] \tag{27}$$

equal to zero.

Hence the optimal control is given by

$$\begin{aligned}\Delta U_{opt\rightarrow}(t) &= \left(\mathbf{E}^T\mathbf{M}_{\rightarrow}\mathbf{E} + \mathbf{\Lambda}_{\rightarrow}\right)^{-1}\mathbf{E}^T\mathbf{M}_{\rightarrow}\left[Y^r_{\rightarrow}(t) - Y^0_{\rightarrow}(t)\right] \\ &= \mathbf{K}_{\rightarrow}\left[Y^r_{\rightarrow}(t) - Y^0_{\rightarrow}(t)\right]\end{aligned} \tag{28}$$

Since only the first component of the computed control vector is used at a given time instant t (according to the principle of the receding horizon), we obtain finally

$$\Delta u_{opt}(t\,|t\,) = [1\ 0 \cdots 0]\,\Delta U_{opt\rightarrow}(t) = \mathbf{k}_{\rightarrow}[Y^r_{\rightarrow}(t) - Y^0_{\rightarrow}(t)] \tag{29}$$

where $\mathbf{k}_{\rightarrow}$ is the first row of the matrix $\mathbf{K}_{\rightarrow}$.

As may be seen, the fractional orders of discrete summation α_1, α_2 have a pronounced effect on the control signal. This confirms the significant potential difference in that how a fractional-order predictive algorithm operates as compared to a constant-order algorithm. On the one hand, it offers the controller designer new possibilities, on the other hand, it impedes significantly the controller tuning, since general guidelines on how to choose fractional orders in the cost function (12) are still lacking. One may find in literature preliminary suggestions only, e.g. those following from having been used to tune genetic algorithms [12]. Connection between the both types of controllers is shown in Table 1.

Remarks

1. As regards the structure, the relationship obtained for the fractional-order predictive controller is identical with the one for the integer-order controller. However, it should be noted that dimensions of individual vectors $\mathbf{k}^T_{\rightarrow}$, $Y^r_{\rightarrow}(t)$, $Y^0_{\rightarrow}(t) \in \mathbb{R}^{N_2}$, as well as those of the weight matrix $\mathbf{M}_{\rightarrow} \in \mathbb{R}^{N_2 \times N_2}$ and the process dynamics matrix $\mathbf{E} \in \mathbb{R}^{N_2 \times N_u}$ are in general greater than those in the integer-order algorithm ($\mathbb{R}^{N_2-N_1+1}, \mathbb{R}^{(N_2-N_1+1)\times(N_2-N_1+1)}$ and $\mathbb{R}^{(N_2-N_1+1)\times N_u}$ respectively).
2. For $\alpha_1 = \alpha_2 = 1$ in (19) and (20), (29) transforms into the equation of the integer-order predictive controller since
$$c_i^{-1} = 1 \text{ for } i \geqslant 0 \tag{30}$$
3. In the integer-order predictive controller the weighting coefficients $\mu(j), \lambda(j)$ are non-negative and are chosen by the designer in accordance with known rules. In the fractional-order predictive controller the values of weighting coefficients $w_\mu(j), w_\lambda(j)$ depend only partly on $\mu(j), \lambda(j)$, and more on fractional orders α_1, α_2, which follows from (19) and (20); they also may assume negative values.

Table 1. Connection between predictive controllers of fractional and integer order

fractional-order MPC		
$0 < \alpha_1, \alpha_2 < 1$ $w_\mu, w_\lambda \in \mathbb{R}$	$\alpha_1, \alpha_2 = 1$ $w_\mu, w_\lambda \in \mathbb{R}^+$ integer-order MPC	$1 < \alpha_1, \alpha_2$ $w_\mu, w_\lambda \in \mathbb{R}^+$

4 Fractional Predictive Control with Few Coincidence Points

In Model Predictive Control algorithms with few coincidence points [3] the cost function (8) is affected only by errors in selected points of the prediction horizon, called coincidence points. These algorithms are very easy to implement and quite readily used in practice. The ease of controller tuning (and intelligibility to service people) is also a contributory factor here. Among classic integer-order predictive algorithms the *Extended Prediction Self Adaptive Control* (EPSAC) presented originally in [13] is the well-known example of such an algorithm. The objective of the algorithm was to minimize the squared prediction errors without attaching weights to control costs. In the case of the fractional-order EPSAC algorithm the weighting coefficients in the cost function (26) are given by

- for $N_1 = 1$

$$w_\mu(j) = \begin{cases} \mu(j) c_{N_2-j}^{-\alpha_1} & \text{for} \quad j \in \Sigma \\ 0 & \text{for} \quad j \notin \Sigma \end{cases} \qquad 1 \leqslant j \leqslant N_2 \tag{31}$$

- for $N_1 > 1$

$$w_\mu(j) = \begin{cases} c_{N_2-j}^{-\alpha_1} - c_{N_1-1-j}^{-\alpha_1} & \text{for} \quad 1 \leqslant j < N_1 \\ \mu(j) c_{N_2-j}^{-\alpha_1} & \text{for} \quad j \in \Sigma \\ 0 & \text{for} \quad j \notin \Sigma \end{cases} \qquad N_1 \leqslant j \leqslant N_2 \tag{32}$$

and

$$w_\lambda(j) = 0 \qquad \text{for} \qquad 0 \leqslant j \leqslant N_u - 1 \tag{33}$$

where Σ means a set of coincidence points.

Another example of an integer-order model predictive control algorithm with few coincidence points is the proposed in [14] the *Extended Horizon Predictive Control* (EHPC) algorithm. Here the output prediction coincidence with the reference trajectory only at the end of the prediction horizon, i.e. $\mu(j) = 0$ for $j = N_1, N_1 + 1, \ldots, N_2 - 1, \mu(N_2) = \mu \neq 0$ has been adopted as a condition to be met. It can easily be shown that for the fractional-order EHPC algorithm we have

– for $N_1 = 1$

$$w_\mu(j) = \begin{cases} 0 & \text{for} \quad 1 \leqslant j < N_2 \\ \mu & \text{for} \quad j = N_2 \end{cases} \tag{34}$$

– for $N_1 > 1$

$$w_\mu(j) = \begin{cases} c_{N_2-j}^{-\alpha_1} - c_{N_1-1-j}^{-\alpha_1} & \text{for} \quad 1 \leqslant j < N_1 \\ 0 & \text{for} \quad N_1 \leqslant j < N_2 \\ \mu & \text{for} \quad j = N_2 \end{cases} \tag{35}$$

and

$$w_\lambda(j) = \lambda(j) c_{N_u-1-j}^{-\alpha_2} \quad \text{for} \quad 0 \leqslant j \leqslant N_u - 1 \tag{36}$$

If all weighting coefficients $\lambda(j)$ are identical, then eq. (36) takes a simpler form

$$w_\lambda(j) = \lambda c_{N_u-1-j}^{-\alpha_2} \quad \text{for} \quad 0 \leqslant j \leqslant N_u - 1 \tag{37}$$

Another example is the *Extended Horizon Adaptive Control* (EHAC) proposed in [14], which represents a special case of the EHPC algorithm. Here a unit prediction horizon $N_u = 1$ has been adopted. From this it follows that (34) or (35) will hold true for the fractional-order EHAC algorithm, whereas eq. (36) will take the following form

$$w_\lambda(j) = w_\lambda(0) = \lambda(0) c_{N_u-1}^{-\alpha_2} = \lambda \tag{38}$$

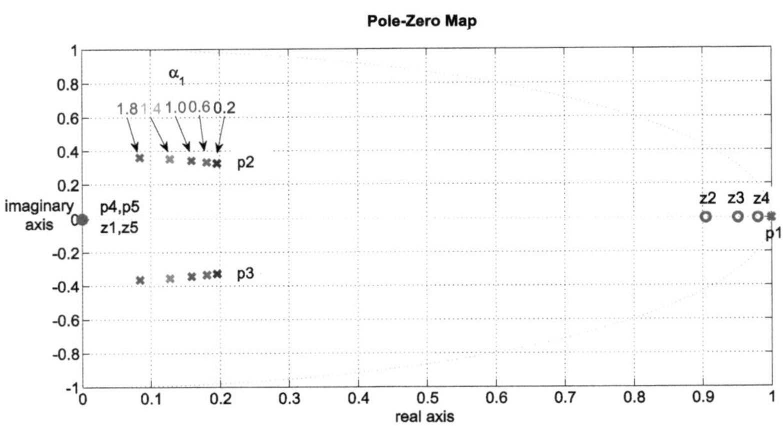

Fig. 1. Pole-zero map for EPSAC algorithm with four coincidence points $\mu_2 = \mu_4 = \mu_6 = 1$, $\alpha_1 = \{0.2 \ldots 1.8\}$

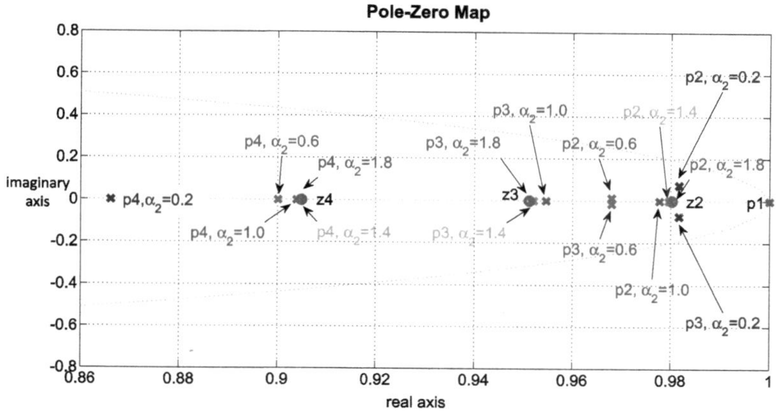

Fig. 2. Pole-zero map for EHPC algorithm for $\mu = \lambda = 1$, $\alpha_1 = 1.8$, $\alpha_2 = \{0.2 \ldots 1.8\}$

Remarks

1. It can be seen from (31) and (32) that the EPSAC algorithm of arbitrary fractional orders α_1, α_2 differs from the integer-order EPSAC algorithm for the same coincidence points,
2. It can be seen from (34) and (38) that the EHAC algorithm of arbitrary fractional orders α_1, α_2 for $N_1 = 1$ does not differ from the integer-order EHAC algorithm of the same prediction horizon length N_2.
3. The matrix $\mathbf{E}^T\mathbf{M}_{\rightarrow}\mathbf{E} + \mathbf{\Lambda}_{\rightarrow}$ in (28) must be non-singular, which is particularly important for EPSAC algorithm.

4.1 Properties of the Proposed Fractional Predictive Control

The principal advantage of the model predictive control with a small set of coincidence points is the ease of implementation and intelligible interpretation of tuning parameters. In the case of fractional-order algorithms their properties are also dependent on fractional orders α_1, α_2 in the cost function (12). Their impact on properties exhibited by the algorithm is not yet well studied, and their selection is thus not simple. To illustrate the impact of orders α_1, α_2 on properties of a fractional predictive control algorithm with few coincidence points, the dependence of the pole-zero map on fractional orders for individual algorithms has been studied. The plant model has been chosen as $\frac{1}{(1+10s)(1+20s)(1+50s)}$ with sampling time equal to 1[s]. In Figures 1 and 2 examples for selected parameters of fractional EPSAC ($N_1 = 1, N_2 = 6, N_u = 2$), and EHPC ($N_1 = 3, N_2 = 8, N_u = 4$) algorithms with a prediction horizon are displayed.

5 Conclusion

In the paper the model predictive control with a small set of coincidence points is proposed and next its properties are analyzed. The fractional orders of summation in the cost function permit affecting the control performance. At the same time, the proposed algorithms are very easy to implement, just as their integer-order counterparts.

References

1. Domek, S.: Switched state model predictive control of fractional-order nonlinear discrete-time systems. In: Pisano, A., Caponetto, R. (eds.) Advances in Fractional Order Control and Estimation, Asian J. Control, Special Issue, vol. 15(3), pp. 658–668 (2013)
2. Lawrynczuk, M.: Computationally efficient model predictive control algorithms; A neural network approach. Studies in System, Decision and Control, vol. 3. Springer (2014)
3. Tatjewski, P.: Advanced control of industrial processes. Springer, London (2007)
4. Dzielinski, A., Sierociuk, D., Sarwas, G.: Some applications of fractional order calculus. Bulletin of the Polish Academy of Sciences, Technical Sciences 58(4), 583–592 (2010)
5. Kaczorek, T.: Selected Problems of Fractional Systems Theory. LNCIS, vol. 411. Springer, Berlin (2011)
6. Podlubny, I.: Fractional Differential Equations. Academic Press, San Diego (1999)
7. Stanislawski, R.: Advances in modeling of fractional difference systems new accuracy, stability and computational results. Oficyna Wydawnicza Politechniki Opolskiej, Opole (2013)
8. Chen, Y.Q., Petras, I., Xue, D.: Fractional order control. In: American Control Conference, St. Louis, pp. 1397–1410 (2009)
9. Domek, S.: Fractional-order differential calculus in model predictive control. West Pomeranian University of Technology Academic Press, Szczecin (2013)
10. Ostalczyk, P., Rybicki, T.: Variable-fractional order dead-beat control of an electromagnetic servo. Journal of Vibration and Control 14(910), 1457–1471 (2008)
11. Romero, M., De Madrid, A.P., Manoso, C., Vinagre, B.M.: Fractional-order generalized predictive control: Formulation and some properties. In: 11th Int. Conf. Control, Automation, Robotics and Vision, Singapore, pp. 1495–1500 (2010)
12. Romero, M., Vinagre, B.M., De Madrid, A.P.: GPC Control of a Fractional Order Plant: Improving Stability and Robustness. In: 17th IFAC World Congress, Seoul, pp. 14266–14271 (2008)
13. De Keyser, R.M.C., Van Cuawenberghe, A.R.: Extended prediction self-adaptive control. In: IFAC Symp. on Identification and System Parameter Estimation, York, pp. 1317–1322 (1985)
14. Ydstie, B.E.: Extended horizon adaptive control. In: 9th IFAC World Congress, Budapest, vol. 7, pp. 133–137 (1984)

Tuning of the Half-Order Robust PID Controller Dedicated to Oriented PV System

Wojciech Mitkowski and Krzysztof Oprzedkiewicz

AGH University of Science and Technology,
Faculty of Electrotechnics, Automatics, Informatics and Biomedical Engineering,
Dept. of Automatics and Biomedical Engineering,
Al. A. Mickiewicza 30, 30059 Krakow, Poland

Abstract. In the paper tuning rules for half - order PID controller dedicated to control an oriented PV system were presented. The plant is described with the use of interval transfer function. Results were by simulations depicted.

Keywords: Fractional order PID controller, PV systems, minimal-energy control, interval systems, robust control.

1 An introduction

Problems of modeling and control for dynamic systems described with the use of non-integer models were presented by many Authors, for example: [23], [24], [2], [4] , [5], [6], [19], [25]. Problems of control fractional order, interval systems were presented for example in [20], [21], [22].

The another area of non-integer order calculus in control are non integer order controllers. They are often applied in many control systems. This problem was presented for example in [24] or [22]. It is caused by a fact, that this class of controllers assures better control performance, than traditional integer order control.

In this paper a minimal-energy control of an uncertain-parameter oriented PV system with the use of half-order PID controller is considered. The term "half order" describes orders of derivative and integral actions: they both are equal 0.5. For this system an analysis of BIBO (Bounded Input Bounded Output) stability is possible with respect to uncertainty of plants parameters.

In the paper the following problems will be presented:

- An oriented PV system and its model,
- Closed-loop control system with fractional order PID controller,
- Stability analysis for closed-loop system,
- ORA approximation for both parts of controller,
- PID tuning methods for the considered system,
- An example.

© Springer International Publishing Switzerland 2015
K.J. Latawiec et al. (eds.), *Advances in Modeling and Control of Non-integer Order Systems*, Lecture Notes in Electrical Engineering 320, DOI: 10.1007/978-3-319-09900-2_14

2 An Oriented PV System and Its Model

Let us consider a moving part of an oriented PV system shown in figure 2. The most simple scheme of this plant is a DC motor with gearbox, considered by many Authors (see [1], p. 453, 459; see [22], p. 121) shown in figure 1.

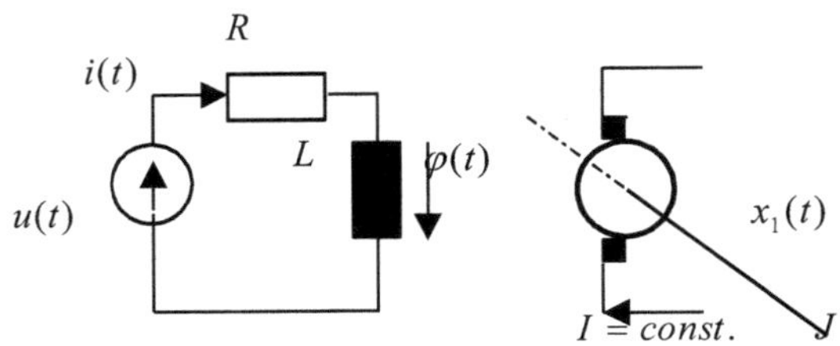

Fig. 1. A DC electric drive as a model of moving part of the oriented PV system

Fig. 2. An oriented PV system

The exact description of the plant we deal with can be found in [15], [12]. The most simple model of the plant shown in figure 1 is a transfer function with interval parameters:

$$G(s, q) = \frac{k}{T_i s} \tag{1}$$

where: $k > 0$ and T_i denote interval parameters of the PV, assembled in vector $q \in Q$ defined as follows:

$$Q = \{q = [k, T_i] : \underline{k} \leq k \geq \overline{k}\ \underline{T_i} \leq T_i \geq \overline{T_i}\} \subset I(\mathbb{R}^{\nvDash}) \tag{2}$$

Vertices (corners) of the set Q are defined as underneath:

$$\begin{aligned} q_{ll} &= [\underline{k}, \underline{T_i}] \\ q_{lh} &= [\underline{k}, \overline{T_i}] \\ q_{hh} &= [\overline{k}, \overline{T_i}] \\ q_{hl} &= [\overline{k}, \underline{T_i}] \end{aligned} \tag{3}$$

Vector $q \in Q$ describes parameters of the plant, changing during work of the system outdoor in extremally different atmospheric conditions (summer and winter, with and without snow, etc.). Additionally - these parameters have different values for moving up and moving down the PV. Exemplary values of these parameters are given in an example.

3 Closed-Loop Control System with Fractional Order PID Controller

The closed-loop control system with half-order PID controller is shown in figure 3. In scheme shown in figure 3 $G(s,q)$ denotes the plant described with the use of transfer function (1), $G_c(s)$ denotes a fractional-order PID controller, described as underneath:

$$G_c(s,p) = k_p + \frac{k_I}{s^\alpha} + k_D s^\beta \tag{4}$$

where $k_p > 0$, $k_I > 0$ and $k_D > 0$ describe the proportional, integrating and derivating actions of the controller, α and β are fractional orders of the integration and derivation actions. The controller parameters are assembled in vector p:

$$p = [k_P, k_I, k_D] \in P \subset \mathbb{R}^{+3} \tag{5}$$

where $P(q)$ is a set of all vectors p possible to technical realization. It can be also described by intervals.
The transfer function of the open loop system shown in figure 1 is equal:

$$G_o(s,p,q) = \frac{kk_D s^{\alpha+\beta} + kk_P s^\alpha + kk_I}{T_i s^{\alpha+1}} \tag{6}$$

The transfer function of the whole closed-loop system defined as: $G_z(s) = \frac{Y(s)}{R(s)}$ is expressed as follows:

$$G_z(s,p,q) = \frac{kk_D s^{\alpha+\beta} + kk_P s^\alpha + kk_I}{T_i s^{\alpha+1} + kk_D s^{\alpha+\beta} + kk_P s^\alpha + kk_I} \tag{7}$$

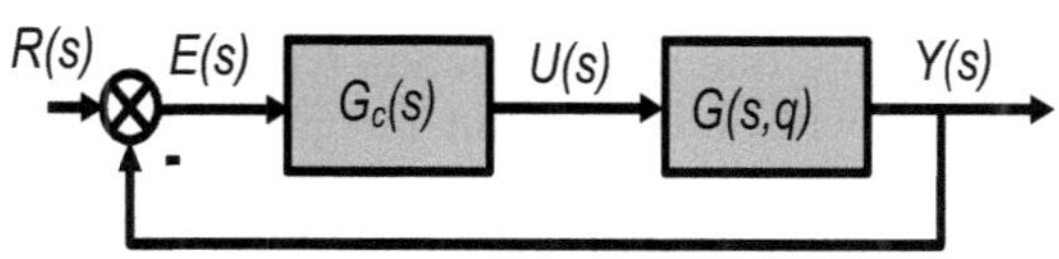

Fig. 3. A closed-loop control system for the considered PV system

The above transfer function will be applied to stability analysis and optimal tuning of the fractional - order PID controller.

4 Robust Stability Analysis of Control System

The characteristic polynomial of the control system we deal with is a function of vectors p and q and it is equal:

$$W(s,p,q) = T_i s^{\alpha+1} + kk_D s^{\alpha+\beta} + kk_P s^{\alpha} + kk_I \tag{8}$$

Now let us make an important assumption about fractional orders of integrating and derivating actions in controller: assume, that $\alpha = 0.5$ and $\beta = 0.5$. This assumption will make the asymptotic stability analysis of the system much more simplier. Additionally, results of simulations show, that stability conditions obtained for those fixed values can be generalized for broad range of fractional orders α and β. It will be shown in the Example.

After the above assumption the polynomial (8) reduces to the form:

$$W(s,p,q) = T_i s^{1.5} + kk_D s + kk_P s^{0.5} + kk_I \tag{9}$$

If we replace: $\lambda = s^{0.5}$, then the polynomial (8) reduces to the following integer order quasi-polynomial:

$$W_\lambda(\lambda,p,q) = T_i \lambda^3 + kk_D \lambda^2 + kk_P \lambda + kk_I \tag{10}$$

The BIBO stability of quasi polynomial $W_\lambda(\lambda,p,q)$ described by (10) determines the BIBO stability of polynomial (9) with respect to the following conditon (see for example [2], p. 21 and 22):

Theorem 1. *The system (10) is BIBO stable (Bounded Input Bounded Output) if and only if:*

$$|arg(\lambda_i)| > \alpha\tfrac{\pi}{2} \ \ \forall i = 1...N$$

where λ_i denotes i-th root of characteristic polynomial $W(s^\alpha)$

Notice, that for $\alpha = 0.5$ the stability area covers also the part of right complex semiplane and it covers the whole stability area for integer order system with $\alpha = 1$.

Next, the PD controller ($k_I = 0$) makes the system structurally stable, but the PI controller (with $k_D = 0$) is not able to assure the asymptotic stability, because it causes the lossing one of parameters the quasi polynonial (10). The system with PI controller will be stable, but not asymptotically stable.

The quasi-polynomial $W_\lambda(\lambda,p,q)$ is an interval polynomial and to test their stability the Charitonov theorem can be applied (see for example [3]). Next - the stability of quasi-polynomial(10) implies the stability of fractional-order polynomial (9).

The Hurwitz array $H(p, q)$ for the quasi-polynomial (10) is also a function of vectors p and q and it has the following simple form:

$$H(p,q) = \begin{bmatrix} kk_D & kk_I & 0 \\ T_i & kk_P & 0 \\ 0 & kk_D & kk_I \end{bmatrix} \tag{11}$$

From Hurwtiz criterion we at once formulate the sufficient and necessary condition of asymptotic stability the quasi-polynomial (10), expressed with the use of both vectors p and q:

$$k\left(kk_D k_P - k_I T_i\right) > 0 \iff k_I < \frac{k}{T_i} k_D k_P \tag{12}$$

With the use of (12) we can express the set $P_{\lambda q}(q)$, for which the condition (12) is met:

$$P_{\lambda q}(q) = \{p_\lambda = [k_P, k_I, k_D] \in \mathbb{R}^3, q \in Q : \forall q \in Q : k_I < \frac{k}{T_i} k_D k_P\} \tag{13}$$

Furthermore, let us introduce the set of controller parameters $P_\lambda(q)$ for which the interval quasi polynomial $W_\lambda(\lambda, p, q)$ will be asymptotically stable. It can be assigned as a common part of sets $P_{\lambda q}$ assigned for all corners the set Q:

$$P_\lambda(q) = P_{\lambda q}(q_{ll}) \cap P_{\lambda q}(q_{lh}) \cap P_{\lambda q}(q_{hh}) \cap P_{\lambda q}(q_{hl}) \tag{14}$$

where q_{ll}, q_{lh},q_{hh},q_{hl} are described by (3). between sets $P(q)$ and $P_\lambda(q)$ there exist the following dependence:

$$P_\lambda(q) \subset P(q) \tag{15}$$

The sets$P(q)$ described by (13) have simple geometric interpretation. For fixed plant parameters (for example corners of set Q) they are solids in $\mathbb{R}^3$ space, limited by ranges of parameters k_P and k_D and plane described by function (12). Next, the set (14) is a common part of all these sets. Its estimating will be shown in an example.

5 The ORA Approximation

The modeling of fractional order controller at MATLAB/SIMULINK platform requires us the use of a finite-dimensional and integer order approximation. It this case the ORA approximation can be applied (see for example [9]). It allows us to estimate the elementary factor s^α as follows (recommended range of α $0 < \alpha < 1$):

$$s^\alpha \cong k_a \prod_{n=1}^{N} \frac{1 + \frac{s}{\mu_n}}{1 + \frac{s}{\nu_n}} \tag{16}$$

In (16) N denotes the order of approximation, μ_n and ν_n denote approximation coefficients equal:

$$\begin{aligned} \mu_1 &= \omega_l \sqrt{\eta} \\ \nu_n &= \mu_n \gamma, \quad n = 1...N \\ \mu_{n+1} &= \nu_n \eta, \quad n = 1...N-1 \end{aligned} \tag{17}$$

where:

$$\begin{aligned} \gamma &= \left(\tfrac{\omega_h}{\omega_l}\right)^{\frac{\alpha}{N}} \\ \eta &= \left(\tfrac{\omega_h}{\omega_l}\right)^{\frac{1-\alpha}{N}} \end{aligned} \tag{18}$$

In (18) ω_l and ω_h denote the range of pulsace, for which the approximation is going to be applied. The estimating of these parameters is presented in paper [8].

If we need to approximate the non integer order β greater than 1, it can be done with the use of the following elementary dependence:

$$s^{\beta} = s^{\alpha} s^{m}, \quad \alpha \in (0;1), \quad m \in \mathbb{Z} \tag{19}$$

Furthermore, if an elementary integral described by the transfer function: $\frac{1}{s^{\alpha}}$ is required to be approximated, it can be done as follows:

$$\frac{1}{s^{\alpha}} = \frac{1}{s} s^{1-\alpha}, \quad \alpha \in (0;1) \tag{20}$$

6 Robust Half-Order PID Tuning for the Considered System

During control an oriented PV system the most typical control stategy is the minimal energy control. Energy consumption and dissipation can be described with the use of the following cost function:

$$I(p,q) = \int_0^{t_k} \left(w_1 u^2(t) + w_2 e^2(t)\right) dt \tag{21}$$

where $e(t)$ denotes the error in the control system, w_1 and w_2 denote normalized weight coefficients: $w_1 + w_2 = 1$, $t_k < +\infty$ denotes the maximal, final time of moving the PV.

The cost function (21) is a function both of plant parameters, described by vector q and controller parameters, described by vector $p_{\lambda q} \in P_{\lambda q}$.

The optimal tuning of the considered robust, half-order PID controller consists in finding such a vector : $p_{\lambda q} \in P_{\lambda q}$ for which the value of cost function (21) will meet the following conditions:

- It should be minimal or close to minimal in the whole set of uncertain plant parameters Q, described by (2) and (3).

- big perturbations of the vector q should cause small perturbations of cost function (21).

To assign the parameters assuring the robustness of controller the following simple algorithm can be applied:

1. Calculate the optimal controller parameters for all corners of set Q. We obtain four vectors $p_0(q)$.
2. Assign set of robust controller parameters as a mean of all sets for all corners

This above algorithm can be easily numerically done with the use of MATLAB and it will be shown in an example.

7 An Example

As an Example let us consider the PV system shown in figure 1 and described with the use of transfer function (1)-(3). Their parameters were assigned with the use of identification experiments at real PV system (see [17]) and for elevation angle they are equal:
$T_i = [0.57; 0.71][s]$, $k = [0.55; 0.64]$
Furthermore, let us consider the following ranges of controller parameters k_P and k_D:
$0.1 < k_P < 10.0$
$0.1 < k_D < 10.0$
For the above ranges of parameters the third parameter k_I can be estimated with the use of stability condition (12). It is easy to see, that the stability set $P_\lambda(q)$ defined by (14) is equal: $P_\lambda(q) = P_{\lambda q}(q_{lh})$. The stability set $P_\lambda(q)$ is shown in figure 4. Analogically the stability set $P_\lambda(q)$ can be estimated. It is achieved also for corner q_{lh} and it is shown in figure 5.

The SIMULINK diagram of the control system shown in the figure 3 is shown in figure 6. To modeling the half order PID controller the ORA approximation

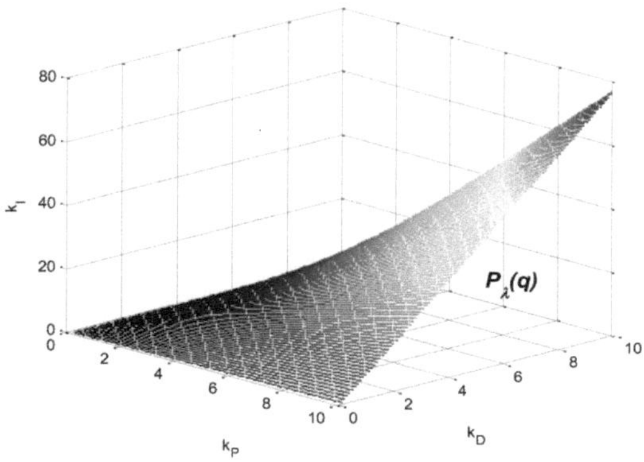

Fig. 4. The stability set $P_\lambda(q)$, moving up

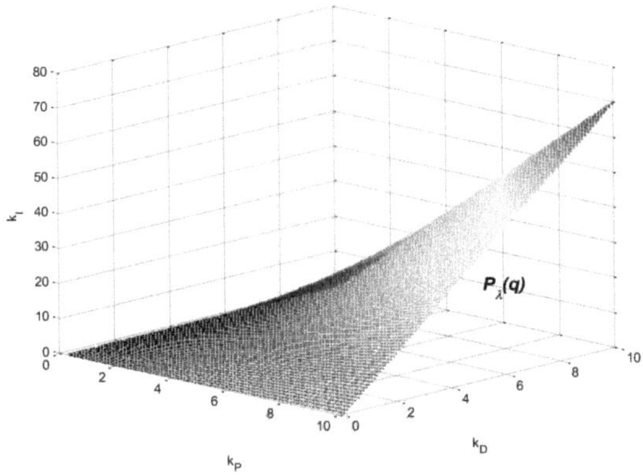

Fig. 5. The stability set P_λ, moving down

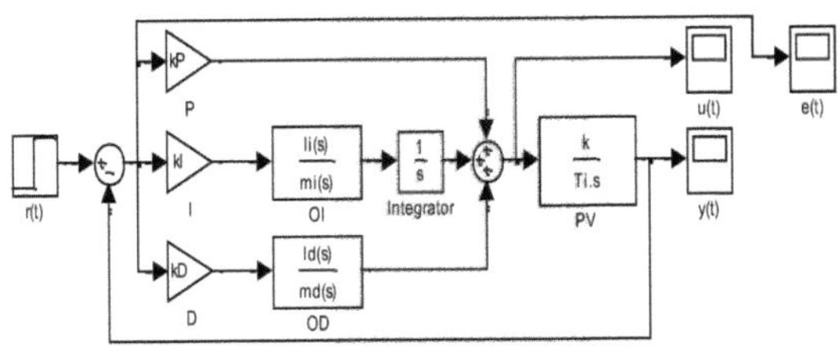

Fig. 6. The SIMULINK model of the considered control system

described by (16) - (20) was applied. The parameters of ORA approximation are shown in table 1, the final time t_k was equal: $t_k = 10[s]$. Model shown in 6 was applied to numerical calculation of coefficients minimizing the cost function (21). Results are shown in tables 2 - 5. Next the parameters of robust controller will be calculated with the use of algorithm presented in the previous section. They can be calculated for all sets of weight coefficient of cost function and they are shown in table 6.

Collection of step responses of the control system with robust half-order PID controller we deal with are shown in figure 7.

From analysis of tables 2 - 6 and figure 7 it turns out, that proposed robust controller described in table 6 assures very similar control performance in each corner of set Q, as optimal controller dedicated especially to this corner.

An another interesting problem is the behaviour of the considered control system in the situation, when orders of the the derivation and integral actions, α

Table 1. Parameters of ORA approximation applied in Example

α	N	ω_l	ω_h
0.5	5	0.1	10.0

Table 2. Optimal parameters of PID controller for different weight coefficients and corner $q_{ll} = [0.55; 0.57]$ of the set Q

weights	PID parameters	$I(p,q)$
w_e=0.2 w_u=0.8	k_P=0.4037 k_I=0.1 k_D=0.1551	0.4137
w_e=0.5 w_u=0.5	k_P=0.9181 k_I=0.1 k_D=0.1104	0.5212
w_e=0.8 w_u=0.2	k_P=1.9071 k_I=0.1 k_D=0.1	0.4186

Table 3. Optimal parameters of PID controller for different weight coefficients and corner $q_{lh} = [0.55; 0.71]$ of the set Q

weights	PID parameters	$I(p,q)$
w_e=0.2 w_u=0.8	k_P=0.3957 k_I=0.1 k_D=0.1744	0.5205
w_e=0.5 w_u=0.5	k_P=0.9125 k_I=0.1 k_D=0.1242	0.6487
w_e=0.8 w_u=0.2	k_P=1.9111 k_I=0.1 k_D=0.1	0.5205

Table 4. Optimal parameters of PID controller for different weight coefficients and corner $q_{hh} = [0.64; 0.71]$ of the set Q

weights	PID parameters	$I(p,q)$
w_e=0.2 w_u=0.8	k_P=0.4014 k_I=0.1 k_D=0.1604	0.4471
w_e=0.5 w_u=0.5	k_P=0.9166 k_I=0.1 k_D=0.1141	0.5578
w_e=0.8 w_u=0.2	k_P=1.9086 k_I=0.1 k_D=0.1	0.4479

Table 5. Optimal parameters of PID controller for different weight coefficients and corner $q_{hl} = [0.64; 0.57]$ of the set Q

weights	PID parameters	$I(p,q)$
w_e=0.2 w_u=0.8	k_P=0.4088 k_I=0.1 k_D=0.1433	0.3590
w_e=0.5 w_u=0.5	k_P=0.9218 k_I=0.1 k_D=0.1006	0.4483
w_e=0.8 w_u=0.2	k_P=1.9036 k_I=0.1 k_D=0.1	0.3603

Table 6. Optimal parameters of robust PID controller and cost function $I(q)$ for all corners of set Q

weights	PID parameters	$I(q_{ll})$	$I(q_{lh})$	$I(q_{hh})$	$I(q_{hl})$
w_e=0.2 w_u=0.8	k_P=0.4024 k_I=0.1 k_D=0.1583	0.4178	0.5205	0.4472	0.3591
w_e=0.5 w_u=0.5	k_P=0.9172 k_I=0.1 k_D=0.1123	0.5731	0.7134	0.5578	0.4483
w_e=0.8 w_u=0.2	k_P=1.9076 k_I=0.1 k_D=0.1	0.4601	0.5721	0.4479	0.3603

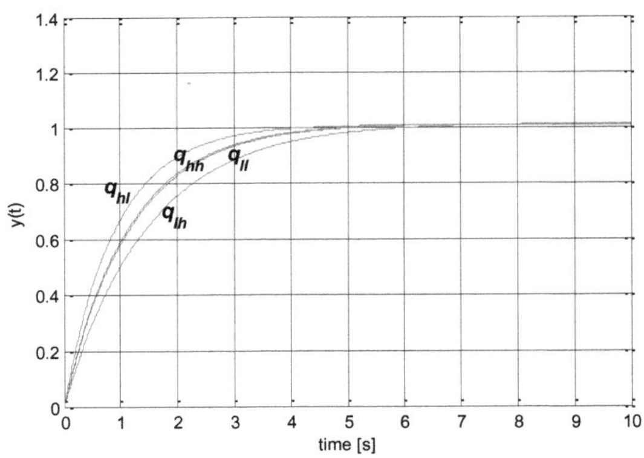

Fig. 7. The step responses of control system with optimal, robust half-order PID controller for all corners of set Q (weights of cost function during tuning: $w_e = w_u = 0.5$)

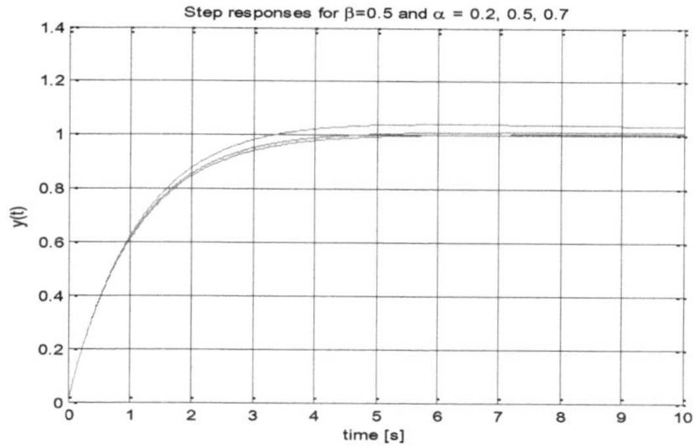

Fig. 8. The step responses of control system with perturbed order of integration action

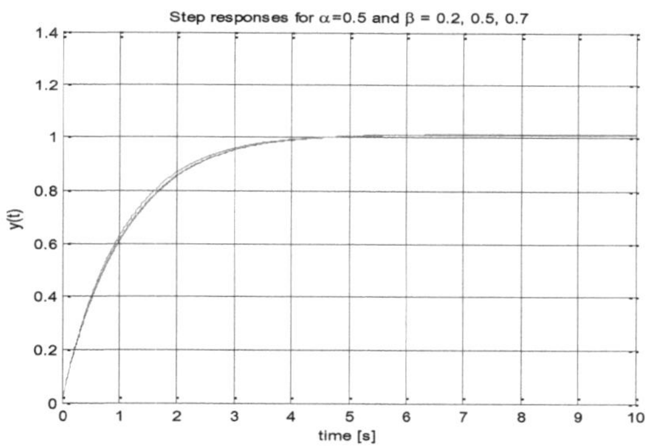

Fig. 9. The step responses of control system with perturbed order of derivation action

and β are not exactly equal 0.5, but they are perturbated in the wide range. The most important property of the system is the asymptotic stability. Exemplary tests of the system behaviour for perturbed values of α and β are shown in figures 8 and 9. The simulations were done for the following parameters of control system: $k = 0.55$, $T_i = 0.57[s]$, $k_P = 0.9166$, $k_I = 0.1$, $k_D = 0.1141$.

From the both figures it can be concluded, that the asymptotic stability is kept and the control performance does not singnificantly depend on values α and β.

8 Final Conclusions

Final conclusions from the paper can be formulated as follows:

- The assumption about fixed orders of both controller actions equal 0.5 allows us to analyze the robust stability analysis with the use of tools dedicated to integer-order systems. This makes this analysis much more simplier.

- The proposed in paper robust half order PID controller assures the good control performance for each corner of the uncertain parameter space Q.
- The tuning of the controller can be done with the use of numerical methods and iterval model of the PV system. In practice it can be done with the use of suitable tools implemented at a SCADA system.
- The optimal parameters calculated for half-order controller (with fixed orders) assures the asymptotic stability and good control performance also for perturbed values of fractional orders α and β.
- An interesting problem for the control system presented in this paper is the analytical estimating of asymptotic stability areas with assumption, that fractional orders of controller α andβ are not constant and equal 0.5, but they can vary in certain range. Solving of this problem should significantly increase the usefulnees of presented results.
- The proposed robust half-order PID controller is recently implemented at controller dedicated to control the considered PV system.

Acknowledgments. This paper was sponsored by NCN grant no 6693/B/T02/ 2011/40.

References

1. Athans, M., Falb, P.L.: Optimal Control. An Introduction to the Theory and Its Applications, pp. 1–858. WNT, Warszawa (1969) (in Polish)
2. Caponetto, R., Dongola, G., Fortuna, L., Petr, I.: Fractional Order Systems: Modeling and Control Applications. Series A, Series on Nonlinear Science, vol. 72, pp. 1–178. World Scientific Publishing Co. Pte. Ltd., New Jersey (2010); Chua, L.O. (Series edn.)
3. Charitonow, W.L.: Ob asimptoticzeskoj ustojciwosti polozenija rawnowie-sija siemiejstwa sistiem liniejnych differencyalnych urawnienij. Diff. Urawnienija 14(11), 2086–2088 (1978)
4. Das, S.: Functional Fractional Calculus for System Identyfication and Control, pp. 1–240. Springer, Berlin (2010)
5. Kaczorek, T.: Selected Problems of Fractional Systems Theory. LNCIS, vol. 411. Springer, Berlin (2011)
6. Mathews, J.H., Fink, K.D.: Numerical Methods Using Matlab, 4th edn., pp. 1–680. Person Education Inc., New Jerssey (2004)
7. Merrikh-Bayat, F., Karimi-Ghartemani, M.: Method for designing $PI^{\lambda}D^{\mu}$ stabilisers for minimum-phase fractional-order systems. IET Control Theory Appl. 4(1), 61–70 (2008, 2010)
8. Merrikh-Bayat, F.: Efficient method for time-domain simulation of the linear feedback systems containing fractional order controllers. ISA Trans. 50, 1706 (2011)
9. Merrikh-Bayat, F.: Rules for selecting the parameters of Oustaloup recursive approximation for the simulation of linear feedback systems containing $PI^{\lambda}D^{\mu}$ controller. Commun. Nonlinear Sci. Numer. Simulat. 17(2012), 1852–1861 (2012)
10. Merrikh-Bayat, F.: General Rules for Optimal Tuning the $PI^{\lambda}D^{\mu}$ Controllers With Application to First-Order Plus Time Delay Processes. The Canadian Journal of Chemical Engineering 90, 1400–14010 (2012)

11. Mitkowski, W., Oprzedkiewicz, K.: Optimal control for an uncertain parameters, second order plant with the use of discrete controller. Journal of Mathematics and System Science 2(4), 258–263 (2012) ISSN 2159-5291
12. Mitkowski, W., Oprzędkiewicz, K.: Fractional-order $P2D^{\beta}$ controller for uncertain parameter DC motor. In: Mitkowski, W., Kacprzyk, J., Baranowski, J. (eds.) Theory & Appl. of Non-integer Order Syst. LNEE, vol. 257, pp. 249–259. Springer, Heidelberg (2013)
13. Mitkowski, W., Oprzedkiewicz, K.: Application of fractional order transfer functions to modeling of high-order systems. In: IFIP TC 7/2013: System modelling and Optimization, September 8-13. programme and abstracts / IFIP, S. 71. AlpenAdria Universitt Klagenfurt, IFIP, Klagenfurt (2013)
14. Oprzedkiewicz, K.: An additional PD-corrector in a PID-control system for a high-order control plant POmiary. Automatyka, Robotyka (PAR) 11, 36–39 (2006) (in Polish)
15. Oprzedkiewicz, K.: Practical control of dynamic systems with point spectrum and interval parameters Krakw: AGH Uczelniane Wydawnictwa Naukowo-Dydaktyczne. s. - (Rozprawy Monografie / Akademia Grniczo-Hutnicza im. Stanisawa Staszica w Krakowie) 176 (2008) ISSN 0867-6631; 186
16. Oprzedkiewicz, K.: A robust suboptimal PD servo controller for an oriented PV system. Automatyka = Automatics 16(1), 59–71 (2012), `http://journals.bg.agh.edu.pl/AUTOMAT/2012.16.1/automat.2012.16.1.59.pdf` ISSN 1429-3447
17. Oprzdkiewicz, K., Teneta, J., Zaczyk, M., Wickowski, W., Gowacz, W.: Metoda pomiarw parametrw orientowanego ogniwa sonecznego Parameters identification of an oriented PV system. PAR Pomiary Automatyka Robotyka 9, 74–79 (2012), ISSN 1427-9126
18. Padula, F., Visioli, A.: Tuning rules for optimal PID and fractional-order PID controllers. Journal of Process Control 21, 69–81 (2011)
19. Pan, I., Das, S.: Intelligent Fractional Order Systems and Control. SCI, vol. 438. Springer, Berlin (2013)
20. Petr, I., Chen, Y.Q., Vinagre, B.M.: A Robust Stability Test Procedure for a Class of Uncertain LTI Fractional-Order Systems. In: Proc. of the ICCC 2002, Malenovice, Czech Republic, May 27 - 30, pp. 247–252 (2002)
21. Petr, I., Chen, Y.Q., Vinagre, B.M., Podlubny, I.: Stability of Linear Time Invariant Systems with Interval fractional Order and Interval Coefficients. In: Proc. of the Second IEEE International Conference on Computational Cybernetics, Vienna, August 30, September- 1, pp. 341–346 (2004)
22. Petr, I.: Fractional-Order Feedback Control of a DC Motor. J. of Electrical Engineering 60(3), 117–128 (2009)
23. Podlubny, I.: Fractional Differential Equations. Academic Press, San Diego (1999a)
24. Podlubny, I.: Fractional-order systems and controllers. IEEE Trans. on Automatic Control 44(1), 208–214 (1999b)
25. Woodford, C., Phillips, C.: Numerical Methods with Worked Examples, pp. 1–273. Chapman and Hall, London (1997)

Order Functions Selection in the Variable-, Fractional-Order PID Controller

Piotr W. Ostalczyk, Piotr Duch, Dariusz W. Brzeziński, and Dominik Sankowski

Institute of Applied Computer Science, Lodz University of Technology
18/22 Stefanowskiego St., 90-924 Łódź, Poland
postalcz@p.lodz.pl

Abstract. In the paper a variable-, fractional-order PID (VFOPID) controller microprocessor realization problems are discussed. In such controllers the variable-, fractional-orders backward differences and sums (VFOBD/S) are used to perform closed-loop system error discrete-time differentiation and integration. In practice all digitally differentiated and integrated signals are noised so there is a necessity of a digital signal pre-filtering. This additionally loads the DSP system. A solution of this problem is proposed. Also the possibilities of the VFOPID controller DSP realizations are presented and compared with the computer simulation results.

Keywords: Variable-, fractional-order PID controller, discrete-time system, real-time system.

1 Introduction

The PID control strategies have been for over 60 years a fundamental structure in the control with feedback [1, 19]. Because of its usefulness it seems to be desirable to develop a digital control algorithms based on the fractional-order backward differences and sums being approximations of the differentiation and integration actions [16, 20] The proposed solution has numerous advantages in comparison to the classical PID and fractional-order PID controllers [2, 9, 10, 14, 16, 17, 22–24] which can be treated as special cases of our VFOPID controller. The VFOPID controller has the characteristics of both the classic and the fractional-order PID controller. The main advantages of the proposed VFOPID control strategies are:

* it contains both classical PID (for constant order functions $\nu_k = 1 = const$, $\mu_k = -1 = const$) and FOPID (for constant order functions $\nu_k \in \mathbb{R}_+ \setminus \mathbb{Z}_+ = const$, $\mu_k \in \mathbb{R}_- \setminus \mathbb{Z}_- = const$) controller properties,
* presented in the paper simple choice of the order functions guarantee both FOPID and PID transient behaviour in two time intervals removing the disadvantage of the FOPID controller related to so called growing calculation tail,

© Springer International Publishing Switzerland 2015
K.J. Latawiec et al. (eds.), *Advances in Modeling and Control of Non-integer Order Systems*,
Lecture Notes in Electrical Engineering 320, DOI: 10.1007/978-3-319-09900-2_15

- ⋆ application of the variable fractional and integer differentiation and integration orders in a finite time interval equal to the duration of the transition process allows richer possibilities of dynamics shaping of the closed-loop system in comparison to the classical PID controller,
- ⋆ smooth order functions e.g. with bounded first-order difference lead to smoother transient states,
- ⋆ appropriate choice of the OF leads to much less ISE performance criterion with comparison to the classical PID control,- application of the differentiation and summation integer orders after a finite period of time close to the closed-loop system steady-state allows the removal of computational problems related to the fractional-order numerical integration and differentiation,
- ⋆ assumption of the simplified forms of the fractional-order difference and sum allows the action of the closed-loop system during the transients with a controller with constant order L (constant value of the calculation tail). In practical applications of the PID controllers there is a need to pre-filter a processed signal [6]. This additionally ballasts the microprocessor due to the growing calculation time and memory occupation in the fractional differentiation and integration.

The paper is organized as follows. In section 2 the Grünwald-Letnikov and the Horner equivalent forms of the variable-, fractional-order differences/sums (VFOBD/S) are introduced. Also, the two PID controller structures are presented: parallel and serial. Next, in section 3 the VFOBD/S numerical evaluation accuracy is considered. After that a problem of signal pre-filtering is shortly discussed. Some proposals of the differentiation and integration order functions are given in section 4.Finally, some chosen step responses of the variable-, fractional orders PI and PID controllers are presented.

2 Mathematical Preliminaries

2.1 The Grünwald-Letnikov Form of the VFOBD/S [13]

For a given discrete-time bounded function f_k and an order function $\nu_k > 0$ the Grünwald-Letnikov variable-, fractional-order backward difference (VFOBD) is defined as a sum

$$ {}_0^{GL}\Delta_k^{(\nu_k)} f_k = \sum_{i=0}^{k} a_i^{(\nu_k)} f_{k-i} \tag{1} $$

where

$$ \begin{aligned} a_i^{(\nu_k)} &= 0 && \text{for } i = -1, -2, -3, \cdots \\ a_0^{(\nu_k)} &= 1 && \\ a_i^{(\nu_k)} &= a_{i-1}^{(\nu_k)} \left(1 - \frac{\nu_k + 1}{i}\right) && \text{for } \quad i = 1, 2, 3, \cdots \end{aligned} \tag{2} $$

2.2 The Horner Form of the VFOBD/S

For a given discrete-time bounded function f_k and an order function $\nu_k > 0$ the Horner form of the VFOBD is defined as follows [12]

$$ {}_0^H\Delta_k^{(\nu_k)} f_k = c_0^{(\nu_k)}[f_k + c_1^{(\nu_k)}[f_{k-1} + \cdots + c_{k-2}^{(\nu_k)}[f_{k-2} + c_{k-1}^{(\nu_k)}[f_{k-1} + c_k^{(\nu_k)} f_0]] \cdots]] \quad (3) $$

with coefficients $c_i^{(\nu_k)}$

$$ c_i^{(\nu_k)} = \begin{cases} 1 & \text{for} \quad i = 0 \\ 1 - \dfrac{\nu_k + 1}{i} & \text{for} \quad i = 1, 2, \cdots \end{cases} \quad (4) $$

one can easily prove that

$$ {}_0^H\Delta_k^{(\nu_k)} f_k = {}_0^{GL}\Delta_k^{(\nu_k)} f_k. $$

2.3 The VFOPID Controller

Two VFOPID controller structures are considered: parallel and serial. The VFOPID controller structure is presented in Figure 1.

$$ u_k = K_P e_k + K_I {}_0^H\Delta_k^{(\nu_k)} e_k + K_D {}_0^H\Delta_k^{(\nu_k)} e_k \quad (5) $$

where K_p, K_I, K_D are the proportional, integral and derivative gains, respectively, e_k is the input to the controller and u_k is the controller output signal. Discrete-variable $\nu_k > 0$ and $\mu_k < 0$ are the differentiation and summation order functions respectively. For constant order functions (constant orders) one immediately gets the FOPID controller structure. As a special case of constant fractional orders one obtains the classical PID controller structure. The principle of the VFOPID controller is an operation during the closed-loop system transients. At the system steady state, the VFOPID controller resets and waits for the appearance of a non-zero error signal, which is further processed according to the VFOPID algorithm. Therefore the closed-loop system dynamics always begins at $k = 0$ and ends when the system reaches its steady-state. This operation allows the application of the VFOPD algorithm proposed by us for instance to the control the robot arm.

The cascade VFOPID controller structure is presented in Figure 2 and described by equation

$$ u_k = K_{PID} {}_0^H\Delta_k^{(\zeta_k)} e_k \quad (6) $$

where K_{PID} is the controller gain and the controller order function satisfies condition

$$ \zeta_k = \begin{cases} \zeta_{D,k} > 0 & \text{for} \quad 0 \le k < k_d \\ \zeta_{P,k} = 0 & \text{for} \quad k_d \le k < k_i \\ \zeta_{I,k} < 0 & \text{for} \quad k_i \le k \end{cases} \quad (7) $$

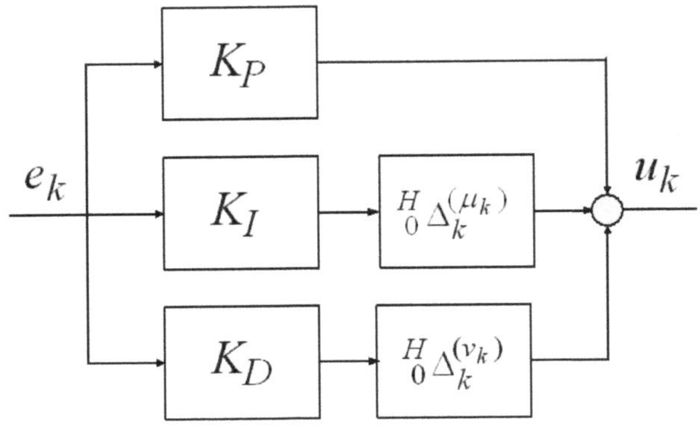

Fig. 1. Block diagram of the parallel VFOPID controller

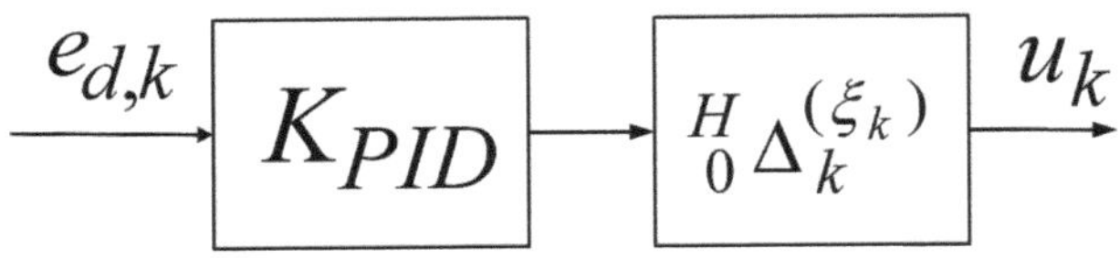

Fig. 2. Block diagram of the serial VFOPID controller

3 VFOPID Controller Order Functions Shape Requirements

There is an immense choice of the order functions (for differentiation and integration) possible in realization. Determining maximal number of samples after which the VFOPID controller transforms itself into the classical PID one stops the increase of the number of processed samples. The VFOPID controller has also characteristics of FOPID (with removed growing number of samples disadvantage). To preserve a closed-loop system with the VFOPID controller and typical plants (stable and without poles at $z = 1$) fundamental properties one should impose the following assumptions on the order functions. The zero steady-state error will be preserved for

$$\lim_{k\to\infty} \mu_k = -1 \tag{8}$$

It is well known that the closed-loop system stability is protected by the controller differentiation action [11]. Hence one may admit

$$\lim_{k\to\infty} \nu_k = 1 \tag{9}$$

Conditions (8) and (9) can be strengthened by assumptions that for some constant integers $0 \leq L_\mu, L_\nu < \infty$

$$\mu_k = \begin{cases} f_k & \text{for} \quad k \leq L_\mu \\ -1 & \text{for} \quad k > L_\mu \end{cases}, \quad \nu_k = \begin{cases} g_k & \text{for} \quad k \leq L_\nu \\ 1 & \text{for} \quad k > L_\nu \end{cases} \tag{10}$$

for $-\infty < f_k \leq 0$ and $0 \leq g_k < +\infty$

Order functions (10) avoid the fundamental problems of the VFOPID practical realizations related to the microprocessor memory and calculation time lack. This is caused by the fact that for $k \geq max\left(L_\mu, L_\nu\right)$ the VFOPID controller achieves properties of the classical PID one.This causes that the VFOPID controller parameters K_p, K_I, K_D can be evaluated due to the commonly known methods [1]. The special choice of the order functions enables also a preservation of the closed-loop system stability. One considers a closed-loop system with the classical PID controller. For a chosen PID controller parameters K_p, K_I, K_D preserving the closed-loop stability.

Theorem 1. *If the closed-loop system is (asymptotically) stable for the PID controller with some K_p, K_I, K_D , it is also (asymptotically) stable for the VFOPID controller with the order functions satisfying* (10) *and the same parameters.*

Proof. The linear, time-invariant closed-loop system stability doee not depend on the initial conditions. The closed-loop system with the VFOPID controller over a time interval $0 \leq k < max\left(L_\mu, L_\nu\right)$ may be stable or unstable. By assumption it is (asymptotically) stable over $k \geq \left(L_\mu, L_\nu\right)$ and the system transient behaviour over $0 \leq k < \left(L_\mu, L_\nu\right)$ is memorized by initial conditions.

Additional requirement imposed on the order functions μ_k, ν_k may be their monotonicity. Below two examples of the VFOPID controllers satisfying conditions (8) and (9) with the monotonicity requirement.

$$\mu_k = \begin{cases} -\dfrac{k}{L_\mu} & \text{for} \quad k < L_\mu \\ -1 & \text{for} \quad k \geq L_\mu \end{cases}, \quad \nu_k = \begin{cases} \dfrac{k}{L_\nu} & \text{for} \quad k < L_\nu \\ 1 & \text{for} \quad k \geq L_\nu \end{cases} \tag{11}$$

$$\mu_k = \begin{cases} \dfrac{1 - e^{\alpha_\mu k}}{e^{\alpha_\mu L_\mu} - 1} & \text{for} \quad k \leq L_\mu \\ -1 & \text{for} \quad k > L_\mu \end{cases}, \quad \nu_k = \begin{cases} \dfrac{e^{\alpha_\nu k} - 1}{e^{\alpha_\nu L_\nu} - 1} & \text{for} \quad k \leq L_\nu \\ 1 & \text{for} \quad k > L_\nu \end{cases} \tag{12}$$

$a_\nu, a_\mu < 0$

From (10) and (11) it is clear that the VFOPID controllers are the VO during the finite time interval $0 \leq k < max\left(L_\mu, L_\nu\right)$ transforming after into the classical PID controller. Regardless of whether the system has reached steady state or not the system memory is cleared and the VFOPID controller waits for the change of the closed-loop system reference signal r_k or disturbance d_k. Considering as a closed-loop system performance criterion a function $J\left(K_p, K_I, K_D, \nu_k, \mu_k\right)$ one may expect that

$$J\left(K_p, K_I, K_D, \nu_k, \mu_k\right) \leq J\left(K_p, K_I, K_D, \nu, \mu\right) \leq J\left(K_p, K_I, K_D, 1, -1\right)$$

what is the main advantage of the VFOPID controller.

It is possible to synthesis the controller having properties of the VFO controller over only few samples (in this paper an opposite case arm is presented in the next Section). To present wide possibilities of the VFOPID controller in closed-loop system transients shaping an example is presented.

Example 1

For the typical in robotics angular path (see Figure 1) one considers two controllers: the PD and VIOPI. All controllers have the same gain coefficients $K_p = K_D = 1.7$. The robot arm has properties of an integrating element and is modeled by a discrete integrator. Such element preserves the output signal value even when the input signal reaches zero value. so is in the robot arm, especially when it is equipped in the self-locking gear. The simulation results (order functions, errors, control and output signals) of a closed-loop system (CLS) are presented in Figs. 4, 5, 6, 7. The controllers order functions are described by formulae given in Table 1. In Fig. 3 values of integral square error (ISE) performance indexes are plotted.

The stability of the CLS is important in the time interval $k > max\,(L_{\mu,L_\nu})$ and it preservation successfully guarantee theorem 1. The presented theorem is entirely sufficient for a design of a working (and safe) real CLS with the VFOPID

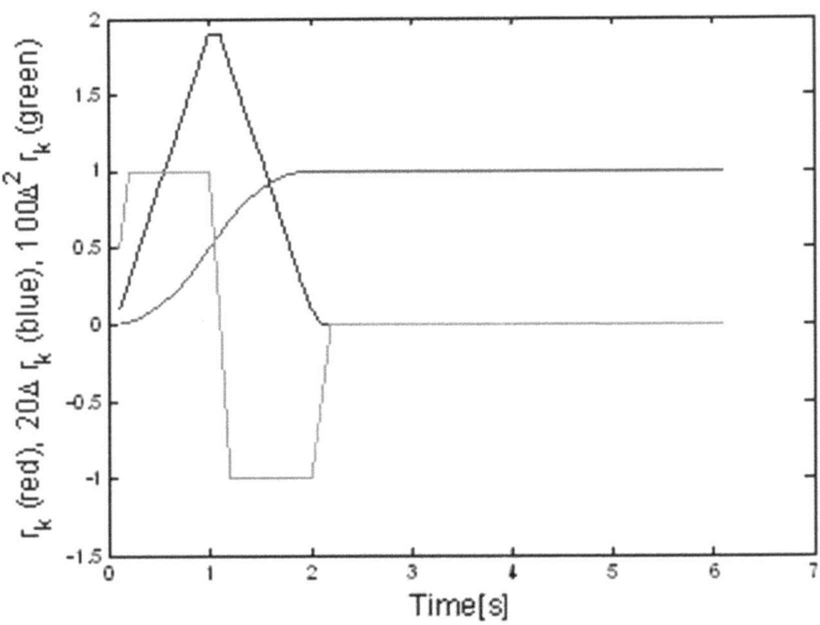

Fig. 3. Robot path reference signal (in red), its angular velocity (in blue) and angular acceleration (in green)

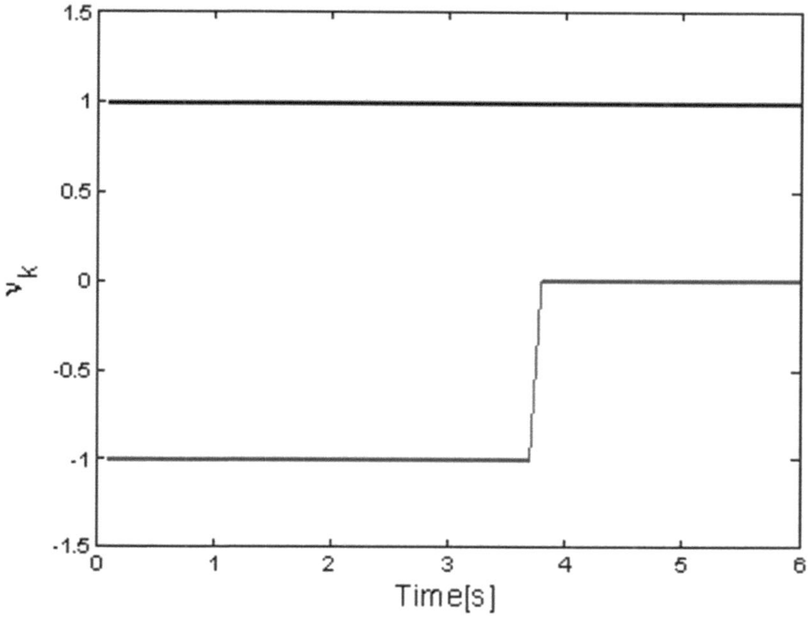

Fig. 4. The controllers 1 and 2 order functions

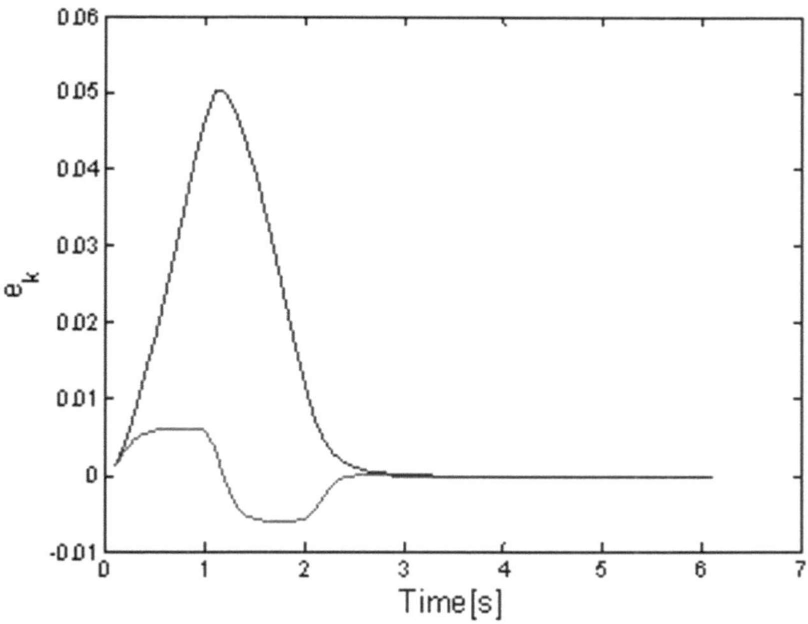

Fig. 5. The CLSs errors with controllers 1 and 2

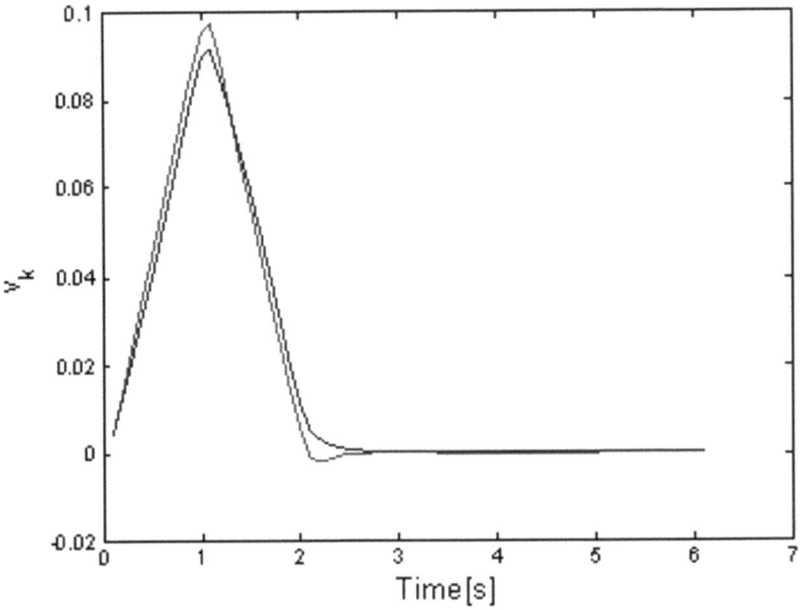

Fig. 6. The controllers 1 and 2 output signals

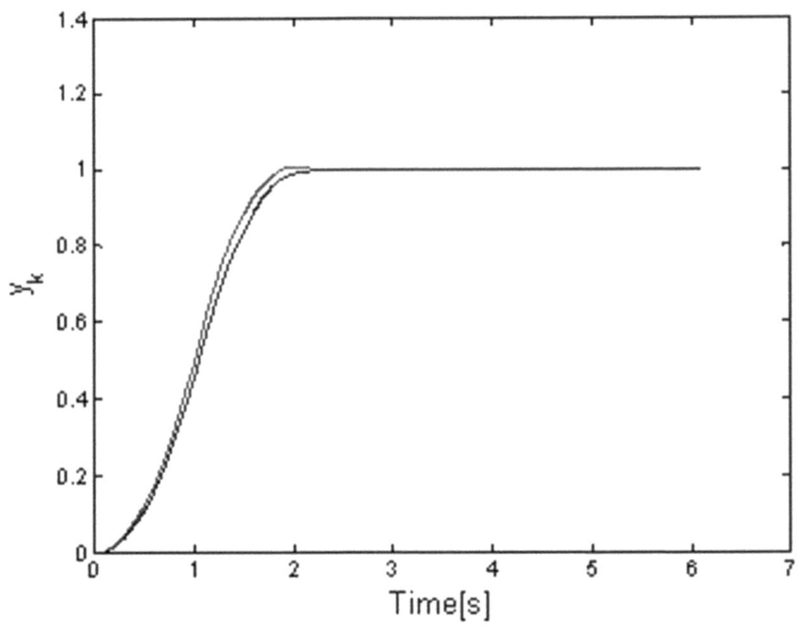

Fig. 7. The CLSs with controllers 1 and 2 output signals

Table 1. The controllers order functions

Controller	OF	CLS stability/unstability	ISE
PD	$\nu_k = 1$ $for\ 0 \leq k \leq L_\nu = +\infty$	$CLS = stable$ for $0 \leq k \leq L_\nu = +\infty$	$2.1048e-004$
VIOPI	$\nu_k = \begin{cases} -1\ for\ 0 \leq k \leq L_\mu \\ 0\ \ for\ \ L_\mu \leq k \end{cases}$	$CLS = \begin{cases} stable\ for\ 0 \leq k \leq L_\nu \\ stable\ for\ \ L_\nu \leq k \end{cases}$	5.6690e-006

Table 2. The OFs defined by formulae

Controller	OF	CLS stability/unstability	ISE
VIOPI	$\mu_k = -3$ $for\ 0 \leq k \leq L_\mu = +\infty$	$CLS = stable$ for $0 \leq k \leq L_\nu = +\infty$	$+\infty$
VIOPI	$\mu_k = \begin{cases} -3\ for\ 0 \leq k \leq L_\mu \\ 0\ \ for\ \ L_\mu \leq k \end{cases}$	$CLS = \begin{cases} unstable\ for\ 0 \leq k \leq L_\nu \\ stable\ \ for\ \ L_\nu \leq k \end{cases}$	1.9189e-005
VFOP	$\mu_k = \begin{cases} -3\ for\ \ 0 \leq k \leq L_{\mu_1} \\ 0\ \ for\ L_{\mu_1} \leq k < L_{\mu_2} \\ 0\ \ for\ \ \ L_{\mu_2} \leq k \end{cases}$	$CLS = \begin{cases} stable\ \ for\ \ 0 \leq k \leq L_{\nu_1} \\ unst./stable\ for\ L_{\nu_1} \leq k \leq L_{\nu_2} \\ stable\ \ for\ \ L_{\nu_1} \leq k \end{cases}$	4.5285e-006

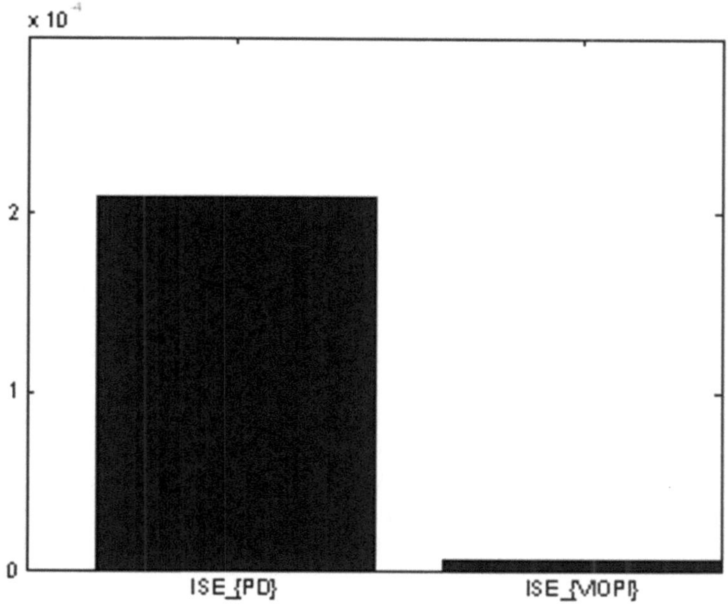

Fig. 8. The CLSs with controllers 1 and 2 ISE performance indexes values

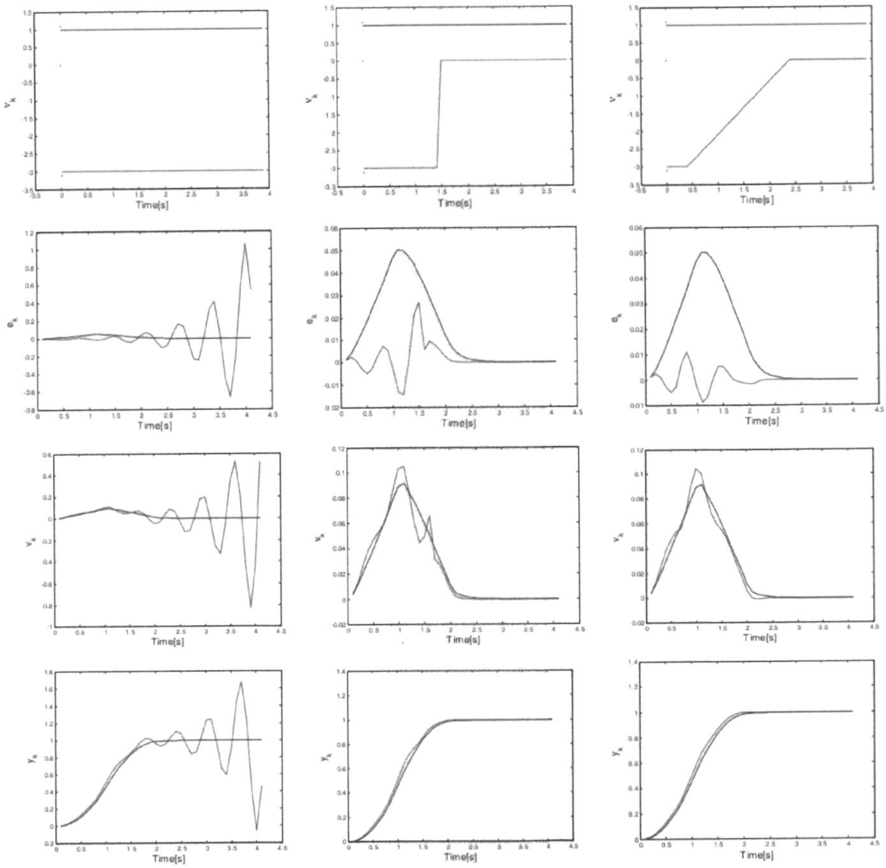

Fig. 9. Transient states of the CLSs with controllers 3,4 and 5

controller. In $0 \leq k \leq max\,(L_{\mu,L_{\nu}})$ stable or even unstable. The CLS system unstability in strictly defined finite time interval may positively influence a system transient-state. Below, in example simulation results of a CLS consisting of the robot arm model and the very simple VIOPI controller are presented. The CLS is intervally unstable. The OFs are defined by formulae given in Table 2. In Fig. 9 the simulation results are presented for controller 3,4 and 5, respectively.

4 Conclusions

The idea of the VFOPID controller shows an immense possibilities of the controller transient characteristic shaping. The proposed special choice of the integrating and differentiation order functions enables to avoid such problems related to fractional differences and sums calculations as the growing calculations tail and the memory insufficiency in the DSP devices. The conditions (8) and (9) imposed on the order functions preserve the closed-loop system stability. This, however, does not solve all of the problems related to the design of the VFOPID controllers. The closed-loop system with the VFOPID controller should be treated as a time-varying system for which one cannot determine the frequency characteristics and to use the commonly known tuning regulators methods. The choice of the VFOPID controller order functions is still an open issue, although one can suggest some solutions relating the order functions with error signal $\nu(e_k)$, $\mu(e_k)$.

References

1. Aštrom̃, K.J., Hägglund, T.: PID controllers: Theory, design and tuning. Instrument Society of America (1995)
2. Chen, Y.Q., Petras, I., Xue, D.: Fractional Order Control - A Tutorial. In: 2009 American Control Conference Hyatt Regency Riverfront. St. Louis, MO (2009)
3. Gene, F.: Digital signal processor trends, pp. 52–55. Texas Instruments (2000)
4. Franklin, G.F., Powell, J.D., Ememi-Naeini, A.: Feedback Control of Dynamic Systems. Prentice Hall, Upper Saddle River (2002)
5. Hussain, Z.M., Sadik, A.Z., Shea, P.: Digital signal Processing. An Introduction with MATLAB and Applications. Springer, Berlin (2011)
6. Ifeachor, E.C., Jervis, B.W.: Digital Signal Processing. Addison- Wesley, Harlow (1998)
7. Jacquot, R.G.: Modern Digital Control Systems. Marcel Dekker, Inc., New York (1994)
8. Karam, L.J., AlKamal, I., Gatherer, A., Frantz, G.A., Anderson, D.V., Evans, B.L.: Trends in Multicore DSP Platforms. IEEE Signal Provessing Magazine (38), 28–49 (2009)
9. Monje, C.A., Chen, Y., Vinagre, B., Xue, D., Feliu, V.: Fractional-order Systems and Controls Fundamentals and Applications. Advances in Industrial Control. Springer London Limited (2010)
10. Monje, C.A., Vinagre, B.M., Chen, Y.Q., Feliu, V., Lanusse, P., Sabatier, J.: Proposals for fractional $PI\lambda D\mu$ tuning. In: Proceedings of Fractional Differentiation and its Applications, Bordeaux, France (2004)
11. Ogata, K.: Discrete-Time Control Systems. Prentice Hall International Editions, Englewood Cliffs (1987)
12. Ostalczyk, P.: Fractional-order backward difference equivalent forms. In: Le Mehaute, Tenreiro Machado, J.A., Trigeassou, J.C., Sabatier, J. (eds.) Fractional differentiation and its Applications. Systems Analysis, Implementation and Simulation, System Identification and Control, pp. 545–556. Ubooks Verlag, Neusä Germany (1995)
13. Ostalczyk, P.: A note on the Grünwald-Letnikov fractional-order backward difference. Physica Scripta (T136), 1–5 (2009)
14. Podlubny, I.: Fractional-order Systems and Fractional-order Controllers. Slovak Academy of Sciences Institute of Experimental Physics, UEF SAV Kosice (1994)
15. Gene, F.: Digital signal processor trends. Texas Instruments, pp. 5–52 (2000)
16. Podlubny, I.: Fractional Differential Equations. Academic Press, London (1999)

17. Podlubny, I.: Fractional-Order Systems and controllers. IEEE Transactions on Automatic Control 44(1), 208–214 (1999)
18. Weeks, M.: Digital Signal Processing. Jones and Bartlett Publishers, Boston (2011)
19. Sankowski, D., Kucharski, J., Łobodziński, W.: Auto-tuning Temperature Control Using Identification by MBS. IEE Proc. Control Theory Appl. 144(3), 233–240 (1997)
20. Shenga, H., Sunb, H., Coopmansc, C., Chenc, Y.Q., Bohannanc, G.: Physical Experimental Study of Variable-order Fractional Integrator and Differentiator. In: Podlubny, I., Vinagre Jara, B.M., Chen, Y., Feliu Batlle, V., Tejado Balsera, I. (eds.) Proceedings of FDA10. The 4th IFAC Workshop Fractional Differentiation and its Applications, Badajoz, Spain, October 18-20 (2010)
21. Silva, M.F., Tenreiro Machado, J.A.: Fractional Order PD-Joint Control of Legged Robots. Journal of Vibration and Control 12(12), 1483–1501 (2006)
22. Vilanova, R., Visioli, A. (eds.): PID Control in the Third Millennium. Advances in Industrial ControlLimited. Springer, London (2012)
23. Xue, D., Chen, Y.: A Comparative Introduction of Four Fractional Order Controllers. In: Proceedings of the 4th World Congress on Intelligent Control and Automation, Shanghai, P.R.China, June 10-14 (2002)
24. Xue, D., Zhao, C., Chen, Y.: Fractional Order PID Control of A DC-Motor with Elastic Shaft: A Case Study. In: Proceedings of the 2006 American Control Conference Minneapolis, Minnesota, USA, June 14-16 (2006)

Minimum Energy Control of Fractional Positive Continuous-Time Linear Systems with Two Different Fractional Orders and Bounded Inputs

Łukasz Sajewski

Bialystok University of Technology
Faculty of Electrical Engineering
Wiejska 45D, 15-351 Bialystok
l.sajewski@pb.edu.pl

Abstract. The minimum energy control problems for fractional positive continuous-time linear systems with two different fractional orders and with bounded input is formulated. Solution to the minimum energy control problem with bounded input is derived. Procedure is proposed and demonstrated on example of electrical circuit.

Keywords: fractional calculus, different order, positive systems, continuous-time, bounded input.

1 Introduction

In positive systems inputs, state variables and outputs take only non-negative values. Examples of positive systems are industrial processes involving chemical reactors, heat exchangers and distillation columns, storage systems, compartmental systems, water and atmospheric pollution models. Positive linear systems are defined on cones and not on linear spaces. An overview of state of art in positive systems theory is given in the monographs [4, 12].

The first definition of the fractional derivative was introduced by Liouville and Riemann in the middle of the 19th century [24, 25] and another on was proposed in 20th century by Caputo [26]. This idea has been used by engineers for modeling different processes [3, 5]. Mathematical fundamentals of fractional calculus are given in the monographs [17, 23-25]. The positive fractional linear systems have been investigated in [7, 8, 13, 14, 17]. Stability of fractional linear 1D discrete-time and continuous-time systems has been investigated in [1, 20] of 2D fractional positive linear systems in [6] and of continuous-time linear systems consisting of n subsystem with different fractional orders [2]. The minimum energy control problem for standard linear systems has been formulated and solved in [18-22], for fractional positive continuous-time linear systems in [10, 11] and for systems with two different fractional orders in [28]. Reachability and observability of fractional positive continuous-time linear systems has been investigated in [16] and reachability of systems with two different fractional orders in [27].

© Springer International Publishing Switzerland 2015
K.J. Latawiec et al. (eds.), *Advances in Modeling and Control of Non-integer Order Systems*
Lecture Notes in Electrical Engineering 320, DOI: 10.1007/978-3-319-09900-2_16

In this paper minimum energy problems for fractional positive continuous-time linear systems with two different fractional orders and bounded input will be formulated and solved.

The paper is organized as follows. In section 2 the basic definitions and theorems of the fractional continuous-time linear systems with two different fractional orders are recalled and the necessary and sufficient conditions for positivity and reachability of the systems are given. Section 3 gives the formulation and solution to the minimum energy control problem. Illustrating example of electrical circuit is given in section 4. Concluding remarks are given in section 5.

The following notation will be used: $\Re$ - the set of real numbers, $\Re^{n\times m}$ - the set of $n\times m$ real matrices, $\Re_+^{n\times m}$ - the set of $n\times m$ matrices with nonnegative entries and $\Re_+^n = \Re_+^{n\times 1}$, M_n - the set of $n\times n$ Metzler matrices (real matrices with nonnegative off-diagonal entries), I_n - the $n\times n$ identity matrix, A^T - the transpose matrix A. A real square matrix is called monomial if each its row and each its column contains only one positive entry and the remaining entries are zero.

2 System with Two Different Fractional Orders

In this paper the following Caputo definition of the fractional derivative will be used [17, 26]

$$\frac{d^\alpha f(t)}{dt^\alpha} = \frac{1}{\Gamma(n-\alpha)} \int_0^\infty \frac{f^{(n)}(\tau)}{(t-\tau)^{\alpha+1-n}} d\tau, \tag{2.1}$$

where $n-1<\alpha<n$, $n\in W=\{1,2,\ldots\}$,

$$\Gamma(x) = \int_0^\infty t^{x-1} e^{-t} dt \tag{2.2}$$

is the Euler gamma function and

$$f^{(n)}(\tau) = \frac{d^n f(\tau)}{d\tau^n}. \tag{2.3}$$

It is well known [17] that the Laplace transform ($\mathcal{L}$) of (2.1) is given by the formula

$$\mathcal{L}\left[\frac{d^\alpha f(t)}{dt^\alpha}\right] = \int_0^\infty \frac{d^\alpha f(t)}{dt^\alpha} e^{-st} dt = s^\alpha F(s) - \sum_{k=1}^n s^{\alpha-k} f^{(k-1)}(0+), \tag{2.4}$$

$n-1<\alpha<n$, $n\in W$, where $F(s)=\mathcal{L}[f(t)]$ and $n-1<\alpha<n$, $n\in W$.

Consider a fractional linear system described by the equation [14]

$$\begin{bmatrix} \dfrac{d^{\alpha} x_1(t)}{dt^{\alpha}} \\ \dfrac{d^{\beta} x_2(t)}{dt^{\beta}} \end{bmatrix} = \begin{bmatrix} A_{11} & A_{12} \\ A_{21} & A_{22} \end{bmatrix} \begin{bmatrix} x_1(t) \\ x_2(t) \end{bmatrix} + \begin{bmatrix} B_1 \\ B_2 \end{bmatrix} u(t), \tag{2.5a}$$

$$y(t) = [C_1 \quad C_2] \begin{bmatrix} x_1(t) \\ x_2(t) \end{bmatrix} \tag{2.5b}$$

and $p-1<\alpha,\beta<p;\quad p,q \in W$ where $x_1(t) \in \Re^{n_1}$, $x_2(t) \in \Re^{n_2}$, $u(t) \in \Re^{m}$ and $y(t) \in \Re^{p}$ are the state, input and output vectors respectively, $A_{ij} \in \Re^{n_i \times n_j}$, $B_i \in \Re^{n_i \times m}$; $i, j = 1,2$.

Initial conditions for (2.5) have the form

$$x_1(0) = x_{10},\ x_2(0) = x_{20} \text{ and } x_0 = \begin{bmatrix} x_{10} \\ x_{20} \end{bmatrix}. \tag{2.6}$$

Theorem 2.1. The solution of the equation (2.5) for $0<\alpha<1;\ 0<\beta<1$ with initial conditions (2.6) has the form

$$x(t) = \begin{bmatrix} x_1(t) \\ x_2(t) \end{bmatrix} = \Phi_0(t)x_0 + \int_0^t M(t-\tau)u(\tau)d\tau, \tag{2.7a}$$

where

$$M(t) = \Phi_1(t)B_{10} + \Phi_2(t)B_{01} = \begin{bmatrix} \Phi_{11}^1(t) & \Phi_{12}^1(t) \\ \Phi_{21}^1(t) & \Phi_{22}^1(t) \end{bmatrix} \begin{bmatrix} B_1 \\ 0 \end{bmatrix} + \begin{bmatrix} \Phi_{11}^2(t) & \Phi_{12}^2(t) \\ \Phi_{21}^2(t) & \Phi_{22}^2(t) \end{bmatrix} \begin{bmatrix} 0 \\ B_2 \end{bmatrix}$$

$$= \begin{bmatrix} \Phi_{11}^1(t)B_1 + \Phi_{12}^2(t)B_2 \\ \Phi_{21}^1(t)B_1 + \Phi_{22}^2(t)B_2 \end{bmatrix} = \begin{bmatrix} \Phi_{11}^1(t) & \Phi_{12}^2(t) \\ \Phi_{21}^1(t) & \Phi_{22}^2(t) \end{bmatrix} \begin{bmatrix} B_1 \\ B_2 \end{bmatrix} \tag{2.7b}$$

and

$$\Phi_0(t) = \sum_{k=0}^{\infty}\sum_{l=0}^{\infty} T_{kl} \frac{t^{k\alpha+l\beta}}{\Gamma(k\alpha+l\beta+1)}, \tag{2.8a}$$

$$\Phi_1(t) = \sum_{k=0}^{\infty}\sum_{l=0}^{\infty} T_{kl} \frac{t^{(k+1)\alpha+l\beta-1}}{\Gamma[(k+1)\alpha+l\beta]},\quad \Phi_2(t) = \sum_{k=0}^{\infty}\sum_{l=0}^{\infty} T_{kl} \frac{t^{k\alpha+(l+1)\beta-1}}{\Gamma[k\alpha+(l+1)\beta]}, \tag{2.8b}$$

$$T_{kl} = \begin{cases} I_n & \text{for} \quad k = l = 0 \\ \begin{bmatrix} A_{11} & A_{12} \\ 0 & 0 \end{bmatrix} & \text{for} \quad k = 1, l = 0 \\ \begin{bmatrix} 0 & 0 \\ A_{21} & A_{22} \end{bmatrix} & \text{for} \quad k = 0, l = 1 \\ T_{10} T_{k-1,l} + T_{01} T_{k,l-1} & \text{for} \quad k + l > 1 \end{cases} \tag{2.8c}$$

Proof is given in [14].

Definition 2.1. The fractional system (2.5) is called positive if $x_1(t) \in \Re_+^{n_1}$ and $x_2(t) \in \Re_+^{n_2}$, $t \geq 0$ for any initial conditions $x_{10} \in \Re_+^{n_1}$, $x_{20} \in \Re_+^{n_2}$ and all input vectors $u \in \Re_+^m$, $t \geq 0$.

Theorem 2.2. The fractional system (2.5) for $0 < \alpha < 1;\ 0 < \beta < 1$ is positive if and only if

$$A = \begin{bmatrix} A_{11} & A_{12} \\ A_{21} & A_{22} \end{bmatrix} \in M_N,\ B = \begin{bmatrix} B_1 \\ B_2 \end{bmatrix} \in \Re_+^{N \times m},\ C = [C_1 \quad C_2] \in \Re_+^{p \times N},\ N = n_1 + n_2. \tag{2.9}$$

Proof is given in [12, 15].

Theorem 2.3. The positive fractional continuous-time linear system with two different fractional orders (2.5a) is reachable if and only if $A \in M_N$ is diagonal and $B \in \Re_+^{N \times N}$ is a monomial matrix.

Proof is given in [28].

3 Minimum Energy Control with Bounded Input

Consider the fractional positive system with two different fractional orders (2.5) with $A \in M_N$, $B \in \Re_+^{N \times N}$ and $C \in \Re_+^{N \times N}$ monomial. If the system is reachable in time $t \in [0, t_f]$, then usually there exists many different inputs $u(t) \in \Re_+^N$ that steers the state of the system from $x_0 = [x_{10} \quad x_{20}]^T = 0$ to $x_f = [x_{1f} \quad x_{2f}]^T \in \Re_+^N$. Among these inputs we are looking for input $u(t) \in \Re_+^N$, $t \in [0, t_f]$ satisfying the condition

$$u(t) \le U \in \Re_+^n,\ t \in [0, t_f], \tag{3.1}$$

that minimizes the performance index [9-11]

$$I(u) = \int_0^{t_f} u^T(\tau) Q u(\tau) d\tau, \tag{3.2}$$

where $Q \in \Re_+^{N\times N}$ is a symmetric, positive defined matrix and $Q^{-1} \in \Re_+^{N\times N}$.

The minimum energy control problem for the fractional positive continuous-time linear systems with two different fractional orders (2.5) can be stated as follows [9-11]: Given the matrices $A \in M_N$, $B \in \Re_+^{N\times N}$, α, β and $Q \in \Re_+^{N\times N}$ of the performance matrix (3.2), $x_f \in \Re_+^N$ and $t_f > 0$, find an input $u(t) \in \Re_+^N$ for $t \in [0, t_f]$ satisfying (3.1), that steers the state vector of the system from $x_0 = 0$ to $x_f \in \Re_+^N$ and minimizes the performance index (3.2).

To solve the problem, following [9-11], we define the matrix

$$W(t_f) = \int_0^{t_f} M(t_f - \tau) Q^{-1} M^T(t_f - \tau) d\tau, \tag{3.3}$$

where $M(t)$ is defined by (2.7b). From (3.3) and Theorem 2.3 it follows that the matrix (3.3) is monomial if and only if the fractional positive system (2.5) is reachable in time $[0, t_f]$. In this case we may define the input

$$\hat{u}(t) = Q^{-1} M^T(t_f - t) W^{-1}(t_f) x_f \text{ for } t \in [0, t_f]. \tag{3.4}$$

Note that the input (3.4) satisfies the condition $u(t) \in \Re_+^N$ for $t \in [0, t_f]$ if

$$Q^{-1} \in \Re_+^{N\times N} \text{ and } W^{-1}(t_f) \in \Re_+^{N\times N}. \tag{3.5}$$

Theorem 3.1. Let $\bar{u}(t) \in \Re_+^N$ for $t \in [0, t_f]$ be an input satisfying (3.1), that steers the state of the fractional positive system (2.5) from $x_0 = 0$ to $x_f \in \Re_+^N$. Then the input (3.4) satisfying (3.1) also steers the state of the system from $x_0 = 0$ to $x_f \in \Re_+^N$ and minimizes the performance index (3.2), i.e. $I(\hat{u}) \le I(\bar{u})$.

The minimal value of the performance index (3.2) is equal to

$$I(\hat{u}) = x_f^T W^{-1}(t_f) x_f. \tag{3.6}$$

Proof is given in [28].

Now following [11], to find $t \in [0, t_f]$ for which $\hat{u}(t) \in \Re_+^n$ reaches its minimal value, using (3.4) we compute the derivative

$$\frac{d\hat{u}(t)}{dt} = Q^{-1} B^T \Psi(t) W^{-1}(t_f) x_f, \; t \in [0, t_f], \tag{3.7}$$

where

$$\Psi(t) = \frac{d}{dt}[M^T(t_f - t)]. \tag{3.8}$$

Knowing $\Psi(t)$ and using the equality

$$\Psi(t) W^{-1}(t_f) x_f = 0 \tag{3.9}$$

we can find $t \in [0, t_f]$ for which $\hat{u}(t)$ reaches its maximal value.

Note that if the system is asymptotically stable $\lim_{t \to \infty} M(t) = 0$ then $\hat{u}(t)$ reaches its maximal value for $t = t_f$ and if it is unstable then for $t = 0$.

From the above considerations we have the following procedure for computation of the optimal inputs:

Procedure 4.1

Step 1. Using (2.7b) and (2.8) compute the matrix $M(t)$.

Step 2. Knowing the matrices A, B, Q and α, β, t_f using (3.3) compute the matrix W.

Step 3. Using (3.4) and (3.9) find t_f for which $\hat{u}(t)$ satisfying (3.1) reaches its maximal value and the desired $\hat{u}(t)$ for given $U \in \Re_+^n$ and $x_f \in \Re_+^n$.

Step 4. Using (3.6) compute the maximal value of the performance index $I(\hat{u})$.

4 Example

Consider the fractional electrical circuit shown on Figure 4.1 [15, 27, 28] with given source voltages e_1, e_2, ultracapacitor C_1 of the fractional order $\alpha = 0.7$, ultracapacitor C_2 of the fractional order $\beta = 0.6$, conductances G_1, G_2, G'_1, G'_2 and $G_{12} = 0$.

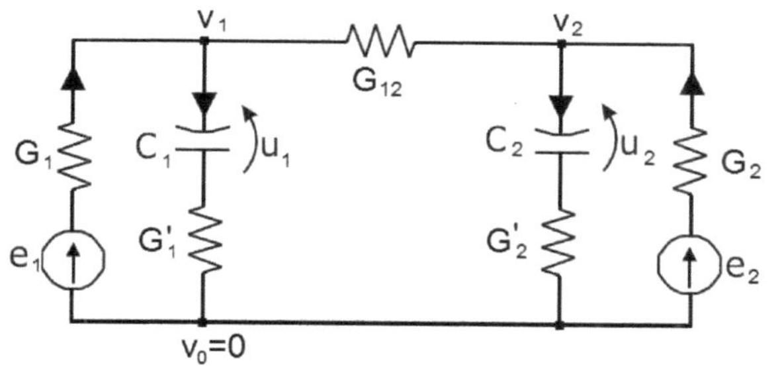

Fig. 1. Fractional electrical circuit

Using the Kirchhoff's laws we can write the equations

$$C_1 \frac{d^\alpha u_1}{dt^\alpha} = G'_1 (v_1 - u_1),$$
$$C_2 \frac{d^\beta u_2}{dt^\beta} = G'_2 (v_2 - u_2) \tag{4.1}$$

and

$$\begin{bmatrix} (G_1 + G'_1) & 0 \\ 0 & (G_2 + G'_2) \end{bmatrix} \begin{bmatrix} v_1 \\ v_2 \end{bmatrix} = \begin{bmatrix} G'_1 & 0 \\ 0 & G'_2 \end{bmatrix} \begin{bmatrix} u_1 \\ u_2 \end{bmatrix} + \begin{bmatrix} G_1 & 0 \\ 0 & G_2 \end{bmatrix} \begin{bmatrix} e_1 \\ e_2 \end{bmatrix}. \tag{4.2}$$

From (4.2) we obtain

$$\begin{aligned} \begin{bmatrix} v_1 \\ v_2 \end{bmatrix} &= \begin{bmatrix} (G_1 + G'_1) & 0 \\ 0 & (G_2 + G'_2) \end{bmatrix}^{-1} \begin{bmatrix} G'_1 & 0 \\ 0 & G'_2 \end{bmatrix} \begin{bmatrix} u_1 \\ u_2 \end{bmatrix} \\ &+ \begin{bmatrix} (G_1 + G'_1) & 0 \\ 0 & (G_2 + G'_2) \end{bmatrix}^{-1} \begin{bmatrix} G_1 & 0 \\ 0 & G_2 \end{bmatrix} \begin{bmatrix} e_1 \\ e_2 \end{bmatrix}. \end{aligned} \tag{4.3}$$

Substitution of (4.3) into

$$\begin{bmatrix} \frac{d^\alpha u_1}{dt^\alpha} \\ \frac{d^\beta u_2}{dt^\beta} \end{bmatrix} = \begin{bmatrix} -\frac{G'_1}{C_1} & 0 \\ 0 & -\frac{G'_2}{C_2} \end{bmatrix} \begin{bmatrix} u_1 \\ u_2 \end{bmatrix} + \begin{bmatrix} \frac{G'_1}{C_1} & 0 \\ 0 & \frac{G'_2}{C_2} \end{bmatrix} \begin{bmatrix} v_1 \\ v_2 \end{bmatrix} \tag{4.4}$$

yields

$$\begin{bmatrix} \frac{d^\alpha u_1}{dt^\alpha} \\ \frac{d^\beta u_2}{dt^\beta} \end{bmatrix} = A \begin{bmatrix} u_1 \\ u_2 \end{bmatrix} + B \begin{bmatrix} e_1 \\ e_2 \end{bmatrix}, \tag{4.5}$$

where

$$A = \begin{bmatrix} \frac{G'_1}{C_1}\left(\frac{G'_1}{(G_1+G'_1)} - 1\right) & 0 \\ 0 & \frac{G'_2}{C_2}\left(\frac{G'_2}{(G_2+G'_2)} - 1\right) \end{bmatrix},$$
$$B = \begin{bmatrix} \frac{G'_1 G_1}{C_1(G_1+G'_1)} & 0 \\ 0 & \frac{G'_2 G_2}{C_2(G_2+G'_2)} \end{bmatrix}. \tag{4.6}$$

From (4.6) it follows that A is a diagonal Metzler matrix and the matrix B is monomial matrix with positive diagonal entries. Therefore, the fractional electrical circuit is positive for all values of the conductances and capacitances.

Without lost of generality, to simplify the notation, wet take for example values of the electrical circuit: $C_1 = 1$, $\alpha = 0.7$, $C_2 = 2$, $\beta = 0.6$, $G_1 = 4$, $G'_1 = 4$, $G_2 = 3$, $G'_2 = 6$, $G_{12} = 0$ and $N = n_1 + n_2 = 2$.

The matrices (4.6) takes the form

$$A = \begin{bmatrix} -2 & 0 \\ 0 & -1 \end{bmatrix} = \begin{bmatrix} A_1 & 0 \\ 0 & A_2 \end{bmatrix}, \quad B = \begin{bmatrix} 2 & 0 \\ 0 & 1 \end{bmatrix} = \begin{bmatrix} B_1 \\ B_2 \end{bmatrix}. \tag{4.7}$$

Find the optimal input (source voltage) $\hat{e}(t) \in \Re^2_+$, $t \in [0, t_f]$ satisfying the condition

$$\hat{e}(t) = \begin{bmatrix} \hat{e}_1(t) \\ \hat{e}_2(t) \end{bmatrix} < \begin{bmatrix} 1 \\ 1 \end{bmatrix} \text{ for } t \in [0, t_f] \tag{4.8}$$

for the performance index (3.2) with $Q = \mathrm{diag}[2,2]$, which steers the system from initial state (voltage drop on capacitances) $u_0 = [0 \quad 0]^T$ to the finite state $u_f = [2 \quad 3]^T$ and minimize the performance index (3.2) with (4.7).

Using (2.8) and (4.7) we obtain

$$M(t) = \begin{bmatrix} \Phi_{11}^{1}(t) & 0 \\ 0 & \Phi_{22}^{2}(t) \end{bmatrix} \begin{bmatrix} B_1 \\ B_2 \end{bmatrix}, \tag{4.9}$$

where

$$\begin{aligned} \Phi_{11}^{1}(t) &= \sum_{k=0}^{\infty}\sum_{l=0}^{\infty} \frac{t^{(k+1)0.7+l0.6-1}}{\Gamma[(k+1)0.7+l0.6]}(-2)^k, \\ \Phi_{22}^{2}(t) &= \sum_{k=0}^{\infty}\sum_{l=0}^{\infty} \frac{t^{k0.7+(l+1)0.6-1}}{\Gamma[k0.7+(l+1)0.6]}(-1)^l. \end{aligned} \tag{4.10}$$

From (2.7b), (4.7) and (3.3) we have

$$\begin{aligned} W(t_f) &= \int_0^{t_f} M(t_f-\tau)Q^{-1}M^T(t_f-\tau)d\tau \\ &= \int_0^{t_f} \begin{bmatrix} \Phi_{11}^{1}(\tau) & 0 \\ 0 & \Phi_{22}^{2}(\tau) \end{bmatrix} \begin{bmatrix} 8 & 0 \\ 0 & 2 \end{bmatrix} \begin{bmatrix} \Phi_{11}^{1}(\tau) & 0 \\ 0 & \Phi_{22}^{2}(\tau) \end{bmatrix}^T d\tau. \end{aligned} \tag{4.11}$$

Note that the electrical circuit is stable. Therefore, $\hat{e}(t)$ reach its maximal value for $t = t_f$.

Now using (3.4) and (4.11) we obtain

$$\begin{aligned} \hat{e}(t) &= Q^{-1}M^T(t_f-t)W^{-1}(t_f)u_f = \begin{bmatrix} 1 & 0 \\ 0 & 0.5 \end{bmatrix} \begin{bmatrix} \Phi_{11}^{1}(t_f-t) & 0 \\ 0 & \Phi_{22}^{2}(t_f-t) \end{bmatrix}^T W^{-1}(t_f) \begin{bmatrix} 2 \\ 3 \end{bmatrix} \\ &= \begin{bmatrix} \Phi_{11}^{1}(t_f-t) & 0 \\ 0 & 0.5\Phi_{22}^{2}(t_f-t) \end{bmatrix}^T \left[\int_0^{t_f} \begin{bmatrix} 8\Phi_{11}^{1}(\tau)[\Phi_{11}^{1}(\tau)]^T & 0 \\ 0 & 2\Phi_{22}^{2}(\tau)[\Phi_{22}^{2}(\tau)]^T \end{bmatrix} d\tau \right]^{-1} \begin{bmatrix} 2 \\ 3 \end{bmatrix} \\ &= \begin{bmatrix} 0.25[\Phi_{11}^{1}(t_f-t)]^T \int_0^{t_f} [\Phi_{11}^{1}(\tau)[\Phi_{11}^{1}(\tau)]^T]^{-1} d\tau \\ 1.5[\Phi_{22}^{2}(t_f-t)]^T \int_0^{t_f} [\Phi_{22}^{2}(\tau)[\Phi_{22}^{2}(\tau)]^T]^{-1} d\tau \end{bmatrix}, \end{aligned} \tag{4.12}$$

The minimal value of t_f satisfying the condition (3.1) can be found from the inequality

$$\begin{bmatrix} 0.25[\Phi_{11}^1(t_f - t)]^T \int_0^{t_f} [\Phi_{11}^1(\tau)[\Phi_{11}^1(\tau)]^T]^{-1} d\tau \\ 1.5[\Phi_{22}^2(t_f - t)]^T \int_0^{t_f} [\Phi_{22}^2(\tau)[\Phi_{22}^2(\tau)]^T]^{-1} d\tau \end{bmatrix} < \begin{bmatrix} 1 \\ 1 \end{bmatrix} \text{ for } t \in [0, t_f] \quad (4.13)$$

and the minimal value of the performance index (3.2) is equal to

$$I(\hat{e}) = u_f^T W^{-1} u_f = [2 \quad 3] \left[\int_0^{t_f} \begin{bmatrix} 8\Phi_{11}^1(\tau)[\Phi_{11}^1(\tau)]^T & 0 \\ 0 & 2\Phi_{22}^2(\tau)[\Phi_{22}^2(\tau)]^T \end{bmatrix} d\tau \right]^{-1} \begin{bmatrix} 2 \\ 3 \end{bmatrix}$$

$$= 0.5 \int_0^{t_f} [\Phi_{11}^1(\tau)[\Phi_{11}^1(\tau)]^T]^{-1} d\tau + 3 \int_0^{t_f} [\Phi_{22}^2(\tau)[\Phi_{22}^2(\tau)]^T]^{-1} d\tau. \quad (4.14)$$

5 Concluding Remarks

The minimum energy control problems for the fractional positive continuous-time linear systems with two different fractional orders and bounded input have been formulated. Solution to the minimum energy control problem has been given. Effectiveness of the proposed considerations has been demonstrated on example of electrical circuit. Extension of this considerations on systems consisting of n subsystems with different fractional orders is also possible.

Acknowledgment. This work was supported under work S/WE/1/11.

References

1. Busłowicz, M.: Stability of linear continuous time fractional order systems with delays of the retarded type. Bull. Pol. Ac.: Tech. 56(4), 319–324 (2008)
2. Busłowicz, M.: Stability analysis of continuous-time linear systems consisting of *n* subsystem with different fractional orders. Bull. Pol. Ac.: Tech. 60(2), 279–284 (2012)
3. Dzieliński, A., Sierociuk, D., Sarwas, G.: Ultracapacitor parameters identification based on fractional order model. In: Proc. of European Control Conference, ECC, Budapest, Hungary (2009)
4. Farina, L., Rinaldi, S.: Positive Linear Systems, Theory and Applications. J. Wiley, New York (2000)
5. Ferreira, N.M.F., Machado, J.A.T.: Fractional-order hybrid control of robotic manipulators. In: Proc. of 11th Int. Conf. Advanced Robotics, ICAR, Coimbra, Portugal, pp. 393–398 (2003)
6. Kaczorek, T.: Asymptotic stability of positive fractional 2D linear systems. Bull. Pol. Ac.: Tech. 57(3), 289–292 (2009)

7. Kaczorek, T.: Fractional positive continuous-time systems and their Reachability. Int. J. Appl. Math. Comput. Sci. 18(2), 223–228 (2008)
8. Kaczorek, T.: Fractional positive linear systems. Kybernetes 38(7/8), 1059–1078 (2009)
9. Kaczorek, T.: Minimum energy control of fractional descriptor positive discrete-time linear systems. Int. J. Appl. Math. Comput. Sci. (in Press 2014)
10. Kaczorek, T.: Minimum energy control of fractional positive continuous-time linear systems. Bull. Pol. Ac.: Tech. 61(4), 803–807 (2013)
11. Kaczorek, T.: Minimum energy control of fractional positive continuous-time linear systems with bounded inputs. Int. J. Appl. Math. Comput. Sci. (in Press 2014)
12. Kaczorek, T.: Positive 1D and 2D Systems. Springer, London (2002)
13. Kaczorek, T.: Positive linear systems consisting of n subsystems with different fractional orders. IEEE Transactions on Circuits and Systems-I: Regular Paper 58(6), 1203–1210 (2011)
14. Kaczorek, T.: Positive linear systems with different fractional orders. Bull. Pol. Acad. Sci. Tech. 58(3), 453–458 (2010)
15. Kaczorek, T.: Positivity and reachability of fractional electrical circuits. Acta Mechanica et Automatica 5(2), 42–51 (2011)
16. Kaczorek, T.: Reachability and observability of fractional positive continuous-time linear systems. In: Proc. of XIV Int. Conf. System Modelling and Control, SMC, Lodz, Poland (2013)
17. Kaczorek, T.: Selected Problems in Fractional Systems Theory. LNCIS, vol. 411. Springer, Berlin (2011)
18. Klamka, J.: Controllability and minimum energy control problem of fractional discrete-time systems, chapter in monograph. In: Baleanu, D., Guvenc, Z.B., Tenreiro Machado, J.A. (eds.) New Trends in Nanotechnology and Fractional Calculus, pp. 503–509. Springer, New York (2010)
19. Klamka, J.: Controllability of dynamical systems: A survey. Bull. Pol. Ac.: Tech. 61(2), 221–229 (2013)
20. Klamka, J.: Controllability of Dynamical Systems. Kluwer Academic Press, Dordrecht (1991)
21. Klamka, J.: Relative controllability and minimum energy control of linear systems with distributed delays in control. IEEE Transactions on Automatic Control 21(4), 594–595 (1976)
22. Klamka, J.: Controllability of dynamical systems-a survey. Bull. Pol. Ac. Tech. 61(2), 335–342 (2013)
23. Miller, K.S., Ross, B.: An Introduction to the Fractional Calculus and Fractional Differenctial Equations. J. Willey, New York (1993)
24. Nishimoto, K.: Fractional Calculus. Decartess Press, Koriama (1984)
25. Oldham, K.B., Spanier, J.: The Fractional Calculus. Academmic Press, New York (1974)
26. Podlubny, I.: Fractional Differential Equations. Academic Press, San Diego (1999)
27. Sajewski, Ł.: Reachability of fractional positive continuous-time linear systems with two different fractional orders. In: Szewczyk, R., Zieliński, C., Kaliczyńska, M. (eds.) Recent Advances in Automation, Robotics and Measuring Techniques. AISC, vol. 267, pp. 239–250. Springer, Heidelberg (2014)
28. Sajewski, Ł.: Reachability, observability and minimum energy control of fractional positive continuous-time linear systems with two different fractional orders. Multidim. Syst. Sign. Process. (2014), doi:10.1007/s11045-014-0287-2

Relative Controllability of Differential-Algebraic Systems with Delay within Riemann-Liouville Fractional Derivatives

Zbigniew Zaczkiewicz

Faculty of Computer Science, Bialystok University of Technology,
15-351 Bialystok, Poland
z.zaczkiewicz@pb.edu.pl

Abstract. The paper deals with a problem of relative controllability for linear fractional differential-algebraic systems with delay (FDAD). FDAD systems consists of fractional differential in the Riemann-Liouville sense and difference equations. We introduce the determining equation systems and their properties. By solution representations into series of their determining equation solutions we obtain effective parametric rank criteria for relative controllability.

Keywords: Fractional differential equations, determining equations, differential-algebraic systems.

1 Introduction

Controllability is one of the fundamental concepts in modern mathematical control theory. This is qualitative property of control systems and is of particular importance in control theory. Systematic study of controllability was started at the beginning of 1960's. Theory of controllability is based on the mathematical description of the dynamical system. Roughly speaking, controllability generally means, that it is possible to steer dynamical control system form an arbitrary initial state to an arbitrary final state using the set of admissible controls.

Infinite-dimensional systems have many different definitions of controllability. One of them is relative controllability. Roughly speaking, relative controllability generally means, that it is possible to steer dynamical control system form infinite-dimensional initial state to an arbitrary finite-dimensional final state. Relative controllability of infinite-dimensional systems was studied through decades from the end of 1960's (e. g. [1]-[10]).

In recent years, much attention has been paid to fractional control systems, see the monograph [11]. During last few years many results concerning theory of fractional control system with delays in control or state variables have been published in the literature (e. g. [12]-[14]). However, it should be pointed out, that the most controllability results are known only for fractional systems with Caputo fractional derivative.

The paper deals with linear fractional differential-algebraic systems with delay (FDAD). FDAD systems consist of some equations being fractional differential

© Springer International Publishing Switzerland 2015
K.J. Latawiec et al. (eds.), *Advances in Modeling and Control of Non-integer Order Systems*, Lecture Notes in Electrical Engineering 320, DOI: 10.1007/978-3-319-09900-2_17

in the Riemann-Liouville sense, the other-difference, with some variables being continuous the other piecewise continuous. We introduce the determining equations the same as for differential-algebraic systems (for example see [15] or [16]). To obtain solutions representations we apply fractional differential calculus especially dealing with the Laplace transform. By this result we obtain effective parametric rank criteria for several types of controllability. Our results can be considered as a generalization of the known corresponding results: for the integer order case [17] and for the fractional differential case with the Caputo fractional derivative [18].

The paper is organized as follows. In section 2 basic definition and fractional differential-algebraic systems with delay FDAD are recalled. Deterministic equations and representation of solutions into series of determining equations solutions are established in section 3. The controllability results are given in section 4.

2 Preliminaries

Let us introduce the following notation:

D_t^α is the left-sided Riemann-Liouville fractional derivatives of order α defined by

$$D_t^\alpha f(t) = \frac{1}{\Gamma(1-\alpha)} \frac{d}{dt} \int_0^t \frac{f(\tau)}{(t-\tau)^\alpha} d\tau,$$

where $0 < \alpha < 1$, $\alpha \in \mathbb{R}$ and $\Gamma(t) = \int_0^\infty e^{-\tau} \tau^{t-1} d\tau$ is the Euler gamma function (see [19] for more details). $T_t = \lim_{\epsilon \to +0} \left[\frac{t-\epsilon}{h}\right]$, where the symbol $[z]$ means entire part of the number z; I_n is the identity n by n matrix.

In this paper, we concentrate on the stationary FDAD system in the following form:

$$D_t^\alpha x_1(t) = A_{11}x_1(t) + A_{12}x_2(t) + B_1 u(t), \ t > 0, \tag{1a}$$
$$x_2(t) = A_{21}x_1(t) + A_{22}x_2(t-h) + B_1 u(t), \ t \geq 0, \tag{1b}$$

where $x_1(t) \in \mathbb{R}^{n_1}$, $x_2(t) \in \mathbb{R}^{n_2}$, $u(t) \in \mathbb{R}^r$, $A_{11} \in \mathbb{R}^{n_1 \times n_1}$, $A_{12} \in \mathbb{R}^{n_1 \times n_2}$, $A_{21} \in \mathbb{R}^{n_2 \times n_1}$, $A_{22} \in \mathbb{R}^{n_2 \times n_2}$, $B_1 \in \mathbb{R}^{r \times n_1}$, $B_2 \in \mathbb{R}^{r \times n_2}$ are constant (real) matrices, $0 < h$ is a constant delay. We regard an absolute continuous n_1-vector function $x_1(\cdot)$ and a piecewise continuous n_2-vector function $x_2(\cdot)$ as a solution of System (1) if they satisfy the equation (1a) for almost all $t > 0$ and (1b) for all $t \geq 0$.

System (1) should be completed with initial conditions:

$$[D_t^{\alpha-1} x_1(t)]_{t=0} = x_{10}, \ x_2(\tau) = \psi(\tau), \ \tau \in [-h, 0), \tag{2}$$

where $x_{10} \in \mathbb{R}^{n_1}$; $\psi \in PC([-h,0), \mathbb{R}^{n_2})$ and $PC([-h,0), \mathbb{R}^{n_2})$ denotes the set of piecewise continuous n_2-vector-functions in $[-h,0]$. Observe that $x_2(t)$ at $t=0$ is determined from Equation (1b).

3 Representation of Solutions into Series of Determining Equations Solutions

Let us introduce the determining equations of System (1) (see [16] for more details).

$$\begin{aligned}
&X_{1,k}(t) = A_{11}X_{1,k-1}(t) + A_{12}X_{2,k-1}(t) + B_1U_{k-1}(t), \\
&X_{2,k}(t) = A_{21}X_{1,k}(t) + A_{22}X_{2,k}(t-h) + B_2U_{k-1}(t),\ k = 0,1,\dots; \\
&\text{with initial conditions} \\
&X_{1,k}(t) = 0, X_{2,k}(t) = 0 \text{ for } t < 0 \ \text{ or } \ k \leq 0; \\
&U_0(0) = I_{n_1},\ U_k(t) = 0 \ \text{ for } \ t^2 + k^2 \neq 0.
\end{aligned} \tag{3}$$

Here, we establish some algebraic properties of $X_{1,k}$, $X_{2,k}$.

Proposition 1. *([20]) The following identities hold :*

$$\left(A_{11} + A_{12}(I_{n_2} - \omega A_{22})^{-1}A_{12}\right)^k \left(B_1 + A_{12}(I_{n_2} - \omega A_{22})^{-1}B_2\right) = \sum_{j=0}^{+\infty} X_{1,k+1}(jh)\omega^j,\ k = 0,1,\dots;$$

$$(I_{n_2} - \omega A_{22})^{-1}A_{12}\left(A_{11} + A_{12}(I_{n_2} - \omega A_{22})^{-1}A_{12}\right)^k \left(B_1 + A_{12}(I_{n_2} - \omega A_{22})^{-1}B_2\right) = \sum_{j=0}^{+\infty} X_{2,k+1}(jh)\omega^j,\ k = 1,2,\dots;$$

$$(I_{n_2} - \omega A_{22})^{-1}B_2 = \sum_{j=0}^{+\infty} X_{2,0}(jh)\omega^j,$$

where $|\omega| < \omega_1$ and ω_1 is a sufficiently small real number.

Let us introduce the determining equations of homogenous System (1).

$$\begin{aligned}
&\widetilde{X}_{1,k}(t) = A_{11}\widetilde{X}_{1,k-1}(t) + A_{12}\widetilde{X}_{2,k-1}(t), \\
&\widetilde{X}_{2,k}(t) = A_{21}\widetilde{X}_{1,k}(t) + A_{22}\widetilde{X}_{2,k}(t-h),\ t \geq 0,\ k = 1,2,\dots; \\
&\text{with initial conditions} \\
&\widetilde{X}_{1,k}(t) = 0, \widetilde{X}_{2,k}(t) = 0 \text{ for } t < 0 \ \text{ or } k \leq 0; \\
&\widetilde{X}_{1,1}(0) = I_{n_1}, \widetilde{X}_{1,1}(\tau) = 0 \text{ if } \ \tau \neq 0.
\end{aligned} \tag{4}$$

Similar to Proposition 1 we can formulate the following.

Proposition 2. *([20]) The following identities hold :*

$$\left(A_{11} + A_{12}(I_{n_2} - \omega A_{22})^{-1}A_{12}\right)^k = \sum_{j=0}^{+\infty} \widetilde{X}_{1,k+1}(jh)\omega^j,\ k = 1,2,\dots;$$

$$(I_{n_2} - \omega A_{22})^{-1}A_{21}\left(A_{11} + A_{12}(I_{n_2} - \omega A_{22})^{-1}A_{12}\right)^k = \sum_{j=0}^{+\infty} \widetilde{X}_{2,k+1}(jh)\omega^j, k = 1,2,\dots;$$

where $|\omega| < \omega_1$ and ω_1 is a sufficiently small real number.

Theorem 3. *A solution to System (1) with initial conditions (2) for $t \geq 0$ exists, is unique and can be represented in the form of a series in power of solutions to determining systems (3), (4) in the following form:*

$$x_1(t, x_1(0), \psi, u) = \sum_{k=0}^{+\infty} \sum_{\substack{i \\ t-ih>0}} X_{1,k+1}(ih) \int_0^{t-ih} \frac{(t-\tau-ih)^{\alpha(k+1)-1}}{\Gamma(\alpha(k+1))} u(\tau)d\tau + s_1(t, x_1(0), \psi),$$

$$x_2(t, x_1(0), \psi, u) = \sum_{k=0}^{+\infty} \sum_{\substack{i \\ t-ih>0}} X_{2,k+1}(ih) \int_0^{t-ih} \frac{(t-\tau-ih)^{\alpha(k+1)-1}}{\Gamma(\alpha(k+1))} u(\tau)d\tau +$$
$$+ \sum_{\substack{i \\ t-ih>0}} X_{2,0}(ih)u(t-ih) + s_2(t, x_{10}, \psi),$$

where $s_1(t, x_{10}, \psi), s_2(t, x_{10}, \psi)$ - functions depending only on the initial data:

$$s_1(t, x_{10}, \psi) = \sum_{k=0}^{+\infty} \sum_{\substack{j \\ t-jh>0}} \frac{(t-jh)^{\alpha(k+1)-1}}{\Gamma(\alpha(k+1))} \widetilde{X}_{1,k+1}(jh)x_{10} +$$

$$\sum_{k=0}^{+\infty} \sum_{\substack{i,j \\ t-(i+j)h>0}} \widetilde{X}_{1,k+1}(ih)A_{12}(A_{22})^{i+1} \int_0^{t-(i+j)h} \frac{(t-\tau-(i+j)h)^{\alpha(k+1)-1}}{\Gamma(\alpha(k+1))} \psi(\tau-h)d\tau,$$

$$s_2(t, x_{10}, \psi) =$$

$$\sum_{k=0}^{+\infty} \sum_{\substack{i,j \\ t-(i+j)h>0}} \widetilde{X}_{2,k+1}(ih)A_{12}(A_{22})^{i+1} \int_0^{t-(i+j)h} \frac{(t-\tau-(i+j)h)^{\alpha(k+1)-1}}{\Gamma(\alpha(k+1))} \psi(\tau-h)d\tau +$$

$$\sum_{k=0}^{+\infty} \sum_{\substack{j \\ t-jh>0}} \frac{(t-jh)^{\alpha(k+1)-1}}{\Gamma(\alpha(k+1))} \widetilde{X}_{2,k+1}(jh)x_{10} + \sum_{i=0}^{+\infty} (A_{22})^{i+1}\psi(t-(i+1)h)),$$

where $\psi(\tau) \equiv 0$ for $\tau \notin [-h, 0)$.

Proof. First we use the classical formula for the Laplace transformation of the fractional derivative of Equation (1a)

$$\int_0^{\infty} e^{-pt} D_t^{\alpha} x_1(t)dt = p^{\alpha}\breve{x}_1(p) - [D_t^{\alpha-1}x_1(t)]_{t=0} = p^{\alpha}\breve{x}_1(p) - x_{10}.$$

We apply the Laplace transform to System (1)

$$p^{\alpha}\breve{x}_1(p) - x_{10} = A_{11}\breve{x}_1(p) + A_{12}\breve{x}_2(p) + B_1\breve{u}(p), \tag{5}$$

$$\breve{x}_2(p) = A_{21}\breve{x}_1(p) + A_{22}e^{-ph}\breve{x}_2(t) + A_{22}e^{-ph} \int_{-h}^{0} e^{-p\tau}\psi(\tau)d\tau + B_2\breve{u}(p), \tag{6}$$

where $\breve{x}_1(p)$, $\breve{x}_2(p)$, $\breve{u}(p)$ are Laplace transforms of functions $x_1(t)$, $x_2(t)$, $u(t)$ respectively. Solving (6), we obtain

$$\breve{x}_2(p) = \left(I_{n_2} - A_{22}e^{-ph}\right)^{-1} A_{21}\breve{x}_1(p)+ \tag{7}$$
$$\left(I_{n_2} - A_{22}e^{-ph}\right)^{-1} A_{22}e^{-ph} \int_{-h}^{0} e^{-p\tau}\psi(\tau)d\tau + \left(I_{n_2} - A_{22}e^{-ph}\right)^{-1} B_2\breve{u}(p),$$

$$\breve{x}_1(p)=\left(p^\alpha I_{n_1}-A_{11}-A_{12}\left(I_{n_2}-A_{22}e^{-ph}\right)^{-1}A_{21}\right)^{-1}\left[A_{12}\left(I_{n_2}-A_{22}e^{-ph}\right)^{-1}\times \right. \tag{8}$$
$$\left. A_{22}e^{-ph}\int_{-h}^{0} e^{-p\tau}\psi(\tau)d\tau + x_{10} + \left(B_1 + A_{12}\left(I_{n_2} - A_{22}e^{-ph}\right)^{-1} B_2\right)\breve{u}(p)\right]=$$
$$\sum_{k=0}^{+\infty}\frac{1}{(p^\alpha)^{k+1}}\left(A_{11} + A_{12}\left(I_{n_2} - A_{22}e^{-ph}\right)^{-1} A_{21}\right)^k \left[A_{12}\left(I_{n_2} - A_{22}e^{-ph}\right)^{-1}\times\right.$$
$$\left. A_{22}e^{-ph}\int_{-h}^{0} e^{-p\tau}\psi(\tau)d\tau + x_{10} + \left(B_1 + A_{12}(I_{n_2} - \omega A_{22})^{-1}B_2\right)\breve{u}(p)\right].$$

Applying Propositions 1 and 2 to (7) and (8) we obtain

$$\breve{x}_1(p) = \sum_{k=0}^{+\infty}\frac{1}{(p^\alpha)^{k+1}}\sum_{j=0}^{+\infty} e^{-jph}X_{1,k+1}(jh)\breve{u}(p) + \int_0^{+\infty} x_1(t, x_1(0), \psi, 0)dt,$$
$$\breve{x}_2(p) = \sum_{k=0}^{+\infty}\frac{1}{(p^\alpha)^{k+1}}\sum_{j=0}^{+\infty} e^{-jph}X_{2,k+1}(jh)\breve{u}(p) + \sum_{j=0}^{+\infty} e^{-jph}\widetilde{X}_{2,0}(jh)\breve{u}(p)+$$
$$\int_0^{+\infty} x_2(t, x_{10}, \psi, 0)dt.$$

By applying inverse Laplace transform the proof is complete. □

4 Controllability

We present algebraic properties of the determining solutions in the following two statements [16].

Proposition 4. *The solutions $X_{1,\gamma}(t), X_{2,\gamma}(t)$, $t \geq 0$ of the determining equations (3) satisfy the following equations:*

$$X_{1,\gamma}(kh) = -\sum_{j=1}^{\Theta_k} r_{0j}X_{1,\gamma}((k-j)h) - \sum_{i=1}^{n_1}\sum_{j=0}^{\Theta_k} r_{ij}X_{1,\gamma-i}((k-j)h),$$
$$X_{2,\gamma}(kh) = -\sum_{j=1}^{\Theta_k} r_{0j}X_{2,\gamma}((k-j)h) - \sum_{i=1}^{n_1}\sum_{j=0}^{\Theta_k} r_{ij}X_{2,\gamma-i}((k-j)h)$$

for $r_{ij} \in \mathbb{R}, i = 1, 2, \ldots, n_1; j = 0, 1, \ldots, n_2$; $\gamma = n_1 + 1, n_1 + 2, \ldots$ and $k = 0, 1, \ldots$; $\Theta_k = \min\{k, n_1n_2\}$.

Similar to Proposition 4 we can formulate the following:

Proposition 5. *The solutions of the determining equations (3) satisfy the following conditions:*

$$X_{1,k}((\nu+1)h)=-\sum_{j=1}^{\min\{k,n_1(n_2)^2\}} p_{0j}X_{1,k-j}((\nu+1)h)-\sum_{i=1}^{n_2}\sum_{j=0}^{\min\{k,n_1(n_2)^2\}} p_{ij}X_{1,k-j}((\nu+1-i)h),$$

$$X_{2,k}(\nu h)=-\sum_{j=1}^{\min\{k,n_1(n_2)^2\}} p_{0j}X_{2,k-j}(\nu h)-\sum_{i=1}^{n_2}\sum_{j=0}^{\min\{k,n_1(n_2)^2\}} p_{ij}X_{2,k-j}((\nu-i)h)$$

for $p_{ij}\in\mathbb{R}, i=1,2,\ldots,n_1; j=0,1,\ldots,n_2;\ k=0,1,\ldots$ *and* $\nu=n_2,n_2+1,\ldots$.

We will consider the following controllability conditions.

Definition 1. *System (1) is called* $Q-t_1$*-controllable for* $t_1>0$ *if for any vector* $\begin{bmatrix}x_{10}\\x_{20}\end{bmatrix}\in\mathbb{R}^{n_1+n_2}$ *and for any initial conditions (2) there exist a piecewise continuous control* $u(\cdot)$*, such that the condition* $Q\begin{bmatrix}x_1(t_1)\\x_2(t_1)\end{bmatrix}=Q\begin{bmatrix}x_{10}\\x_{20}\end{bmatrix}$ *holds for the corresponding solution* $x_1(t),x_2(t)$ *of System (1).*

In the case $Q=I_{n_1+n_2}$*, System (1) is called relatively* t_1*-controllable; for* $Q=[I_{n_1},0]$*, it is relatively* t_1*-controllable with respect to* x_1*; and, for* $Q=[0,I_{n_2}]$*, it is relatively* t_1*-controllable with respect to* x_2*.*

For simplicity, we put

$$x_{10}=0,\psi(\tau)=0,\ \tau\in[-h,0]. \tag{9}$$

Then the corresponding solution of System (1), (9) can be represented in the form:

$$x_1(t)=\sum_{k=0}^{+\infty}\left(\sum_{j=0}^{T_t-1}\int_{t-(j+1)h}^{t-jh}\sum_{i=0}^{j}+X_{1,k+1}(ih)\frac{(t-\tau-ih)^{\alpha k+\alpha-1}}{\Gamma(\alpha(k+1))}u(\tau)d\tau\right.$$
$$\left.\int_0^{t-T_th}\sum_{i=0}^{T_t}X_{1,k+1}(ih)\frac{(t-\tau-ih)^{\alpha k+\alpha-1}}{\Gamma(\alpha(k+1))}u(\tau)d\tau\right),\ t>0,$$

$$x_2(t)=\sum_{k=0}^{+\infty}\left(\sum_{j=0}^{T_t-1}\int_{t-(j+1)h}^{t-jh}\sum_{i=0}^{j}X_{2,k+1}(ih)\frac{(t-\tau-ih)^{\alpha k+\alpha-1}}{\Gamma(\alpha(k+1))}u(\tau)d\tau+\right.$$
$$\left.\int_0^{t-T_th}\sum_{i=0}^{T_t}X_{2,k+1}(ih)\frac{(t-\tau-ih)^{\alpha k+\alpha-1}}{\Gamma(\alpha(k+1))}u(\tau)d\tau\right)+\sum_{i=0}^{T_t}X_{2,0}(ih)u(t-ih),\ t>0.$$

The $Q-$attainability set $K(t_1)$ of System (1), (9) at the moment t_1 is described as follows:

$$K(t_1)=\left\{\begin{array}{l}\gamma\in\mathbb{R}^{n_3}:\gamma=\\=Q\sum\limits_{k=0}^{+\infty}\sum\limits_{j=0}^{T_{t_1}-1}\int\limits_{t_1-(j+1)h}^{t_1-jh}\sum\limits_{i=0}^{j}\begin{bmatrix}X_{1,k+1}(ih)\\X_{2,k+1}(ih)\end{bmatrix}\frac{(t_1-\tau-ih)^{\alpha k+\alpha-1}}{\Gamma(\alpha(k+1))}u(\tau)d\tau\\ \\+Q\sum\limits_{k=0}^{+\infty}\int\limits_{0}^{t_1-T_{t_1}h}\sum\limits_{i=0}^{T_{t_1}}\begin{bmatrix}X_{1,k+1}(ih)\\X_{2,k+1}(ih)\end{bmatrix}\frac{(t_1-\tau-ih)^{\alpha k+\alpha-1}}{\Gamma(\alpha(k+1))}u(\tau)d\tau\\ \\+Q\sum\limits_{j=0}^{T_{t_1}}\begin{bmatrix}0\\X_{2,0}(jh)\end{bmatrix}(t_1-jh),\ \forall u(\cdot)\in U(\cdot)\end{array}\right\}.$$

Here $U(\cdot)$ is the set of piecewise continuous $r-$vector-function in the interval $[0,t_1]$ and $K_0=\{Q\mu:\forall\mu\in\mathbb{R}^{n_1+n_2}\}$ is the linear span of the columns of matrix Q. Then $Q-t_1-$controllability of System (1) is equivalent to the inclusion $K(t_1)\supset K_0$ or $(K(t_1))^{\perp}\subset K_0^{\perp}$ for orthogonal complements. Then we state the following:

Theorem 6. *System (1) is $Q-t_1-$controllable if and only if for every vector $w\in\mathbb{R}^{n_3}$ such that :*

$$\begin{aligned}&w'Q\sum_{k=0}^{+\infty}\sum_{i=0}^{j}\frac{(t_1-\tau-ih)^{\alpha k+\alpha-1}}{\Gamma(\alpha(k+1))}\begin{bmatrix}X_{1,k}(ih)\\X_{2,k}(ih)\end{bmatrix}=0,k=0,1,\ldots;\\&\tau\in(t_1-(j+1)h,t_1-jh),t>0,\\&w'Q\sum_{j=0}^{T_{t_1}}\begin{bmatrix}0\\X_{2,0}(ih)\end{bmatrix}=0,\ j=0,\ldots,T_{t_1},\end{aligned}$$

the condition $w'Q=0$ also takes place.

Proof. We have:

$$\begin{aligned}&(\forall w\in\mathbb{R}^{n_3},w'K(r_1)=0,\Rightarrow w'K_0=0)\Leftrightarrow\\&(\forall w\in\mathbb{R}^{n_3}\quad w'Q\sum_{j=0}^{T_{t_1}-1}\int_{t_1-(j+1)h}^{t_1-jh}\sum_{k=0}^{+\infty}\sum_{i=0}^{j}\begin{bmatrix}X_{1,k+1}(ih)\\X_{2,k+1}(ih)\end{bmatrix}\frac{(t_1-\tau-ih)^{\alpha k+\alpha-1}}{\Gamma(\alpha(k+1))}u(\tau)d\tau\\&+w'Q\int_0^{t_1-T_{t_1}h}\sum_{k=0}^{+\infty}\sum_{i=0}^{T_{t_1}}\begin{bmatrix}X_{1,k+1}(ih)\\X_{2,k+1}(ih)\end{bmatrix}\frac{(t_1-\tau-ih)^{\alpha k+\alpha-1}}{\Gamma(\alpha(k+1))}u(\tau)d\tau\\&\left.+w'Q\sum_{j=0}^{T_{t_1}}\begin{bmatrix}0\\X_{2,0}(jh)\end{bmatrix}u(t_1-jh),\ \forall u(\cdot)\in U(\cdot),\forall u(\cdot)\in U(\cdot),\Rightarrow w'K_0=0\right).\end{aligned}$$

Setting

$$u(\tau) = \left(w'Q \sum_{k=0}^{+\infty} \sum_{i=0}^{j} \frac{(t_1 - \tau - ih)^{\alpha k+\alpha-1}}{\Gamma(\alpha(k+1))} \begin{bmatrix} X_{1,k+1}(ih) \\ X_{2,k+1}(ih) \end{bmatrix} \right)', \ j = 0, \dots, T_{t_1},$$

$$\tau \in (t_1 - (j+1)h, t_1 - jh), \tau > 0, \ u(t_1 - jh) = \left(w'Q \begin{bmatrix} 0 \\ X_{2,0}(ih) \end{bmatrix} \right)',$$

we obtain

$$\left(\forall w \in \mathbb{R}^{n_3}, \ \sum_{j=0}^{T_{t_1}-1} \int_{t_1-(j+1)h}^{t_1-jh} \left\| w'Q \sum_{k=0}^{+\infty} \sum_{i=0}^{j} \frac{(t_1-\tau-ih)^{\alpha k+\alpha-1}}{\Gamma(\alpha(k+1))} \begin{bmatrix} X_{1,k+1}(ih) \\ X_{2,k+1}(ih) \end{bmatrix} \right\|^2 d\tau \right.$$

$$+ \int_0^{t_1 - T_{t_1} h} \left\| w'Q \sum_{k=0}^{+\infty} \sum_{i=0}^{T_{t_1}} \frac{(t_1 - \tau - ih)^{\alpha k+\alpha-1}}{\Gamma(\alpha(k+1))} \begin{bmatrix} X_{1,k+1}(ih) \\ X_{2,k+1}(ih) \end{bmatrix} \right\|^2 d\tau$$

$$\left. + \sum_{j=0}^{T_{t_1}} \left\| w'Q \begin{bmatrix} 0 \\ X_{2,0}(jh) \end{bmatrix} \right\|^2 = 0, \Rightarrow w'K_0 = 0 \right).$$

This finishes the proof.

For the sequel, we need the following result:

Lemma 7. *Functions* $f_{kj}(t) = \frac{(t-jh)^{\alpha k+\alpha-1}}{\Gamma(\alpha(k+1))}$ *for* $t - jh > 0$ *and* $f_{kj}(t) = 0$ *for* $t - jh \leq 0$, *where* $k = 0, 1, \dots; j = 0, 1, \dots,$ *are linearly independent for* $t > 0$.

The proof of the lemma can be performed similarly to [16], [18] using the method of mathematical induction.

Now we can formulate an explicit criterion of controllability.

Theorem 8. *System (1) is* t_1-*controllable if and only if*

$$rank \left[\begin{bmatrix} X_{1,k}(ih) \\ X_{2,k}(ih) \end{bmatrix}, k = 0, 1, \dots, n_1; \ i = 0, 1, \dots, \min\{T_{t_1}, n_2\} \right] = n_1 + n_2;$$

Proof. Taking into account Proposition 4 and Lemma 7, we deduce from Theorem 6 that System (1) is t_1-controllable if and only if the condition

$$w' \begin{bmatrix} X_{1,k}(t) \\ X_{2,k}(t) \end{bmatrix} = 0, \ w' \begin{bmatrix} 0 \\ X_{2,0}(ih) \end{bmatrix} = 0$$

$$w \in \mathbb{R}^{n_3}, \ t \in [0, t_1), k = 0, 1, \dots, n_1, \ i = 0, 1, \dots, T_{t_1}$$

implies the equality $w' = 0$. Then by Proposition 5 it is easy seen that $X_{1,k}(ih)$, $X_{2,k}(ih)$, where $k > n_1$, $j > n_2$, are a linear combination of $X_{1,\eta}(\xi h)$, $X_{2,\eta}(\xi h)$,$\eta = 1, 2, \dots, n_1$; $\xi = 0, 1, \dots, n_2$, which completes the proof. □

Corollary 9

System (1) is

(i) relatively t_1-controllable with respect to x_1 if and only if

$$rank\,[X_{1,k}(ih), k=0,1,\dots,n_1;\ i=0,1,\dots,\min\{T_{t_1},n_2\}]=n_1;$$

(ii) relatively t_1-controllable with respect to x_2 if and only if

$$rank\,[X_{2,k}(ih), k=0,1,\dots,n_1;\ i=0,1,\dots,\min\{T_{t_1},n_2\}]=n_2.$$

5 Example 1

Let us consider the following system:

$$\begin{aligned} D_t^{\alpha}x_1(t) =&[1]x_1(t)+\begin{bmatrix}0 & -1\end{bmatrix}x_2(t)+[1]u(t),\ t>0,\\ x_2(t) =& \begin{bmatrix}0\\1\end{bmatrix}x_1(t)+\begin{bmatrix}0&1\\0&0\end{bmatrix}x_2(t-h)+\begin{bmatrix}1\\0\end{bmatrix}u(t),\ t\geq 0.\end{aligned} \tag{10}$$

First compute the solutions of the determining systems:

$$\begin{bmatrix}X_{1,0}(0)\\X_{2,0}(0)\end{bmatrix}=\begin{bmatrix}0\\1\\0\end{bmatrix};\ \begin{bmatrix}X_{1,1}(0)\\X_{2,1}(0)\end{bmatrix}=\begin{bmatrix}1\\0\\1\end{bmatrix};\ \begin{bmatrix}X_{1,k}(0)\\X_{2,k}(0)\end{bmatrix}=\begin{bmatrix}0\\0\\0\end{bmatrix},k\geq 2;\ \begin{bmatrix}X_{1,0}(h)\\X_{2,0}(h)\end{bmatrix}=\begin{bmatrix}0\\0\\0\end{bmatrix};$$

$$\begin{bmatrix}X_{1,1}(h)\\X_{2,1}(h)\end{bmatrix}=\begin{bmatrix}0\\1\\0\end{bmatrix};\ \begin{bmatrix}X_{1,k}(h)\\X_{2,k}(h)\end{bmatrix}=\begin{bmatrix}0\\0\\0\end{bmatrix},\ k\geq 2;\ \begin{bmatrix}X_{1,i}(jh)\\X_{2,i}(jh)\end{bmatrix}=\begin{bmatrix}0\\0\\0\end{bmatrix},i\geq 0,\ j\geq 2;$$

$$\begin{bmatrix}\widetilde{X}_{1,1}(0)\\\widetilde{X}_{2,1}(0)\end{bmatrix}=\begin{bmatrix}1\\0\\1\end{bmatrix};\ \begin{bmatrix}\widetilde{X}_{1,2}(0)\\\widetilde{X}_{2,2}(0)\end{bmatrix}=\begin{bmatrix}0\\0\\0\end{bmatrix};\ \begin{bmatrix}\widetilde{X}_{1,k}(0)\\\widetilde{X}_{2,k}(0)\end{bmatrix}=\begin{bmatrix}0\\0\\0\end{bmatrix},k\geq 3;\ \begin{bmatrix}\widetilde{X}_{1,1}(h)\\\widetilde{X}_{2,1}(h)\end{bmatrix}=\begin{bmatrix}0\\1\\0\end{bmatrix};$$

$$\begin{bmatrix}\widetilde{X}_{1,2}(h)\\\widetilde{X}_{2,2}(h)\end{bmatrix}=\begin{bmatrix}0\\0\\0\end{bmatrix};\ \begin{bmatrix}\widetilde{X}_{1,k}(h)\\\widetilde{X}_{2,k}(h)\end{bmatrix}=\begin{bmatrix}0\\0\\0\end{bmatrix},\ k\geq 2;\ \begin{bmatrix}\widetilde{X}_{1,i}(jh)\\\widetilde{X}_{2,i}(jh)\end{bmatrix}=\begin{bmatrix}0\\0\\0\end{bmatrix},i\geq 1,\ j\geq 2;$$

According to representation of Theorem 3 , we have

$$\begin{bmatrix}x_1(t)\\x_2(t)\end{bmatrix}=\begin{bmatrix}\int_0^t \frac{(t-\tau)^{\alpha-1}}{\Gamma(\alpha)}u(\tau)d\tau+\frac{t^{\alpha-1}}{\Gamma(\alpha)}x_{10}\\ u(t)+\phi(t-h)\\ \int_0^t \frac{(t-\tau)^{\alpha-1}}{\Gamma(\alpha)}u(\tau)d\tau+\frac{t^{\alpha-1}}{\Gamma(\alpha)}x_{10}\end{bmatrix};\ 0\leq t<h;$$

$$\begin{bmatrix}x_1(t)\\x_2(t)\end{bmatrix}=\begin{bmatrix}\int_0^t \frac{(t-\tau)^{\alpha-1}}{\Gamma(\alpha)}u(\tau)d\tau+\frac{t^{\alpha-1}}{\Gamma(\alpha)}x_{10}\\ u(t)+\int_0^{t-h} \frac{(t-\tau-h)^{\alpha-1}}{\Gamma(\alpha)}u(\tau)d\tau+\frac{(t-h)^{\alpha-1}}{\Gamma(\alpha)}x_{10}\\ \int_0^t \frac{(t-\tau)^{\alpha-1}}{\Gamma(\alpha)}u(\tau)d\tau+\frac{t^{\alpha-1}}{\Gamma(\alpha)}x_{10}\end{bmatrix};\ h\leq t;$$

We have:

$$\text{rank}\big[\left[X_{1,k}(ih)\right],k=0,1,\dots,n_1;i=0,1,\dots,\min\{T_{t_1},n_2\}\big]=\text{rank}\begin{bmatrix}0 & 1\end{bmatrix}=n_1=1;$$

Thus the system (10) is relatively t_1-controllable with respect to x_1 for $t_1 > 0$. Moreover

$$\text{rank}\left[\left[X_{2,k}(ih)\right], k = 0, 1, \dots, n_1;\ i = 0, 1, \dots, \min\{T_{t_1}, n_2\}\right] =$$
$$\text{rank}\begin{bmatrix}1 & 0\\ 0 & 1\end{bmatrix} = n_2 = 2;$$

Therefore the system (10) is relatively t_1-controllable with respect to x_2 for $t_1 > 0$.

The following holds:

$$\text{rank}\left[\begin{bmatrix}X_{1,k}(ih)\\ X_{2,k}(ih)\end{bmatrix}, k = 0, 1, \dots, n_1; i = 0, 1, \dots, \min\{T_{t_1}, n_2\}\right] =$$
$$\text{rank}\begin{bmatrix}0 & 1 & 0\\ 1 & 0 & 1\\ 0 & 1 & 0\end{bmatrix} = 2 \neq n_1 + n_2 = 3;$$

We conclude that the system (10) is not relatively t_1-controllable.

6 Example 2

Consider the next example in the following form:

$$\begin{aligned} D_t^{\alpha} x_1(t) =& x_1(t) + \begin{bmatrix}1 & 0 & 0\\ 0 & 1 & 0\end{bmatrix} x_2(t) + \begin{bmatrix}1\\ 0\end{bmatrix} u(t),\ t > 0, \\ x_2(t) =& \begin{bmatrix}0 & 1 & 0\\ 0 & 0 & 1\\ 0 & 0 & 0\end{bmatrix} x_2(t-h) + \begin{bmatrix}0\\ 0\\ 1\end{bmatrix} u(t),\ t \geq 0, \\ &\text{where } x_{10} \in \mathbb{R}^2,\ \psi(\cdot) = \begin{bmatrix}\psi(\cdot)\\ \psi(\cdot)\\ \psi(\cdot)\end{bmatrix}. \end{aligned} \tag{11}$$

Solving the determining systems, we have:

$$X_{1,0}(jh) = \begin{bmatrix}0\\ 0\end{bmatrix},\ j \geq 0;\ X_{1,k}(0) = \begin{bmatrix}1\\ 0\end{bmatrix},\ X_{1,k}(h) = \begin{bmatrix}0\\ 1\end{bmatrix},\ X_{1,k}(2h) = \begin{bmatrix}1\\ 0\end{bmatrix},$$

$$X_{1,k}(ih) = \begin{bmatrix}0\\ 0\end{bmatrix},\ k = 1, 2, \dots; i \geq 3;$$

$$X_{2,0}(0) = \begin{bmatrix}0\\ 0\\ 1\end{bmatrix},\ X_{2,0}(h) = \begin{bmatrix}0\\ 1\\ 0\end{bmatrix},\ X_{2,0}(2h) = \begin{bmatrix}1\\ 0\\ 0\end{bmatrix}, X_{2,0}(jh) = \begin{bmatrix}0\\ 0\\ 0\end{bmatrix},\ j \geq 3;$$

$$X_{2,k}(ih) = \begin{bmatrix}0\\ 0\\ 0\end{bmatrix},\ k = 1, 2, \dots; i \geq 0;\ \widetilde{X}_{1,k}(0) = \begin{bmatrix}1 & 0\\ 0 & 1\end{bmatrix},\ k \geq 1;$$

$$\widetilde{X}_{1,k}(jh) = \begin{bmatrix}0 & 0\\ 0 & 0\end{bmatrix}, k \geq 1;\ j \geq 1; \widetilde{X}_{2,k}(ih) = \begin{bmatrix}0 & 0\\ 0 & 0\\ 0 & 0\end{bmatrix},\ k \geq 1,\ j \geq 0;$$

We obtain representation of solutions of system (11) in the following form:

$$x_1(t) = \begin{bmatrix}1\\0\end{bmatrix} \int_0^t \sum_{k=1}^{+\infty} \frac{(t-\tau)^{\alpha(k+1)-1}}{\Gamma(\alpha(k+1))} u(\tau) d\tau + \sum_{k=1}^{+\infty} \frac{(t)^{\alpha(k+1)-1}}{\Gamma(\alpha(k+1))} x_{10}$$
$$+ \int_0^t \sum_{k=1}^{+\infty} \frac{(t-\tau)^{\alpha(k+1)-1}}{\Gamma(\alpha(k+1))} \begin{bmatrix}\psi_2(\tau)\\ \psi_3(\tau)\end{bmatrix} d\tau,$$
$$x_2(t) = \begin{bmatrix}\psi_1(t-h)\\ \psi_2(t-h)\\ \psi_3(t-h)+u(t)\end{bmatrix}, \text{ for } 0 \le t < h.$$
$$x_1(t) = \begin{bmatrix}1\\0\end{bmatrix} \int_{t-h}^t \sum_{k=1}^{+\infty} \frac{(t-\tau)^{\alpha(k+1)-1}}{\Gamma(\alpha(k+1))} u(\tau) d\tau$$
$$+ \int_0^{t-h} \sum_{k=1}^{+\infty} \left(\begin{bmatrix}1\\0\end{bmatrix} \frac{(t-\tau)^{\alpha(k+1)-1}}{\Gamma(\alpha(k+1))} + \begin{bmatrix}0\\1\end{bmatrix} \frac{(t-\tau-h)^{\alpha(k+1)-1}}{\Gamma(\alpha(k+1))} \right) u(\tau) d\tau$$
$$+ \sum_{k=1}^{+\infty} \frac{(t)^{\alpha(k+1)-1}}{\Gamma(\alpha(k+1))} x_{10} + \int_0^{t-h} \sum_{k=1}^{+\infty} \frac{(t-\tau-h)^{\alpha(k+1)-1}}{\Gamma(\alpha(k+1))} \begin{bmatrix}\psi_3(\tau)\\ 0\end{bmatrix} d\tau$$
$$+ \int_0^t \sum_{k=1}^{+\infty} \frac{(t-\tau)^{\alpha(k+1)-1}}{\Gamma(\alpha(k+1))} \begin{bmatrix}\psi_2(\tau)\\ \psi_3(\tau)\end{bmatrix} d\tau,$$
$$x_2(t) = \begin{bmatrix}0\\ u(t-h)\\ u(t)\end{bmatrix}, \text{ for } h \le t < 2h.$$
$$x_1(t) = \begin{bmatrix}1\\0\end{bmatrix} \int_{t-h}^t \sum_{k=1}^{+\infty} \frac{(t-\tau)^{\alpha(k+1)-1}}{\Gamma(\alpha(k+1))} u(\tau) d\tau$$
$$+ \int_0^{t-h} \sum_{k=1}^{+\infty} \left(\begin{bmatrix}1\\0\end{bmatrix} \frac{(t-\tau)^{\alpha(k+1)-1}}{\Gamma(\alpha(k+1))} + \begin{bmatrix}0\\1\end{bmatrix} \frac{(t-\tau-h)^{\alpha(k+1)-1}}{\Gamma(\alpha(k+1))} \right) u(\tau) d\tau$$
$$+ \sum_{j=2}^{T_t} \int_{max\{0,t-(j+1)h\}}^{t-h} \sum_{k=1}^{+\infty} \left(\begin{bmatrix}1\\0\end{bmatrix} \frac{(t-\tau)^{\alpha(k+1)-1}}{\Gamma(\alpha(k+1))} + \begin{bmatrix}0\\1\end{bmatrix} \frac{(t-\tau-h)^{\alpha(k+1)-1}}{\Gamma(\alpha(k+1))} \right.$$
$$\left. + \begin{bmatrix}1\\0\end{bmatrix} \frac{(t-\tau-2h)^{\alpha(k+1)-1}}{\Gamma(\alpha(k+1))} \right) u(\tau) d\tau + \int_0^{t-h} \sum_{k=1}^{+\infty} \frac{(t-\tau-h)^{\alpha(k+1)-1}}{\Gamma(\alpha(k+1))} \begin{bmatrix}\psi_3(\tau)\\ 0\end{bmatrix} d\tau$$
$$+ \sum_{k=1}^{+\infty} \frac{(t)^{\alpha(k+1)-1}}{\Gamma(\alpha(k+1))} x_{10} + \int_0^t \sum_{k=1}^{+\infty} \frac{(t-\tau)^{\alpha(k+1)-1}}{\Gamma(\alpha(k+1))} \begin{bmatrix}\psi_2(\tau)\\ \psi_3(\tau)\end{bmatrix} d\tau,$$
$$x_2(t) = \begin{bmatrix}u(t-2h)\\ u(t-h)\\ u(t)\end{bmatrix}, \text{ for } 2h \le t.$$

We have:

$$\operatorname{rank}\left[\left[X_{1,k}(ih)\right], k=0,1,\ldots,n_1; i=0,1,\ldots,\min\{T_{t_1},n_2\}\right]=$$
$$\operatorname{rank}\left[X_{1,1}(0)\ X_{1,1}(h)\right] \operatorname{rank}\begin{bmatrix}1&0\\0&1\end{bmatrix}=n_1=2$$

and, hence, system (11) is relatively t_1-controllable with respect to x_1 for $t_1 > h$. Next we have:

$$\operatorname{rank}\left[\left[X_{2,k}(ih)\right], k=0,1,\ldots,n_1; i=0,1,\ldots,\min\{T_{t_1},n_2\}\right]=$$
$$\operatorname{rank}\left[X_{2,0}(0)\ X_{2,0}(h)\ X_{2,0}(2h)\right]=\operatorname{rank}\begin{bmatrix}0&0&1\\0&1&0\\1&0&0\end{bmatrix}=n_2=3;$$

Thus the system (11) is relatively t_1-controllable with respect to x_2 for $t_1 > 2h$. Moreover

$$\operatorname{rank}\left[\begin{bmatrix}X_{1,k}(ih)\\X_{2,k}(ih)\end{bmatrix}, k=0,1,\ldots,n_1;\ i=0,1,\ldots,\min\{T_{t_1},n_2\}\right]=$$
$$\operatorname{rank}\left[\begin{bmatrix}X_{1,0}(0)\\X_{2,0}(0)\end{bmatrix}\begin{bmatrix}X_{1,0}(h)\\X_{2,0}(h)\end{bmatrix}\begin{bmatrix}X_{1,0}(2h)\\X_{2,0}(2h)\end{bmatrix}\begin{bmatrix}X_{1,01}(0)\\X_{2,1}(0)\end{bmatrix}\begin{bmatrix}X_{1,1}(h)\\X_{2,1}(h)\end{bmatrix}\right]=$$
$$\operatorname{rank}\begin{bmatrix}0&0&0&1&0\\0&0&0&0&1\\0&0&1&0&0\\0&1&0&0&0\\1&0&0&0&0\end{bmatrix}=n_1+n_2=5.$$

Therefore the system (11) is relatively t_1-controllable for $t_1 > 2h$.

7 Conclusions

Representations of solutions for for linear fractional differential-algebraic systems with delay (FDAD) has been presented (Theorem 3). Effective parametric rank criteria for relative controllability has been established (Theorem 8). These considerations can be extended to systems with many delays.

Acknowledgements. This research was supported by Bialystok University of Technology (grant no. S/WI/2/11).

References

1. Gabasov, R., Curakova, S.V.: The theory of controllability of linear systems with lag. Izv. Akad. Nauk SSSR Tehn. Kibernet 4, 17–28 (1969) (Russian)

2. Manitius, A.: On the controllability conditions for systems with distributed delays in state and control. Arch. Automat. i Telemech 17, 363–377 (1972)
3. Shklyar, B.: On the problem of the relative controllability of systems with deviating argument of neutral type. Differencial'nye Uravnenija 10, 1443–1450 (1974) (Russian)
4. Klamka, J.: Relative controllability and minimum energy control of linear systems with distributed delays in control. IEEE Trans. Automatic Control AC-21(4), 594–595 (1976)
5. Klamka, J.: Relative controllability of linear systems with varying delay. Systems Sci. 2(3), 17–24 (1976)
6. Klamka, J.: Relative controllability of nonlinear systems with delays in control. Automatica J. IFAC 12(6), 633–634 (1976)
7. Dacka, C.: Relative controllability of perturbed nonlinear systems with delay in control. IEEE Trans. Automat. Control 27(1), 268–270 (1982)
8. Balachandran, K., Dauer, J.P.: Relative controllability of nonlinear neutral Volterra integrodifferential systems. J. Austral. Math. Soc. Ser. B 37(3), 346–353 (1996)
9. Respondek, J.: Controllability of second order infinite dimensional dynamical systems with delays. Ann. Univ. Mariae Curie-Skodowska Sect. AI Inform. 8(1), 133–149 (2008)
10. Marchenko, V.M., Zaczkewicz, Z.: Duality in controllability and observability problems for nonstationary hybrid impulsive differential-difference systems. Differ. Equ. 46(12), 1767–1778 (2010)
11. Valerio, D., Sa da Costa, J.: An Introduction to Fractional Control. Institution of Engineering and Technology (2013)
12. Vijayakumar, V., Selvakumar, A., Murugesu, R.: Controllability for a class of fractional neutral integro-differential equations with unbounded delay. Appl. Math. Comput. 232, 303–312 (2014)
13. Balachandran, K., Kokila, J., Trujillo, J.J.: Relative controllability of fractional dynamical systems with multiple delays in control. Comput. Math. Appl. 64(10), 3037–3045 (2012)
14. Balachandran, K., Zhou, Y., Kokila, J.: Relative controllability of fractional dynamical systems with distributed delays in control. Comput. Math. Appl. 64(10), 3201–3209 (2012)
15. Marchenko, V.M., Poddubnaya, O.N., Zaczkiewicz, Z.: On the observability of linear differential-algebraic systems with delays. IEEE Trans. Automat. Control 51(8), 1387–1392 (2006)
16. Marchenko, V.M., Poddubnaya, O.N.: Relative controllability of stationary hybrid systems. In: Proc. IEEE Methods and Models in Automation and Robotics (MMAR 2004), Miedzyzdroje, Poland, August/ September 267-272 (2004)
17. Zaczkiewicz, Z.: Observability of linear differential-algebraic systems with delay. Ph.D. Thesis. Warsaw University of Technology: Poland (2009)
18. Zaczkiewicz, Z.: Controllability, observability, duality for fractional differential-algebraic systems with delay. JAND 1, 195–205 (2012)
19. Kilbas, A.A., Srivastava, H.M., Trujillo, J.J.: Theory and Applications of Fractional Differential Equations, 4th edn. Springer, Berlin (2007)
20. Zaczkiewicz, Z.: Representation of solutions for fractional differential-algebraic systems with delays. Bull. Pol. Acad. Sci. Tech. Sci. 58(4), 607–612 (2010)

Part V
Stability Analysis

Robust Stability Check of Fractional Discrete-Time Linear Systems with Interval Uncertainties

Mikołaj Busłowicz and Andrzej Ruszewski

Bialystok University of Technology, Faculty of Electrical Engineering,
Wiejska 45D, 15-351 Białystok, Poland
{busmiko,andrusz}@pb.edu.pl
http://www.pb.edu.pl

Abstract. The paper presents the problems of robust practical stability and robust asymptotic stability of fractional-order discrete-time linear systems with uncertainty. It is supposed that the system matrix is the interval matrix which elements are the convex combinations of the elements of specified bounded matrices and the fractional order α satisfies $0 < \alpha < 1$. Using the matrix measure the robust stability conditions are given. The considerations are illustrated by numerical examples.

Keywords: Linear system, discrete-time, fractional-order, robust stability, interval matrix.

1 Introduction

Fractional calculus and its application in many areas in science and engineering have been presented in many monographs and papers (see, e.g. [13, 14, 16, 19, 21]).

The stability problem is the fundamental matter in the dynamical systems theory. This problem for linear continuous-time fractional systems has been considered in many publications (see, e.g. [4, 5, 13, 16, 18, 20]), whereas a stability problem of linear discrete-time fractional systems is more complicated and less advanced. In this case beside the asymptotic stability it is also considered the so-called practical stability defined for the length of practical implementation. The problem of practical stability of fractional discrete-time systems has been considered in [6, 7, 12, 13] for positive systems and in [6, 11] for non-positive (standard) systems. Recently, the stability regions in the complex plane of fractional discrete-time linear systems were presented in [8, 17, 22]. The parametric descriptions of boundaries of these regions have been given.

The robust asymptotic stability problem of the fractional continuous-time interval systems has been studied among others in [1, 9, 15]. For the fractional discrete-time interval systems this problem has been analyzed in [3] for positive systems and in [20] for standard systems.

© Springer International Publishing Switzerland 2015
K.J. Latawiec et al. (eds.), *Advances in Modeling and Control of Non-integer Order Systems*, Lecture Notes in Electrical Engineering 320, DOI: 10.1007/978-3-319-09900-2_18

2 Problem Formulation

Consider an uncertain discrete-time linear system of fractional order described by the homogeneous state equation

$$\Delta^{\alpha} x_{i+1} = A_I x_i, \quad \alpha \in (0,\ 1), \tag{1}$$

with the initial condition x_0, where $x_i \in \Re^n$. The system matrix $A_I \in \Re^{n\times n}$ is the interval matrix in which all elements are known only to within a specific closed intervals defined as follows

$$A_I = \{A = [a_{ij}],\ \ b_{ij} \le a_{ij} \le c_{ij},\quad i,j = 1,2,...,n\}, \tag{2}$$

where b_{ij} and c_{ij} are elements of matrices B and C.

A special case of the interval matrix is the matrix of the form

$$A_I = (1-r)B + rC, \qquad \forall r \in [0,\ 1]. \tag{3}$$

Every element a_{ij} of the interval matrix (3) is the convex combination of the elements b_{ij} and c_{ij} of the matrices B and C. Assumption $b_{ij} \le a_{ij}$ $(i,j = 1,2,...,n)$ does not hold. The system (1) with matrix (2) or (3) is called the interval system.

The following definition of the fractional difference [12, 13] will be used

$$\Delta^{\alpha} x_i = \sum_{k=0}^{i} (-1)^k \binom{\alpha}{k} x_{i-k} \tag{4}$$

where $\alpha \in \Re$ is the order of the fractional difference and

$$\binom{\alpha}{k} = \frac{\alpha!}{k!(\alpha-k)!}. \tag{5}$$

Using definition (4) we may write the equation (1) in the form

$$x_{i+1} = (A_I + I\alpha)x_i + \sum_{k=1}^{i} p_k(\alpha)x_{i-k}, \quad k = 1,2,... \tag{6}$$

where I is the $n \times n$ identity matrix and

$$p_k(\alpha) = (-1)^k \binom{\alpha}{k+1}, \quad k = 1,2,.... \tag{7}$$

The coefficients (7) can be easily calculated using the following formula [7]

$$p_{k+1}(\alpha) = p_k(\alpha)\frac{k+1-\alpha}{k+2}, \quad k = 1,2,... \tag{8}$$

with $p_1(\alpha) = 0.5\alpha(1-\alpha)$.

Note that the equation (6) represents a linear discrete-time system with increasing number of delays in state. From (8) it follows that the coefficients $p_k(\alpha)$ are positive for $\alpha \in (0,\ 1)$ and decrease rapidly with an increase of k. Therefore, we can assume that the value of k in the equation (6) may be limited by some natural number L. This number is called the length of the practical implementation [12]. In this case the equation (6) can be written in the form

$$x_{i+1} = (A_I + I\alpha)x_i + \sum_{k=1}^{L} p_k(\alpha)x_{i-k}, \quad k = 1, 2, ... \tag{9}$$

The equation (9) represents the interval linear discrete-time system with L delays in state. Moreover, the system (9) is called the practical realization of the interval fractional system (1).

The definition of practical and asymptotic stability for fractional discrete-time systems has been introduced in the work [12]. With regard to the interval system (1) this definitions take the following forms.

Definition 1. The fractional interval system (1) is called robust practically stable if the system (9) is asymptotically stable for every matrix $A \in A_I \in \Re^{n \times n}$.

Definition 2. The fractional interval system (1) is called robust asymptotically stable if the system (9) is asymptotically stable with $L \to \infty$ for every matrix $A \in A_I \in \Re^{n \times n}$.

In the paper [20] the problems of robust practical stability and robust asymptotic stability of fractional-order discrete-time linear system (1) with the interval matrix of the form (2) has been considered.

The aim of this paper is to give the robust stability conditions for the practical and asymptotic stability for the discrete-time linear interval system (1) with the interval matrix of the form (3).

3 Practical and Asymptotic Stability Regions

In paper [8] (see also [17, 23]) the practical stability and the asymptotic stability of system

$$\Delta^{\alpha} x_{i+1} = Ax_i, \quad \alpha \in (0,\ 1), \tag{10}$$

with precisely known matrix A has been considered. Necessary and sufficient conditions for the practical stability and for the asymptotic stability for system (10) have been established. The conditions have been given in terms of eigenvalues of the matrix $A_0 = A + I\alpha$ for the practical stability and in terms of eigenvalues of the matrix A and A_0 for the asymptotic stability. In particular, it has been shown that location of all eigenvalues of the matrix A_0 or A in the stability regions is necessary and sufficient for practical and asymptotic stability.

Theorem 1. [8] The fractional system (10) with given length L of practical implementation is practically stable if and only if all eigenvalues $\lambda_i(A)$

($i = 1, 2, ..., n$) are located in the stability region $S(\alpha, L)$, i.e. $\lambda_i(A) \in S(\alpha, L)$ for all $i = 1, 2, ..., n$, where

$$\rho(\omega) = e^{j\omega} - \alpha - \sum_{k=1}^{L} p_k(\alpha) e^{-jk\omega}, \quad \omega \in [0,\ 2\pi]. \tag{11}$$

is the parametric description of boundary of stability region $S(\alpha, L)$, in the complex ρ-plane.

Practical stability regions $S(\alpha, L)$, for $L = 50$ and a few values of fractional order $\alpha \in (0,\ 1)$ are shown in Fig. 1a) on the plane of eigenvalues of A.

Similarly as in paper [8] we have the following lemma for the practical stability in the case of real eigenvalues of matrix A.

Lemma 1. If all eigenvalues $\lambda_i(A)$ are real, then the fractional system (10) with given length L of practical implementation is practically stable if and only if

$$-1 - \alpha - \sum_{k=1}^{L} p_k(\alpha)(-1)^k < \lambda_i(A) < 1 - \alpha - \sum_{k=1}^{L} p_k(\alpha), \quad i = 1, 2, ..., n. \tag{12}$$

According to the asymptotic stability of system (10) we have the following theorem and lemma.

Theorem 2. [8] The fractional system (10) is asymptotically stable if and only if all eigenvalues of the matrix A are located in the stability region , where

$$\eta(\omega) = (e^{j\omega} - 1)^{\alpha}(e^{j\omega})^{1-\alpha}, \quad \omega \in [0,\ 2\pi]. \tag{13}$$

is the parametric description of boundary of the stability region $S(\alpha)$ in the complex η-plane.

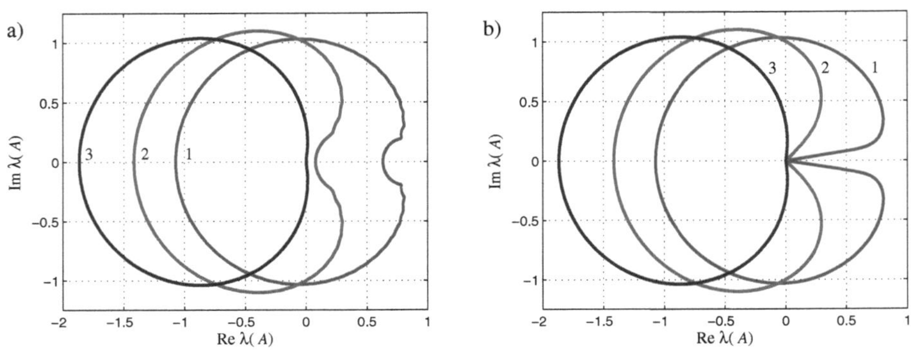

Fig. 1. Stability regions; a) regions $S(\alpha, L)$ for $L = 50$ and $\alpha = 0.1$ (boundary 1), $\alpha = 0.5$ (boundary 2) and $\alpha = 0.9$ (boundary 3), b) regions $S(\alpha)$ for $\alpha = 0.1$ (boundary 1), $\alpha = 0.5$ (boundary 2) and $\alpha = 0.9$ (boundary 3)

Lemma 2. [8] If all eigenvalues $\lambda_i(A)$ are real, then the fractional system (10) is asymptotically stable if and only if

$$-2^{\alpha} < \lambda_i(A) < 0, \quad i = 1, 2, ..., n. \tag{14}$$

Asymptotic stability regions $S(\alpha)$ for a few values of α are shown in Fig. 1b).

4 Main Results

The eigenvalues regions of the interval matrix (3) based on the matrix measure has been presented in the paper [2]. In this paper, using the results of [2] and the parametric description of boundaries of stability regions (11) and (13), the conditions of robust practical stability and robust asymptotic stability of the fractional discrete-time interval systems (1) with the matrix A_I of the form (3) will be given.

Lemma 3. The interval system (1), (3) is robust practically stable if and only if all the eigenvalues of the matrix A_I lie in the open region $S(\alpha, L)$.

Lemma 4. The interval system (1), (3) is robust asymptotically stable if and only if all the eigenvalues of the matrix A_I lie in the open region $S(\alpha)$.

Proof. All eigenvalues of the matrix A_I are in the open region $S(\alpha, L)(S(\alpha))$ if and only if all the eigenvalues of every matrix $A \in A_I$ are in this region. The proof follows from Theorem 1 (Theorem 2).

The matrix $A \in \Re^{n\times n}$ has exactly n eigenvalues, but the interval matrix A_I has an infinite number of eigenvalues. Calculating the eigenvalues of the interval matrix (3) for $\lambda \in [0,\ 1]$ with a sufficiently small step $\Delta\lambda$ we obtain the eigenvalue-loci of this matrix. They are symmetric with respect to the real axis. From Lemmas 3 and 4 it follows that the system (1), (3) is robust practically (asymptotically) stable if and only if these eigenvalue-loci are in the open region $S(\alpha, L)(S(\alpha))$.

Every element a_{ij} of the interval matrix (3) depends on the elements b_{ij} and c_{ij} of the matrices B and C. For $\lambda = 0$ and $\lambda = 1$ we have $A_I = B$ and $A_I = C$, respectively. Hence, we have the following simple necessary conditions of robust practical and robust asymptotic stability of interval system (1), (3). Denote by $\lambda_i(B)$ and $\lambda_i(C)$ i-th eigenvalue of the matrices B and C $(i = 1, 2, ..., n)$, respectively.

Lemma 5. If all eigenvalues $\lambda_i(B)$ or $\lambda_i(C)$ do not lie in the open region $S(\alpha, L)$ then the interval system (1), (3) is not robust practically stable.

Lemma 6. If all eigenvalues $\lambda_i(B)$ or $\lambda_i(C)$do not lie in the open region $S(\alpha)$ then the interval system (1), (3) is not robust asymptotically stable.

From the above and lemma 1 and 2 we have the following lemmas.

Lemma 7. If the condition (12) does not hold for the real eigenvalues $\lambda_i(B)$ or $\lambda_i(C)$ then the interval system (1), (3) is not robust practically stable.

Lemma 8. If the real eigenvalues $\lambda_i(B)$ or $\lambda_i(C)$ are positive then the interval system (1), (3) is not robust asymptotically stable.

For the interval matrix A_I of the form (3) we can determine the eigenvalues region, i.e. the region which consists the eigenvalues of a matrix A_I. This region can be determined by the matrix measure [2].

For a matrix $X = [x_{ij}] \in \mathbb{C}^{n\times n}$ ($\mathbb{C}$ – field of complex numbers), the measures $\mu_k(X)$ $(k = 1, 2, \infty)$ are calculated as follows [10]

$$\mu_1(X) = \max_j [Re(x_{jj}) + \sum_{i=1, i\neq j}^{n} |x_{ij}|], \tag{15}$$

$$\mu_2(X) = 0.5 \max_i \lambda_i(X + X^*), \tag{16}$$

$$\mu_\infty(X) = \max_i [Re(x_{ii}) + \sum_{j=1, j\neq i}^{n} |x_{ij}|]. \tag{17}$$

where (*) denotes the conjugate transpose symbol and $\lambda_i(X)$ is the i-th eigenvalue of matrix X.

From the paper [2] we have the following lemma.

Lemma 9. [2] The eigenvalues of the interval matrix A_I of the form (3), lie in the region (rectangle) determined by the following inequalities

$$u_l \leq Re\lambda_i(A_I) \leq u_r, \tag{18}$$

$$-v \leq Im\lambda_i(A_I) \leq v, \tag{19}$$

where

$$u_l = -\max\{\mu_k(-B), \mu_k(-C)\}, \tag{20}$$

$$u_r = \max\{\mu_k(B), \mu_k(C)\}, \tag{21}$$

$$v = \max\{\mu_k(jB), \mu_k(jC)\}, \quad j^2 = -1, \tag{22}$$

for $k = 1, 2, \infty$.

The eigenvalues region of the interval matrix (3) is determined by the inequalities (18) and (19).

From above we have the following theorems.

Theorem 3. If rectangle determined by the inequalities (18) and (19) lies in the open region $S(\alpha, L)$, then the interval system (1), (3) is robust practically stable for a given α and L.

Theorem 4. If rectangle determined by the inequalities (18) and (19) lies in the open region $S(\alpha)$, then the interval system (1), (3) is robust asymptotically stable for a given α.

5 Illustrative Examples

Example 1. Check robust practical stability of the interval system (1), (3) with length $L = 50$ of practical implementation and the boundary matrices B and C of the form

$$B = \begin{bmatrix} -0.02 & -0.07 & 0.02 \\ -0.15 & -0.05 & -0.25 \\ -0.07 & 0.02 & -0.1 \end{bmatrix}, \quad C = \begin{bmatrix} 0 & 0.01 & 0.1 \\ 0 & 0.05 & -0.15 \\ 0.1 & 0.15 & -0.05 \end{bmatrix}. \tag{23}$$

Eigenvalue-loci of the matrix (3), (23) determined with the step $\Delta\lambda = 0.01$ are shown in Fig. 2. In this figure, stability regions $S(\alpha, L)$ for $L = 50$, $\alpha = 0.7$ and $\alpha = 0.2$ are also shown. It is easy to see that these eigenvalue-loci lie in the open region $S(\alpha, L)$ for $\alpha = 0.7$ and $\alpha = 0.2$. Hence, the sufficient condition given in Lemma 3 is satisfied for $\alpha = 0.7$ and $\alpha = 0.2$. From Figs 2 and 1a) it follows that this holds for all $\alpha \leq 0.7$. Hence, the system (1), (3), (23) is robust practically stable for $L = 50$ and any $\alpha \leq 0.7$.

Calculating the measures of the suitable matrices from (15), (16) and (17) and u_l , u_r and v from (20), (21) and (22), we obtain $u_l = -0.45$, $u_r = 0.35$, $v = 0.4$.

Hence, from Lemma 9 we have

$$-0.45 \leq Re\lambda_i(A_I) \leq 0.35, \quad -0.4 \leq Im\lambda_i(A_I) \leq 0.4. \tag{24}$$

The eigenvalues region, determined by (24), is shown in Fig. 2. From Figs 2 and 1a) it follows that eigenvalues region lie in stability regions $S(\alpha, L)$ for the value $\alpha = 0.2$ and less. It means, according to Theorem 3, that interval matrix (3), (23) is robust practically stable for $L = 50$ and $\alpha \leq 0.2$.

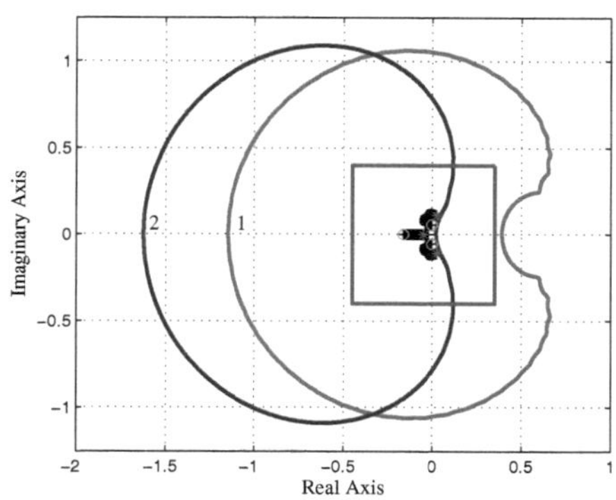

Fig. 2. Regions $S(\alpha, L)$ for $L = 50$, $\alpha = 0.2$ (boundary 1) and $\alpha = 0.7$ (boundary 2), eigenvalues region of matrix (3) (rectangle) and eigenvalue-loci of the matrix (3), (23)

Example 2. Check robust practical stability and robust asymptotic stability of the interval system (1), (3) with the boundary matrices B and C of the form

$$B = \begin{bmatrix} -0.5 & -1 & 0 \\ 0 & 0 & 1 \\ 0.1 & -1 & -1.8 \end{bmatrix}, \quad C = \begin{bmatrix} 0 & 1 & 0 \\ 0 & 0.2 & 3.5 \\ -0.1 & -0.1 & 0.1 \end{bmatrix}. \tag{25}$$

In Fig. 3a) stability regions $S(\alpha, L)$ for $L = 50$, $\alpha = 0.1$, $\alpha = 0.3$ and eigenvalue-loci of the matrix (3), (25) are shown. From Fig. 3a) it follows that the eigenvalues of the matrix B and C lie in stability regions for $\alpha = 0.1$ and $\alpha = 0.3$. Eigenvalue-loci lie completely in stability region for $\alpha = 0.3$, while for $\alpha = 0.1$ part of eigenvalue-loci lie outside the stability region. It means that the convex combination of the matrices (25) is robust practically stable for $L = 50$ and $\alpha = 0.3$ and is not robust practically stable for $L = 50$ and $\alpha = 0.1$. Note, that eigenvalues of B and C lie in stability region in all cases.

From above we have the following important remarks.

Remark 1. Location of the eigenvalues of the matrices B and C in the stability region $S(\alpha, L)$ for a given L and α does not mean that the convex combination of the matrices B and C is robust practically stable.

Remark 2. The robust practical stability of the convex combination of the matrices B and C for a given value α does not mean that the convex combination is robust practically stable for less values of α.

In Fig. 3b) asymptotic stability regions $S(\alpha)$ for $\alpha = 0.1$, $\alpha = 0.3$, $\alpha = 0.5$ and eigenvalue-loci of the matrix (3), (25) are shown. From this figure it follows that for $\alpha = 0.1$ part of eigenvalue-loci lie outside the asymptotic stability region although the eigenvalues of the matrix B and C lie in this region. It means that the convex combination of the matrices (25) is not robust asymptotically stable

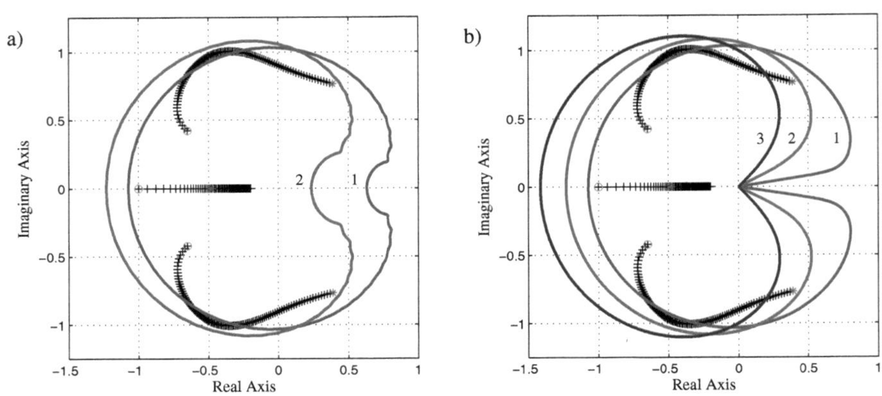

Fig. 3. Eigenvalue-loci of the matrix (3), (25), eigenvalues of B (o), eigenvalues of C (*); a) regions $S(\alpha, L)$ for $L = 50$, $\alpha = 0.1$ (boundary 1) and $\alpha = 0.3$ (boundary 2), b) regions $S(\alpha)$ for $\alpha = 0.1$ (boundary 1), $\alpha = 0.3$ (boundary 2) and $\alpha = 0.5$ (boundary 3)

for $\alpha = 0.1$. The convex combination of the matrices (25) is robust asymptotically stable for $\alpha = 0.3$, i.e. eigenvalue-loci lie completely in the stability region $S(\alpha)$. For $\alpha = 0.5$ the convex combination of the matrices (25) is not robust asymptotically stable. From the above we have that the fractional interval system (1) with the matrices (3), (25) is robust asymptotically stable for $\alpha = 0.3$. Hence, remarks given for the practical stability also applies to the asymptotic stability.

Remark 3. Location of all eigenvalues of the matrices B and C in the asymptotic stability region $S(\alpha)$ for a given α is not equivalent to robust asymptotic stability of the convex combination of the matrices B and C.

Remark 4. The robust asymptotic stability of the convex combination of the matrices B and C for a given value α does not mean that this combination is robust asymptotically stable for less values α.

6 Concluding Remarks

The problems of robust practical stability and robust asymptotic stability of discrete-time linear system (1) of fractional order $\alpha \in (0, 1)$ with the system matrix as the interval matrix of the form (3) have been addressed. It has been shown that location of all eigenvalues of the interval matrix (3) in the stability region is necessary and sufficient for the robust practical stability and robust asymptotic stability. Using the matrix measure and the parametric description of boundaries of stability regions, the sufficient conditions for the robust practical stability (Theorem 3) and for the robust asymptotic stability (Theorem 4) have been established. Also the simple necessary condition of the robust practical and robust asymptotic stability of the interval system (1), (2) (or (1), (3)) has been given in Lemmas 5 and 6, respectively.

It has been shown on example that for given α, location of all eigenvalues of B and C in practical (asymptotic) stability region is not equivalent to the robust practical stability (robust asymptotic stability) of the interval matrix (3) (Remarks 1-4).

Acknowledgement. This work was supported by the Polish Ministry of Science and Higher Education under the work S/WE/1/11.

References

1. Ahn, H.-S., Chen, Y.Q.: Necessary and Sufficient Stability Condition of Fractional-order Interval Linear Systems. Automatica 44, 2985–2988 (2008)
2. Busłowicz, M.: Asymptotic Stability of Dynamical Interval Systems with Pure Delay. Scientific Journal Bialystok University of Technology, Technical Sciences 83, Electricity 11, 61–77 (1992)
3. Busłowicz, M.: Robust Stability of Positive Discrete-time Linear Systems of Fractional Order. Bulletin of the Polish Academy of Sciences, Technical Sciences 58, 567–572 (2010)

4. Busłowicz, M.: Stability of State-space Models of Linear Continuous-time Fractional Order Systems. Acta Mechanica et Automatica 5, 15–22 (2011)
5. Busłowicz, M.: Stability Analysis of Continuous-time Linear Systems Consisting of n Subsystems with Different Fractional Orders. Bulletin of the Polish Academy of Sciences, Technical Sciences 60, 279–284 (2012)
6. Busłowicz, M.: Simple Analytic Conditions for Stability of Fractional Discrete-time Linear Systems with Diagonal State Matrix. Bulletin of the Polish Academy of Sciences, Technical Sciences 60, 809–814 (2012)
7. Busłowicz, M., Kaczorek, T.: Simple Conditions for Practical Stability of Linear Positive Fractional Discrete-time Linear Systems. International Journal of Applied Mathematics and Computer Science 19, 263–269 (2009)
8. Busłowicz, M., Ruszewski, A.: Necessary and Sufficient Conditions for Stability of Fractional Discrete-time Linear State-space Systems. Bulletin of the Polish Academy of Sciences, Technical Sciences 61, 779–786 (2013)
9. Chen, Y.Q., Ahn, H.-S., Podlubny, I.: Robust Stability Check of Fractional Order Linear Time Invariant Systems with Interval Uncertainties. Signal Processing 86, 2611–2618 (2006)
10. Desoer, C.A., Vidyasagar, M.: Feedback Systems: Input-output Properties. Acad. Press, New York (1975)
11. Dzieliński, A., Sierociuk, D.: Stability of Discrete Fractional State-space Systems. Journal of Vibration and Control 14, 1543–1556 (2008)
12. Kaczorek, T.: Practical Stability of Positive Fractional Discrete-time Systems. Bulletin of the Polish Academy of Sciences, Technical Sciences 56, 313–317 (2008)
13. Kaczorek, T.: Selected Problems of Fractional Systems Theory. LNCIS, vol. 411. Springer, Berlin (2011)
14. Kilbas, A.A., Srivastava, H.M., Trujillo, J.J.: Theory and Applications of Fractional Differential Equations. Elsevier, Amsterdam (2006)
15. Liao, Z., Peng, C., Li, W., Wang, Y.: Robust Stability Analysis for a Class of Fractional Order Systems with Uncertain Parameters. Journal of the Franklin Institute 348, 1101–1113 (2011)
16. Monje, C., Chen, Y., Vinagre, B., Xue, D., Feliu, V.: Fractional-order Systems and Controls. Springer, London (2010)
17. Ostalczyk, P.: Equivalent Descriptions of a Discrete-time Fractional-order Linear System and its Stability Domains. Int. J. Applied Mathematics and Computer Science 22, 533–538 (2012)
18. Petras, I.: Stability of Fractional-order Systems with Rational Orders: A Survey. Fractional Calculus & Applied Analysis. An International Journal for Theory and Applications 12, 269–298 (2009)
19. Podlubny, I.: Fractional Differential Equations. Academic Press, San Diego (1999)
20. Ruszewski, A.: Practical Stability and Asymptotic Stability of Interval Fractional Discrete-time Linear State-space System. In: Szewczyk, R., Zieliński, C., Kaliczyńska, M. (eds.) Recent Advances in Automation, Robotics and Measuring Techniques. AISC, vol. 267, pp. 217–228. Springer, Heidelberg (2014)
21. Sabatier, J., Agrawal, O.P., Machado, J.A.T. (eds.): Advances in Fractional Calculus, Theoretical Developments and Applications in Physics and Engineering. Springer, London (2007)
22. Stanisławski, R., Latawiec, K.J.: Stability Analysis for Discrete-time Fractional-order LTI State-space Systems. Part I: New Necessary and Sufficient Conditions for Asymptotic Stability. Bulletin of the Polish Academy of Sciences 61, 353–361 (2013)

On Mittag–Leffler Stability of Fractional Order Difference Systems

Małgorzata Wyrwas and Dorota Mozyrska

Faculty of Computer Science
Bialystok University of Technology
Wiejska 45A, 15-351 Białystok, Poland
{m.wyrwas,d.mozyrska}@pb.edu.pl

Abstract. The definition of Mittag–Leffler stability of the fractional order difference systems is introduced. The conditions for Mittag–Leffler stability of such systems with both the Riemann–Liouville– and Caputo–type operators are studied.

Keywords: fractional order difference systems, Riemann–Liouville–type operator, Caputo–type operator, Mittag–Leffler stability.

1 Introduction

During the last decade the fractional calculus has been used in both theoretical and applied problems of several branches of science and engineering, see for instance [14, 22, 23, 2, 3, 10, 12, 13] and others. Recently, the stability problem of nonlinear fractional systems is investigated, for example in [4, 11, 5, 9, 15–17, 13, 20, 24, 25, 6]. Due to the lack of geometry interpretation of the fractional derivatives, it is difficult to find a valid tool to analyze the stability of fractional equations, and there are few works on the stability of solutions for either fractional differential equations, see for instance [15–17, 5, 24, 25] or fractional difference equations, see [11, 20, 9, 13, 7, 8]. In this paper we propose the definition of Mittag–Leffler stability of nonlinear fractional order difference systems and prove the condition that guarantees the Mittag–Leffler stability. The problem of asymptotic stability of discrete-time fractional-order state-space systems with the Grünwald-Letnikow–type operator can be found in [20, 24, 25, 6], where the authors give the stability condition that coincides with the proposed one in the paper.

The paper is organized as follows. In Section 2 we recall some definitions, notations and results needed in the sequel. In particular, we study Mittag–Leffler discrete type functions and their properties that later on are used to prove the condition for Mittag–Leffler stability. Section 3 contains the definition of Mittag–Leffler stability and the sufficient condition for Mittag–Leffler stability of systems with both the Riemann–Liouville– and Caputo–type operator. Additionally, an illustrative example that describes our results is presented. Finally, Section 4 provides the brief conclusions.

© Springer International Publishing Switzerland 2015
K.J. Latawiec et al. (eds.), *Advances in Modeling and Control of Non-integer Order Systems*,
Lecture Notes in Electrical Engineering 320, DOI: 10.1007/978-3-319-09900-2_19

2 Preliminaries

Firstly, we recall some necessary definitions and properties in fractional discrete calculus connected to two types of operators: Caputo– and Riemann–Liouville–type. Let $a \in \mathbb{R}$, then we denote the set that it is used in notations of domains $\mathbb{N}_a := \{a, a+1, a+2, ...\}$, $\sigma(t) := t+1$, for any $t \in \mathbb{N}_a$. Hence $t = a+n$ for $n \in \mathbb{N}_0$. For a function $x : \mathbb{N}_a \to \mathbb{R}$ the *forward difference operator* is defined as $(\Delta x)(t) = x(\sigma(t)) - x(t)$, where $t \in \mathbb{N}_a$.

Definition 1. *For a function $x : \mathbb{N}_a \to \mathbb{R}$ the* fractional sum of order $\alpha > 0$ *is given by*

$$\left({}_a\Delta^{-\alpha}x\right)(t) := \sum_{s=0}^{n} \binom{n-s+\alpha-1}{n-s} \overline{x}(s)\,, \tag{1}$$

where $\overline{x}(s) := x(a+s)$ and $t = a+\alpha+n$, $n \in \mathbb{N}_0$. Additionally $\left({}_a\Delta^{0}x\right)(t) := x(t)$.

For $a = 0$ we write shortly $\Delta^{-\alpha}$ instead of ${}_0\Delta^{-\alpha}$. Note that ${}_a\Delta^{-\alpha}x : \mathbb{N}_{a+\alpha} \to \mathbb{R}$. Observe that formula (1) has the form of the convolution of sequences. In order to show it let us introduce the family of binomial functions on $\mathbb{Z}$ parameterized by $\mu > 0$ and given by values: $\tilde{\varphi}_\mu(n) = \binom{n+\mu-1}{n}$ for $n \in \mathbb{N}_0$ and $\tilde{\varphi}_\mu(n) = 0$ for $n < 0$. Then using $\tilde{\varphi}_\alpha(n) := \binom{n+\alpha-1}{n} = (-1)^n\binom{-\alpha}{n}$ formula (1) can be rewritten in the form of the convolution of $\tilde{\varphi}_\alpha$ and $\overline{x}$, namely

$$\left({}_a\Delta^{-\alpha}x\right)(t) = \left(\tilde{\varphi}_\alpha * \overline{x}\right)(n)\,, \tag{2}$$

where "$*$" denotes a convolution operator, i.e. $\left(\tilde{\varphi}_\alpha * \overline{x}\right)(n) := \sum\limits_{s=0}^{n} \binom{n-s+\alpha-1}{n-s}\overline{x}(s)$.

We need the following simple but important facts.

Proposition 1.

1. *For $\alpha \in (0,1]$ the values $\widetilde{\varphi}_\alpha(s)$ are nonnegative, where $s \in \mathbb{Z}$.*
2. *The fractional sum operator is positive, i.e. for a nonnegative function $x \not\equiv 0$ we have that $({}_a\Delta^{-\alpha}x)(t) \neq 0$, where $t = a+n+\alpha$, $n \in \mathbb{N}_0$.*

Proposition 2. *Let $\alpha \in (0,1]$. Then $|\,({}_a\Delta^{-\alpha}x)(t)| \le ({}_a\Delta^{-\alpha}|x|)(t)$ holds for $t = a+n+\alpha$.*

Proof. By Proposition 1 we have $|\,({}_a\Delta^{-\alpha}x)(t)| \le (\widetilde{\varphi}_\alpha * |\overline{x}|)(n) = ({}_a\Delta^{-\alpha}|x|)(t)$.

Let us recall that the $\mathcal{Z}$-transform of a sequence $\{y(n)\}_{n\in\mathbb{N}_0}$ is a complex function given by $Y(z) := \mathcal{Z}[y](z) = \sum_{k=0}^{\infty}\frac{y(k)}{z^k}$, where $z \in \mathbb{C}$ is a complex number for which this series converges absolutely. Note that $\mathcal{Z}\left[\tilde{\varphi}_\alpha\right](z) = \sum\limits_{k=0}^{\infty}\frac{1}{z^k}\binom{k+\alpha-1}{k} = \sum\limits_{k=0}^{\infty}(-1)^k\binom{-\alpha}{k}z^{-k} = \left(\frac{z}{z-1}\right)^{\alpha}$.

Proposition 3 ([19]). *For $t = a+\alpha+n \in \mathbb{Z}_{a+\alpha}$ let us denote $y(n) := \left({}_a\Delta^{-\alpha}x\right)(t)$ and $\overline{x}(n) = x(a+n)$. Then*

$$\mathcal{Z}\left[y\right](z) = \left(\frac{z}{z-1}\right)^{\alpha} X(z)\,, \tag{3}$$

where $X(z) := \mathcal{Z}\left[\overline{x}\right](z)$.

2.1 Mittag–Leffler Functions and Their Properties

Let us define a discrete type of *two-parametric Mittag–Leffler functions* that agree with those used in [19], but here we use more general definition. Let

$$E_{(\alpha,\beta)}(\lambda,n) := \sum_{k=0}^{\infty} \lambda^k \widetilde{\varphi}_{k\alpha+\beta}(n-k) = \sum_{k=0}^{n} \lambda^k \widetilde{\varphi}_{k\alpha+\beta}(n-k)\,, \tag{4}$$

where the second equation only claims that for $k > n$ we have values of $\widetilde{\varphi}_{k\alpha+\beta}(n-k) = 0$. This is not in contradiction with the definition of Mittag–Leffler discrete type functions stated in [1] or used in [19]. In the paper we will use in fact three of them, namely

$$E_{(\alpha,\alpha)}(\lambda,n) = \sum_{k=0}^{\infty} \lambda^k \widetilde{\varphi}_{k\alpha+\alpha}(n-k) = \sum_{k=0}^{\infty} \lambda^k \binom{n-k+(k+1)\alpha-1}{n-k},$$

$$E_{(\alpha)}(\lambda,n) := E_{(\alpha,1)}(\lambda,n) = \sum_{k=0}^{\infty} \lambda^k \widetilde{\varphi}_{k\alpha+1}(n-k) = \sum_{k=0}^{\infty} \lambda^k \binom{n-k+k\alpha}{n-k},$$

$$E_{(\alpha,0)}(\lambda,n) = \sum_{k=0}^{\infty} \lambda^k \widetilde{\varphi}_{k\alpha}(n-k) = \sum_{k=0}^{\infty} \lambda^k \binom{n-k+k\alpha-1}{n-k}.$$

For $\alpha = 1$ we have that $E_{(1,1)}(\lambda,n) = E_{(1)}(\lambda,n) = (1+\lambda)^n$, while $E_{(1,0)}(\lambda,n) = (1+\lambda)^n - (1+\lambda)^{n-1}$, for $n \geq 1$.

An important tool in our consideration plays the image of $E_{(\alpha,\beta)}(\lambda,\cdot)$ with respect to the $\mathcal{Z}$-transform.

Proposition 4. *Let $E_{(\alpha,\beta)}(\lambda,\cdot)$ be defined by (4). Then*

$$\mathcal{Z}\left[E_{(\alpha,\beta)}(\lambda,\cdot)\right](z) = \left(\frac{z}{z-1}\right)^{\beta} \left(1 - \frac{\lambda}{z}\left(\frac{z}{z-1}\right)^{\alpha}\right)^{-1},$$

where $|z| > 1$ and $|z-1|^{\alpha}|z|^{1-\alpha} > |\lambda|$.

Proof. By basic calculations we have

$$\begin{aligned}
\mathcal{Z}\left[E_{(\alpha,\beta)}(\lambda,\cdot)\right](z) &= \sum_{k=0}^{\infty}\sum_{n=0}^{\infty} \lambda^k \binom{n-k+k\alpha+\beta-1}{n-k} z^{-n} \\
&= \sum_{k=0}^{\infty} \lambda^k z^{-k} \sum_{s=0}^{\infty} \binom{s+k\alpha+\beta-1}{s} z^{-s} \\
&= \sum_{k=0}^{\infty} \left(\frac{\lambda}{z}\right)^k \sum_{s=0}^{\infty} (-1)^s \binom{-k\alpha-\beta}{s} z^{-s} \\
&= \left(\frac{z}{z-1}\right)^{\beta} \sum_{k=0}^{\infty} \left(\frac{\lambda}{z}\right)^k \left(\frac{z}{z-1}\right)^{k\alpha} \\
&= \left(\frac{z}{z-1}\right)^{\beta} \left(1 - \frac{\lambda}{z}\left(\frac{z}{z-1}\right)^{\alpha}\right)^{-1},
\end{aligned}$$

where the summation exists for $|z| > 1$ and $|z-1|^{\alpha}|z|^{1-\alpha} > |\lambda|$.

Proposition 5. *Let $\alpha \in (0,1]$ and $\beta \leq \alpha + 1$. Let R be the set of all roots of the equation*

$$(z-1)^{\alpha} = \lambda z^{\alpha-1}. \tag{5}$$

If all elements from R are strictly inside the unit circle, then

$$\lim_{n\to\infty} E_{(\alpha,\beta)}(\lambda, n) = 0.$$

Proof. If all roots of (5) are strictly inside the unit circle, then using theorem of final value for $\mathcal{Z}$-transform we easily get the thesis.

By Proposition 5 we get that for some order α there is a "good" λ such that all elements from R are in the unit circle.

Corollary 1. *Let $\lambda \in \mathbb{R}$. If all elements from R (Proposition 5) are inside the unit circle, then $-2^{\alpha} < \lambda < 0$.*

Now let us introduce the Caputo–type difference operator.

Definition 2 ([18]). *The* Caputo–type h-difference operator ${}_a\Delta_*^{\alpha}$ of order α for a function $x : \mathbb{N}_a \to \mathbb{R}$ *is defined by* $({}_a\Delta_*^{\alpha}x)(t) := \left({}_a\Delta^{-(1-\alpha)}(\Delta x)\right)(t)$, *where $t \in \mathbb{N}_{a+1-\alpha}$ and $\alpha \in (0,1]$.*

For the Caputo–type fractional difference operator there exists the inverse operator that is the tool in recurrence and direct solutions of fractional difference equations.

Proposition 6 ([18]). *Let $\alpha \in (0,1]$, $a = \alpha - 1$ and x be a real valued function defined on $\mathbb{N}_a$. Then $\left(\Delta^{-\alpha}({}_a\Delta_*^{\alpha}x)\right)(t) = x(t) - x(a), \quad t \in \mathbb{N}_{\alpha}$.*

Proposition 7 ([19]). *Let $a \in \mathbb{R}$, $\alpha \in (0,1]$ and define $y(n) := ({}_a\Delta_*^{\alpha}x)(t)$, where $t \in \mathbb{N}_{a+1-\alpha}$, i.e. $t = a + 1 - \alpha + n$. Then*

$$\mathcal{Z}[y](z) = \left(\frac{z}{z-1}\right)^{1-\alpha}\left((z-1)X(z) - zx(a)\right), \tag{6}$$

where $X(z) = \mathcal{Z}[\overline{x}](z)$ and $\overline{x}(n) := x(a+n)$.

In [19] we define the family of functions $\varphi_{k,\alpha}^* : \mathbb{Z} \to \mathbb{R}$, parameterized by $k \in \mathbb{N}_0$ and by $\alpha \in (0,1]$ with the following values $\varphi_{k,\alpha}^*(n) = \begin{cases} \binom{n-k+k\alpha}{n-k}, & \text{for } n \in \mathbb{N}_k \\ 0, & \text{for } n < k \end{cases}$. Note that $\varphi_{k,\alpha}^*(n) = \widetilde{\varphi}_{k\alpha+1}(n-k)$. Therefore these two families of functions can be used alternatively. Moreover, using the $\mathcal{Z}$-transform of the Caputo–type operator for $\varphi_{k,\alpha}^*$ we can calculate that for $k \in \mathbb{N}_1$ the following relation: $\left({}_0\Delta_*^{\alpha}\varphi_{k,\alpha}^*\right)(n+1-\alpha) = \varphi_{k-1,\alpha}^*(n)$ holds. For $\varphi_{k,\alpha}^*$ we have also the following $\mathcal{Z}\left[\varphi_{k,\alpha}^*\right](z) = \frac{1}{z^k}\left(\frac{z}{z-1}\right)^{k\alpha+1}$ for z such that $|z| > 1$. Using the family of functions $\varphi_{k,\alpha}^*$ the one-parameter Mittag–Leffler function can be written as $E_{(\alpha)}(\lambda, n) = \sum_{k=0}^{\infty}\lambda^k\varphi_{k,\alpha}^*(n)$. For $\alpha = 1$ we have particular exponential function $E_{(1)}(\lambda, n) = (1+\lambda)^n$.

The next Proposition is the particular case for the multi-indexed functions.

Proposition 8 ([21]). *Let* $\alpha \in (0,1]$ *and* $a = \alpha - 1$. *Then for* $n \in \mathbb{N}_0$ *we have* $\left(\Delta^{-\alpha}\varphi^*_{k,\alpha}\right)(n+a) = \varphi^*_{k+1,\alpha}(n)$.

Now, let us recall the definition of Riemann–Liouville–type operator, see for instance [3], and consider initial value problems of fractional order systems with this operator.

Definition 3. *Let* $\alpha \in (0,1]$. *The* Riemann–Liouville–type fractional h-difference operator ${}_a\Delta^\alpha x$ of order α for a function $x : \mathbb{N}_a \to \mathbb{R}$ *is defined by*

$$\left({}_a\Delta^\alpha x\right)(t) := \left(\Delta\left({}_a\Delta^{-(1-\alpha)}x\right)\right)(t), \tag{7}$$

where $t \in \mathbb{N}_{a+1-\alpha}$.

Let us define the family of functions $\varphi_{k,\alpha} : \mathbb{Z} \to \mathbb{R}$ parameterized by $k \in \mathbb{N}_0$ and by $\alpha \in (0,1]$ with the following values $\varphi_{k,\alpha}(n) := \widetilde{\varphi}_{(k+1)\alpha}(n-k) = \begin{cases} \binom{n-k+k\alpha+\alpha-1}{n-k}, & \text{for } n \in \mathbb{N}_k \\ 0, & \text{for } n < k \end{cases}$. Then for z such that $|z| > 1$ we have $\mathcal{Z}\left[\varphi_{k,\alpha}\right](z) = \frac{1}{z^k}\left(\frac{z}{z-1}\right)^{k\alpha+\alpha}$. Moreover, we can write that $E_{(\alpha,\alpha)}(\lambda, n) = \sum_{k=0}^{\infty} \lambda^k \varphi_{k,\alpha}(n)$.

Proposition 9 ([19]). *Let* $\alpha \in (0,1]$ *and* $a = \alpha - 1$. *Then for* $n \in \mathbb{N}_0$ *we have* $\left(\Delta^{-\alpha}\varphi_{k,\alpha}\right)(n+a) = \varphi_{k+1,\alpha}(n)$.

The operators presented in this section can be extended to vector valued functions in a componentwise manner.

3 Mittag–Leffler Stability

Let us introduce the stability notions for fractional difference systems involving both the Caputo– and Riemann–Liouville–type (or equivalently, Grünwald-Letnikov–type) operators.

Consider the difference systems of fractional orders $\alpha \in (0,1]$

$$\left({}_a\Delta^\alpha_* x\right)(t) = f(t, x(a+t)), \tag{8a}$$

$$\left({}_a\Delta^\alpha x\right)(t) = f(t, x(a+t)), \tag{8b}$$

with the initial condition

$$x(a) = x_0 \in \mathbb{R}^k, \tag{9}$$

where $a = \alpha - 1 \in (-1, 0] \subset \mathbb{R}$, $t \in \mathbb{N}_0$, $f : \mathbb{N}_0 \times \mathbb{R}^k \to \mathbb{R}^k$ is a continuous function.

The constant vector x_{eq} is an *equilibrium point* from time $t_0 = 0$ of fractional difference system (8) if and only if $\left({}_a\Delta^\alpha_* x_{\text{eq}}\right)(t) = f(t, x_{\text{eq}}) = 0$ (and $\left({}_a\Delta^\alpha x_{\text{eq}}\right)(t) = f(t, x_{\text{eq}})$ in the case of the Riemann–Liouville difference systems) for all $t \in \mathbb{N}_0$.

For simplicity, we state all definitions and theorems for the case when the equilibrium point is the trivial solution, i.e. $x_{\text{eq}} = 0$. There is no loss of generality in doing so because any equilibrium point can be shifted to the origin via a change of variables.

We study the relationship between the fact that the function f that appears on the right hand side of (8) is Lipschitz with respect to x and solutions of (8).

Proposition 10. *If $x = 0$ is an equilibrium point of system* (8a)*, f is Lipschitz with respect to x with Lipschitz constant L, then the solution of* (8a) *satisfies $\|x(a+n)\| \leq E_{(\alpha,0)}(L,n)\|x(a)\|$, where $\alpha \in (0,1]$.*

Proof. Let $\overline{f}(n) := f(n, x(a+n))$ and $\overline{x}(n) := x(a+n)$ for $n \in \mathbb{N}_0$. By applying the fractional integral operator ${}_0\Delta^{-\alpha}$ to both sides of (8a) and using Proposition 2 and Lipschitz condition one gets

$$\begin{aligned} |\|x(a+n)\| - \|x(a)\|| \leq & \|x(a+n) - x(a)\| = \| \left({}_0\Delta^{-\alpha}\overline{f}\right)(n+\alpha-1)\| \\ & \leq \left({}_0\Delta^{-\alpha}\|\overline{f}\|\right)(n+\alpha)\| \leq L\left({}_0\Delta^{-\alpha}\|\overline{x}\|\right)(n+\alpha-1)\| . \end{aligned}$$

There exists a nonnegative function $M(\cdot)$ satisfying

$$\|x(a+n) - x(a)\| = L\left({}_0\Delta^{-\alpha}\|\overline{x}\|\right)(n+\alpha-1) - M(n) . \tag{10}$$

Using $\mathcal{Z}$–transform for (10), we receive that $\|X(z)\| = \frac{z}{z-1}W(z)\|x_0\| - W(z)\overline{M}(z)$, where $\|X(z)\| = \mathcal{Z}[\|\overline{x}\|](z)$, $\overline{M}(z) = \mathcal{Z}[M](z)$ and $W(z) = \left(1 - L\frac{1}{z}\left(\frac{z}{z-1}\right)^{\alpha}\right)^{-1}$. Applying $\mathcal{Z}^{-1}$ we get $\overline{x}(n) = E_{(\alpha)}(L,n)\|x(a)\| - \left(E_{(\alpha,0)}(L,\cdot) * M\right)(n)$. And as for positive L and $M(\cdot)$ also: $\left(E_{(\alpha,0)}(L,\cdot) * M\right)(n) > 0$, then $\|\overline{x}(n)\| \leq E_{(\alpha)}(L,n)\|x(a)\|$.

Observe that for $\alpha = 1$ we have $\|\overline{x}(n)\| \leq (1+L)^n\|x_0\|$.

Definition 4. *Let $x(x_0, \cdot)$ denote the solution of the initial value problem* (8)–(9)*. The trivial solution of* (8) *is said to be*

(i) stable *if, for each $\epsilon > 0$ and $n_0 \in \mathbb{N}_0$, there exists $\delta = \delta(\epsilon, n_0) > 0$ such that $\|x_0\| < \delta$ implies $\|x(x_0,k)\| < \epsilon$, for all $k \in \mathbb{N}_{n_0}$.*

(ii) uniformly stable *if it is stable and δ depends solely on ϵ, i.e. for each $\epsilon > 0$ there exists $\delta = \delta(\epsilon) > 0$ such that $\|x_0\| < \delta$ implies $\|x(x_0,k)\| < \epsilon$, for all $k \in \mathbb{N}_0$.*

(iii) attractive *if there exists $\delta(t_0) > 0$ such that $\|x_0\| < \delta$ implies*

$$\lim_{k\to\infty} x(x_0,k) = 0 .$$

(iv) asymptotically stable *if it is stable and attractive.*

(v) uniformly asymptotically stable *if it is uniformly stable and, for each $\epsilon > 0$, there exists $T = T(\epsilon) \in \mathbb{N}_0$ and $\delta_0 > 0$ such that $\|x_0\| < \delta_0$ implies $\|x(x_0,k)\| < \epsilon$ for all $k \in \mathbb{N}_T$.*

Let us define the stability in the sense of Mittag–Leffler.

Definition 5. *The solution of the initial value problem* (8)–(9) *is said to be* Mittag–Leffler stable *if*

$$\|x(x_0,n)\| \le \left\{m(x_0)E_{(\alpha)}(-\lambda,n)\right\}^b \tag{11}$$

where $\alpha \in (0,1)$, $\lambda > 0$, $b > 0$, $m(0) = 0$, $m(x) \geqslant 0$, *and* m *is locally Lipschitz on* $x \in \mathbb{B} \in \mathbb{R}^n$ *with Lipschitz constant* m_0.

Proposition 11. *Mittag–Leffler stability implies asymptotic stability.*

Proof. It follows from Proposition 5.

Now, let us extend the Lyapunov direct method to the case of fractional order systems, which leads to the Mittag–Leffler stability. Let $\overline{V} : \mathbb{N}_0 \to \mathbb{R}$ and $\overline{V}(n) := V(n, x(a+n))$ for $n \in \mathbb{N}_0$.

Theorem 1. *Let* $x = 0$ *be an equilibrium point of the system* (8a) *and* $\mathcal{D} \subseteq \mathbb{R}^n$ *be a domain containing the origin. Let* $V : \mathbb{N}_0 \times \mathcal{D} \to \mathbb{R}$ *be a function that is locally Lipschitz with respect to* x *and such that*

$$\alpha_1\|x\|^a \le \overline{V}(n) \le \alpha_2\|x\|^{ab}, \tag{12a}$$

$$\left({}_0\Delta_*^\beta \overline{V}\right)(n) \le -\alpha_3\|x\|^{ab}, \tag{12b}$$

where $n \in \mathbb{N}_0$, $x \in \mathcal{D}$, $\beta \in (0,1)$, α_1, α_2, α_3, a *and* b *are arbitrary positive constants. Then* $x = 0$ *is Mittag–Leffler stable.*

Proof. It follows from equations (12a) and (12b) that $\left({}_0\Delta_*^\beta \overline{V}\right)(n) \le -\frac{\alpha_3}{\alpha_2}\overline{V}(n)$ There exists a nonnegative function M satisfying

$$\left({}_0\Delta_*^\beta \overline{V}\right)(n) + M(n) = -\frac{\alpha_3}{\alpha_2}\overline{V}(n) \tag{13}$$

Taking the $\mathcal{Z}$-transform of (13) and using Proposition 7 we get

$$\left(\frac{z}{z-1}\right)^{1-\beta}\left((z-1)\mathbf{V}(z) - z\overline{V}(0)\right) + \mathbf{M}(z) = -\frac{\alpha_3}{\alpha_2}\mathbf{V}(z), \tag{14}$$

where nonnegative constant $\overline{V}(0) = V(0, x(a))$ and $\mathbf{V}(z) = \mathcal{Z}[\overline{V}](z)$, $\mathbf{M}(z) = \mathcal{Z}[M](z)$. Then it follows that

$$\begin{aligned}\mathbf{V}(z) =& \frac{z}{z-1}\left(1 + \frac{\alpha_3}{\alpha_2}\frac{1}{z}\left(\frac{z}{z-1}\right)^\beta\right)^{-1}\overline{V}(0) \\ &- \frac{1}{z}\left(\frac{z}{z-1}\right)^\beta\left(1 + \frac{\alpha_3}{\alpha_2}\frac{1}{z}\left(\frac{z}{z-1}\right)^\beta\right)^{-1}\mathbf{M}(z)\end{aligned}$$

If $x(0) = 0$, namely $\overline{V}(0) = 0$, then the solution of (8a) is $x = 0$. If $x(0) \neq 0$, then $\overline{V}(0) > 0$. Then taking the inverse $\mathcal{Z}$-transform one gets that the solution of (13) is as follows:

$$\overline{V}(n) = E_{(\beta)}\left(-\frac{\alpha_3}{\alpha_2}, n\right)\overline{V}(0) - E_{(\beta,\beta)}\left(-\frac{\alpha_3}{\alpha_2}, n-1\right) * M(n).$$

Since $E_{(\beta,\beta)}\left(-\frac{\alpha_3}{\alpha_2}, n-1\right) * M(n) > 0$ for $n \in \mathbb{N}_0$, it follows that

$$\overline{V}(n) \leq E_{(\beta)}\left(-\frac{\alpha_3}{\alpha_2}, n\right)\overline{V}(0). \tag{15}$$

Then substituting equation (15) into (12a) we get

$$\|x(x_0, n)\| \leq \left[\frac{\overline{V}(0)}{\alpha_1}E_{(\beta)}\left(-\frac{\alpha_3}{\alpha_2}, n\right)\right]^{\frac{1}{a}},$$

where $\frac{\overline{V}(0)}{\alpha_1} > 0$ for $x(a) \neq 0$. Let $m = \frac{\overline{V}(0)}{\alpha_1} = \frac{V(0,x(a))}{\alpha_1} \geq 0$, then we have

$$\|x(x_0, n)\| \leq \left[mE_{(\beta)}\left(-\frac{\alpha_3}{\alpha_2}, n\right)\right]^{\frac{1}{a}},$$

where $m = 0$ if and only if $x(0) = 0$. Since V is locally Lipschitz with respect to x and $V(0, x(a)) = 0$ if and only if $x(0) = 0$, it follows that $m = \frac{V(0,x(a))}{\alpha_1}$ is also Lipschitz with respect to $x(a)$ and $m(0) = 0$, which implies the Mittag–Leffler stability of system (8a).

Lemma 1. *If $\beta \in (0,1]$ and $y(b) \geq 0$, then $\left({}_b\Delta_*^\beta y\right)(t) \leq \left({}_b\Delta^\beta y\right)(t)$, where $t \in \mathbb{N}_{b+1-\beta}$, ${}_a\Delta^\beta$ and ${}_a\Delta_*^\beta$ are the Riemann-Liouville- and the Caputo-type fractional operators, respectively.*

Proof. Note that for $\beta = 1$ we have equality. Now, let us consider the case for $\beta \in (0,1)$. From [18] we get

$$\begin{aligned}\left({}_b\Delta_*^\beta y\right)(n+b+1-\beta) &= \left({}_b\Delta^\beta y\right)(n+b+1-\beta) - y(b)\binom{n+1-\beta}{-\beta} \\ &= \left({}_b\Delta^\beta y\right)(n+b+1-\beta) - y(b)\binom{n+1-\beta}{n+1},\end{aligned} \tag{16}$$

for $\beta \in (0,1)$ and $n \in (h\mathbb{N})_0$. Since $y(b) \geq 0$ and $\binom{n+1-\beta}{n+1} > 0$, then we get

$$\left({}_b\Delta_*^\beta y\right)(n+b+1-\beta) \leq \left({}_b\Delta^\beta y\right)(n+b+1-\beta),$$

for $n \in \mathbb{N}_0$.

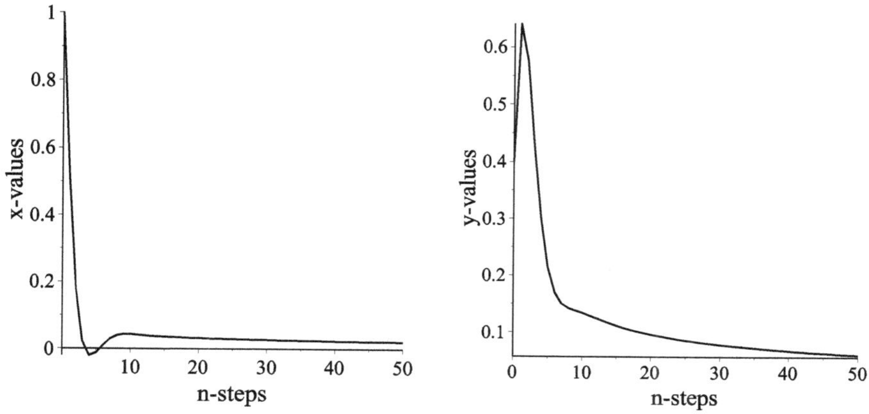

(a) the graph of x (b) the graph of y

Fig. 1. The graphs for the initial value problem (17)-(18) for $n = 0, \ldots, 50$

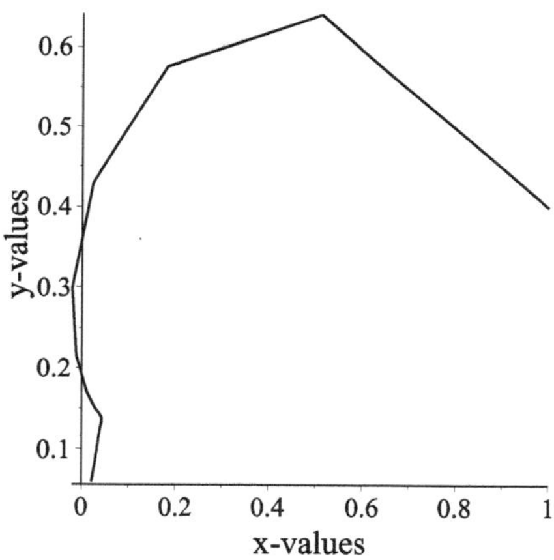

Fig. 2. The phase trajectory for the initial value problem (17)-(18) for $n = 0, \ldots, 50$

Theorem 2. *Assume that the assumptions in Theorem 1 are satisfied except replacing* ${}_a\Delta_*^\beta$ *by* ${}_a\Delta^\beta$*, then we have* $\|x(x_0, n)\| \leq \left[\frac{\overline{V}(0)}{\alpha_1} E_{(\beta)}\left(-\frac{\alpha_3}{\alpha_2}, n\right)\right]^{\frac{1}{a}}$*, where* $x(x_0, \cdot)$ *is the solution of* (8b) *with the initial condition* (9).

Proof. It follows from Lemma 1 and $\overline{V}(n) \geq 0$ for $n \in \mathbb{N}_0$ that $\left({}_0\Delta_*^\beta\right)(\overline{V}) \leq \left({}_0\Delta^\beta\right)(\overline{V})$. Then $\left({}_0\Delta_*^\beta\right)(\overline{V}) \leq \left({}_0\Delta^\beta\right)(\overline{V}) < -\alpha_3\|x\|^{ab}$. Following the same proof in Theorem 1 we get $\|x(x_0, n)\| \leq \left[\frac{\overline{V}(0)}{\alpha_1} E_{(\beta)}\left(-\frac{\alpha_3}{\alpha_2}, n\right)\right]^{\frac{1}{a}}$.

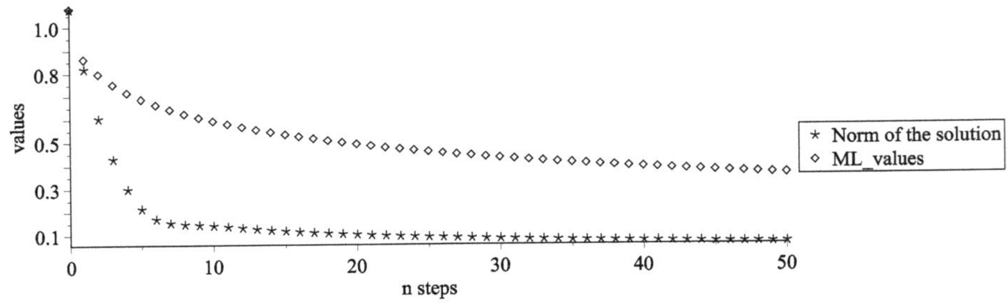

Fig. 3. The comparison of the norm of the solution to the initial value problem (17)-(18) with $\sqrt{1.16} \cdot E_{(0.5)}(-0.2, n)$ for $n = 0, \ldots, 50$

Now, let us give an example that illustrates our results.

Example 1. Consider the following system

$$\left({}_a\Delta_*^{0.5} x\right)(t) = -0.9 \cdot x(a+t) - y(a+t)\,, \tag{17a}$$

$$\left({}_a\Delta_*^{0.5} y\right)(t) = x(a+t) - 0.9 \cdot y(a+t)\,, \tag{17b}$$

with initial values

$$x(a) = 1, \quad y(a) = 0.4\,, \tag{18}$$

where $a = -0.5$.

Then the point $(0, 0)$ is the equilibrium point of the system (17). The values of x and y for $n = 0, \ldots, 50$ are displayed in Figures 1(a) and 1(b), respectively. Figure 2 shows the phase portrait of (x, y) for $n = 0, \ldots, 50$. Note that the points at graphs at Figures 1 and 2 are connected in order to have the better resolution of the behaviour of functions and trajectories. In fact, all graphs are discrete as we are dealing with discrete case of functions and operators.

One can choose the function $V(x, y) = x^2 + y^2$ that is positive definite and decrescent. Then function $\overline{V}(n) = V(x(n-0.5), y(n-0.5))$ satisfies the assumptions of Theorem 1, so the trivial solution of the considered system is Mittag-Leffler stable and consequently, by Proposition 11 asymptotically stable.

The comparison of the norm of the solution to initial value problem (17)-(18) with the Mittag-Leffler function $E_{(0.5)}(-0.2, n)$ multiplied by $\|(x(a), y(a))\| = \sqrt{1.16}$ is presented in Figure 3 for $n = 0, \ldots, 50$. The solution of (17)-(18) satisfies the following inequality

$$\|(x(a+n), y(a+n))\| = \sqrt{(x(a+n))^2 + (y(a+n))^2} \leq \sqrt{1.16} \cdot E_{(0.5)}(-0.2, n)\,.$$

4 Conclusions

The sufficient condition for Mittag–Leffler stability of the fractional difference systems is presented. We show that the Lyapunov direct method can be extended to the case of fractional order systems and it leads to the Mittag–Leffler stability. Our future work will be devoted to investigations of comparison the velocity of convergence for different orders.

Acknowledgment. The project was supported by the founds of National Science Centre granted on the bases of the decision number DEC-2011/03/B/ST7/03476. The work was supported by Bialystok University of Technology grant G/WM/3/2012.

References

1. Abdeljawad, T.: On Riemann and Caputo fractional differences. Computers and Mathematics with Applications 62(3), 1602–1611 (2011), doi:10.1016/j.camwa.2011.03.036.
2. Abdeljawad, T., Baleanu, D.: Fractional differences and integration by parts. Journal of Computational Analysis and Applications 13(3), 574–582 (2011)
3. Atıcı, F.M., Eloe, P.W.: A transform method in discrete fractional calculus. International Journal of Difference Equations 2, 165–176 (2007)
4. Axtell, M., Bise, E.M.: Fractional calculus applications in control systems. In: Proc. of the IEE 1990 Int. Aerospace and Electronics Conf., New York, vol. 311(7), pp. 536–566 (1990)
5. Busłowicz, M.: Stability of continuous-time linear systems described by state equation with fractional commensurate orders of derivatives. In: Przegląd Elektroniczby (Electrical Review), vol. 88(4b) (2012)
6. Busłowicz, M., Ruszewski, A.: Necessary and sufficient conditions for stability of fractional discrete-time linear state-space systems. Bulletin of the Polish Academy of Sciences. Technical Sciences 61, 779–786 (2013)
7. Chen, F.: Fixed points and asymptotic stability of nonlinear fractional difference equations. Electronic Journal of Qualitative Theory of Differential Equations 39, 1–18 (2011)
8. Chen, F., Liu, Z.: Asymptotic stability results for nonlinear fractional difference equations. Journal of Applied Mathematics 2012, 14 (2012), doi:10.1155/2012
9. Guermah, S., Djennoune, S., Bettayeb, M.: Asymptotic stability and practical stability of linear discrete-time fractional order systems. In: 3rd IFAC Workshop on Fractional Differentiation and its Applications, Ankara, Turkey (2008)
10. Holm, M.T.: The theory of discrete fractional calculus: Development and application. Ph.D. dissertation, University of Nebraska - Lincoln (2011)
11. Jarad, F., Abdeljawad, T., Baleanu, D., Biçen, K.: On the stability of some discrete fractional nonautonomous systems. Abstract and Applied Analysis 2012, Article ID 476581, 9 (2012), doi:10.1155/2012
12. Kaczorek, T.: Fractional positive continuous–time linear systems and their reachability. International Journal of Applied Mathematics and Computer Science 18(2), 223–228 (2008)
13. Kaczorek, T.: Practical stability of positive fractional discrete-time linear systems. Bulletin of the Polish Academy of Sciences. Technical Sciences 56(4), 313–317 (2008)
14. Kilbas, A.A., Srivastava, H.M., Trujillo, J.J.: Theory and applications of fractional differential equations. North–Holland Mathematics Studies, vol. 204. Elsevier Science B. V., Amsterdam (2006)

15. Li, Y., Chen, Y.Q., Podlubny, I.: Mittag-Leffler stability of fractional order nonlinear dynamic systems. Automatica 45, 1965–1969 (2009)
16. Li, Y., Chen, Y., Podlubny, I.: Stability of fractional-order nonlinear dynamic systems: Lyapunov direct method and generalized Mittag-Leffler stability. Computers and Mathematics with Applications 59, 1810–1821 (2010)
17. Matignon, D., d'Andréa Novel, B.: Some results on controllability and obseravbility of finite–dimensional fractional differential systems. In: IMACS, IEEE-SMC Proceedings Conference, Lille, France, pp. 952–956 (1996)
18. Mozyrska, D., Girejko, E.: Overview of the fractional h-difference operators. In: Operator Theory: Advances and Applications, vol. 229, pp. 253–267. Birkhäuser (2013), doi:10.1007/978-3-0348-0516-2.
19. Mozyrska, D., Wyrwas, M.: Z-transforms of delta type fractional difference operators. submitted to publication (2014)
20. Ostalczyk, P.: Equivalent descriptions of a discretetime fractionalorder linear system and its stability domains. Int. J. Appl. Math. Comput. Sci. 22(3), 533–538 (2012)
21. Pawłuszewicz, E., Mozyrska, D.: Constrained controllability of h-difference linear systems with two fractional orders. In: Mitkowski, W., Kacprzyk, J., Baranowski, J. (eds.) Theory & Appl. of Non-integer Order Syst. LNEE, vol. 257, pp. 67–75. Springer, Heidelberg (2013)
22. Podlubny, I.: Fractional Differential Equations. Academic Press, San Diego (1999)
23. Samko, S.G., Kilbas, A.A., Marichev, O.I.: Fractional Integrals and Derivatives: Theory and Applications. Gordon and Breach Science Publishers S.A., Yverdon (1993)
24. Stanisławski, R., Latawiec, K.: Stability analysis for discrete-time fractional-order LTI state-space systems. Part I: New necessary and sufficient conditions for the asymptotic stability. Bulletin of the Polish Academy of Sciences. Technical Sciences 61(2), 353–361 (2013)
25. Stanisławski, R., Latawiec, K.: Stability analysis for discrete-time fractional-order LTI state-space systems. Part II: New stability criterion for FD-based systems. Bulletin of the Polish Academy of Sciences. Technical Sciences 61(2), 363–370 (2013)

Lyapunov Direct Method for Non-integer Order Systems

Marta Zagórowska, Jerzy Baranowski, Waldemar Bauer, Tomasz Dziwiński, Paweł Piątek, and Wojciech Mitkowski

AGH University of Science and Technology,
Department of Automatics and Biomedical Engineering,
Al. Mickiewicza 30, 30-059 Kraków
{zagor,jb,bauer,tdz,ppi,wojciech.mitkowski}@agh.edu.pl

Abstract. In this paper an extension of Lyapunov direct method for non-integer order systems is presented. It allows to analyse a special case of classic stability theory - the Mittag-Leffler stability. However, there are some differences that are worth considering. Some of them are analysed in presented examples.

1 Introduction

Non-integer order systems (often called fractional systems) are a rapidly developing field in technical and mathematical sciences. Most focus is oriented on their properties (see for example [2,5]) and applications (see for example [3,4,6,10]). The goal of this paper is to highlight one of the interesting results from the first group.

Lyapunov direct method provides a way to analyse the stability of dynamical systems without solving their differential equations. It is especially advantageous when the solution is difficult or even impossible to find with classical methods. Some basic analysis can be found in [7,8] and [1].

Therefore, it is interesting to investigate extension of the method for non-integer order systems. Such extension relies heavily on a notion of Mittag-Leffler stability, which is presented along with the theorem, after which this paper is titled. Application of the theorem will be illustrated with an example of RC circuit with a supercapacitor, an example of system with diagonal matrix and an example of system with Metzler matrix.

2 Non-integer Order Calculus

Before analysing of non-integer order systems stability, there are some definitions that must be introduced.

In this paper, systems described with the Caputo fractional derivative (CFD) are considered. CFD is given by the following formula

$$ {}_{t_0}^{C}\mathrm{D}_t^p f(t) = \frac{1}{\Gamma(p-n)} \int_{t_0}^{t} \frac{f^{(n)}(\tau)\mathrm{d}\tau}{(t-\tau)^{p-n+1}}, \tag{1} $$

© Springer International Publishing Switzerland 2015

K.J. Latawiec et al. (eds.), *Advances in Modeling and Control of Non-integer Order Systems*,
Lecture Notes in Electrical Engineering 320, DOI: 10.1007/978-3-319-09900-2_20

where

$$\Gamma(x) = \int_0^\infty s^{x-1} e^{-s} \mathrm{d}s$$

is the Gamma function and $n-1 < p < n$. Also the one parameter Mittag-Leffler function will be used given by

$$E_\alpha(z) = \sum_{k=0}^{\infty} \frac{z^k}{\Gamma(k\alpha + 1)},$$

where $\alpha > 0$ and $z \in \mathbb{C}$.

It is worth notice that for $\alpha = 1$ the Mittag-Leffler function becomes exponential function.

3 Mittag-Leffler Stability

The definition of equlibrium point is the first element necessary for stability analysis. For Caputo systems it does not differ from that for integer order systems.

Definition 1 (Equilibrium of non-integer order system). *The solution of equation* (2) *such that* $\mathbf{x}(t) \equiv \mathbf{x_0} = const$ *is called the equilibrium.*

The Mittag-Leffler stability can be formulated as follows [7].

Definition 2 (Mittag-Leffler stability). *The solution of system*

$$ {}_{t_0}^{C}\mathrm{D}_t^\alpha \mathbf{x}(t) = \mathbf{f}(t, x), \quad \mathbf{x}(t_0) = \mathbf{x_0} \in \Omega \subset \mathbb{R}^n \tag{2}$$

where $\alpha \in (0,1)$, $\mathbf{f}\colon [t_0, \infty] \times \Omega \to \mathbb{R}^n$, $\mathbf{f}$ *is piecewise continuous in* t *and locally Lipschitz with respect to* $\mathbf{x}$ *in its domain and* $\Omega \subset \mathbb{R}^n$ *contains the origin, is Mittag-Leffler stable if*

$$\|\mathbf{x}\| \leq \Big(m(\mathbf{x_0}) E_\alpha\big(-\lambda(t-t_0)^\alpha\big)\Big)^\beta$$

where t_0 *is the initial time,* $\lambda \geq 0$, $\beta > 0$, $m(0) = 0$, $m(\mathbf{x}) > 0$ *and* $m(\mathbf{x})$ *is locally Lipschitz.*

Usually the stability of equilibria is investigated. It can be observed from formula (1) that x_0 is an equilibrium of equation (2) iff $\mathbf{f}(t, x_0) = 0$, $\forall t > t_0$. It can be easily observed that Mittag-Leffler stability implies attractivity of equilibrium.

The direct Laypunov method can be extended to verify the Mittag-Leffler stability. The following theorem [8] presents this extension.

Theorem 1 (Lyapunov direct method). *Let* $\mathbf{x} = 0$ *be an equilibrium of* (2), *let* $D \subset \mathbb{R}^n$ *be the domain containing the origin. Let* $V(t, \mathbf{z}) : [0, \infty) \times D \to \mathbb{R}$

be a continuously differentiable of order β and locally Lipschitz with respect to $\mathbf{z}$ *function such that:*

$$\alpha_1\|\mathbf{z}\|^a \leq V(t,\mathbf{z}) \leq \alpha_2\|\mathbf{z}\|^{ab}, \tag{3}$$

$$\left.{}_0^C\mathrm{D}_t^\beta V(t,\mathbf{x}(t))\right|_{\mathbf{x}(t)=\mathbf{z}} \leq -\alpha_3\|\mathbf{z}\|^{ab}, \tag{4}$$

where $\mathbf{x}(t)$ *is given by* (2), $t \geq 0$, $\mathbf{z} \in D$, $\beta \in (0,1)$, α_1, α_2, α_3, a, b *are some positive constants. Then* $\mathbf{x} = 0$ *is Mittag Leffler stable.*

The assumptions of theorem 1 can be relaxed in order to verify only attractivity of the zero solution.

Theorem 2. *Let* $\mathbf{x} = 0$ *be an equilibrium of* (2)*, let* $D \subset \mathbb{R}^n$ *be the domain containing the origin. Let* $V(t,\mathbf{z}) : [0,\infty) \times D \to \mathbb{R}$ *be a continuously differentiable of order* β *and locally Lipshitz with respect to* $\mathbf{z}$ *function such that:*

$$\alpha_1\|\mathbf{z}\|^a \leq V(t,\mathbf{z}) \leq \alpha_2 {}_0^C\mathrm{D}_t^{-\eta}\|\mathbf{z}\|^{ab},$$

$$\left.{}_0^C\mathrm{D}_t^\beta V(t,\mathbf{x}(t))\right|_{\mathbf{x}(t)=\mathbf{z}} \leq -\alpha_3\|\mathbf{z}\|^{ab},$$

where $\mathbf{x}(t)$ *is given by* (2), $t \geq 0$, $\mathbf{z} \in D$, $\beta \in (0,1)$, $\eta \neq \beta$,$\eta > 0$, $|\beta - \eta| < 1$, α_1, α_2, α_3, a, b *are some positive constants. Then* $\mathbf{x} = 0$ *is attractive.*

Proofs of theorems 1 and 2 along with further modifications (including for example analysis with $\mathcal{K}$-class functions) can be found in [8].

4 Examples

In this section three examples are considered. There are used to demonstrate theorem 1 but also to present some faults of the method concerning its practical usage.

4.1 Non-integer RC System

The RC system shown in figure 1 consists of a resistor and a supercapacitor [11] and can be described with the following non-integer order equation

$${}_0^C\mathrm{D}_t^\alpha x(t) = -\frac{1}{RC}x(t),\ x(0) = x_0.$$

It can be easily observed, that the solution is given by (see for example [5, 10])

$$x(t) = E_\alpha\left(-\frac{1}{RC}t^\alpha\right)x_0$$

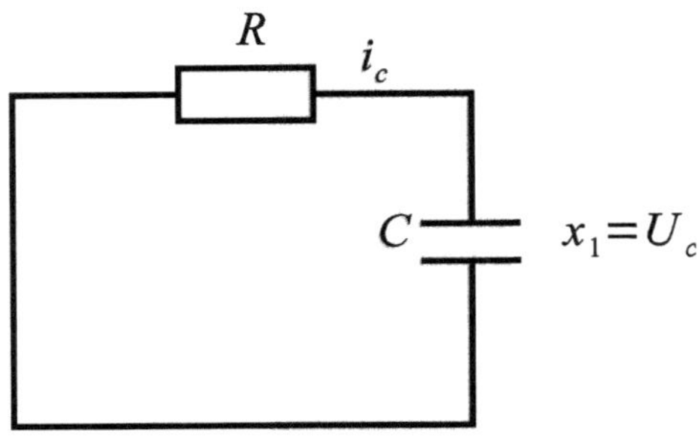

Fig. 1. RC circuit

It will be shown that $V(t,z) = |z|$ is the Lyapunov function for the system. It satisfies the inequality (3) with $\alpha_1 = 1$, $\alpha_2 = 1$, $a = 1$, $b = 1$. If $\beta = \alpha$ then

$$\begin{aligned} {}_0^C\mathrm{D}_t^\alpha V(t, x(t)) &= {}_0^C\mathrm{D}_t^\alpha |x(t)| = \\ &= {}_0^C\mathrm{D}_t^\alpha \left| E_\alpha \left(-\frac{1}{RC} t^\alpha \right) x_0 \right| = \\ &= {}_0^C\mathrm{D}_t^\alpha E_\alpha \left(-\frac{1}{RC} t^\alpha \right) |x_0| = \\ &= -\frac{1}{RC} E_\alpha \left(-\frac{1}{RC} t^\alpha \right) |x_0| = -\frac{1}{RC} |x(t)| . \end{aligned}$$

It fulfils inequality (4) with $\alpha_3 = 1/RC$.

4.2 Non-integer Order with Diagonal Matrix

The following system with Caputo derivative is considered

$${}_0^C\mathrm{D}_t^\alpha \mathbf{x}(t) = \mathbf{A}\mathbf{x}(t),\ \mathbf{x}(0) = \mathbf{x^0} \in \mathbb{R}^n,\ \alpha \in (0,\ 1],$$

where $\mathbf{A}$ as a stable diagonal matrix $\mathbf{A} = (a_{ii})$, where $a_{ii} < 0,\ \forall i \in \{1, ..., n\}$. The exact solution has the form

$$\mathbf{x}(t) = \mathbf{E}_\alpha(\mathbf{A}t^\alpha)\mathbf{x^0}, \tag{5}$$

where $\mathbf{x^0}$ denotes the vector of initial conditions. It should be noted that (5) can be written as

$$\mathbf{x}(t) = [E_\alpha(a_{ii}t^\alpha)x_i^0]^T.$$

The ℓ_1 norm of $\mathbf{x}$ is considered as a Lyapunov function i.e.

$$V(\mathbf{x}) = \|\mathbf{x}\|,$$

$$\|\mathbf{x}\| = \sum_{i=1}^{n} |x_i|.$$

It fulfils the inequality (3) with $\alpha_2 = \alpha_3 = a = b = 1$.

$$\|\mathbf{x}\| \leq V(\mathbf{x}) \leq \|\mathbf{x}\|.$$

The function $V(\mathbf{x})$ can be written as

$$V(\mathbf{x}) = \sum_{i=1}^{n} |E_\alpha(a_{ii}t^\alpha)x_i^0|.$$

Calculating the derivative of order $\beta = \alpha$, the following result is obtained

$$\begin{aligned}
{}_0^C\mathrm{D}_t^\alpha V(x) &= \sum_{i=1}^{n} {}_0^C\mathrm{D}_t^\alpha E_\alpha(a_{ii}t^\alpha)|x_i^0| \\
&= \sum_{i=1}^{n} a_{ii}E_\alpha(a_{ii}t^\alpha)|x_i^0| \\
&\leq \max_{i=1..n} a_{ii} \sum_{i=1}^{n} E_\alpha(a_{ii}t^\alpha)|x_i^0| \\
&= \max_{i=1..n} a_{ii} \sum_{i=1}^{n} \left|E_\alpha(a_{ii}t^\alpha)x_i^0\right| \\
&= \max_{i=1..n} a_{ii}\|\mathbf{x}(t)\|.
\end{aligned}$$

Therefore, in inequality (4), there is $\alpha_3 = |\max_{i=1..n} a_{ii}|$. Hence, the origin is also stable for systems with diagonal matrix and every kind of initial conditions.

4.3 Non-integer Order Positive System

The following positive system with Caputo derivative is considered

$${}_0^C\mathrm{D}_t^\alpha \mathbf{x}(t) = \mathbf{A}\mathbf{x}(t),\ \mathbf{x}(0) = \mathbf{x}^0 \in \mathbb{R}_+^n,\ \alpha \in (0,\ 1], \tag{6}$$

where $\mathbb{R}_+^n$ is the space of nonnegative vectors in $\mathbb{R}^n$ and the matrix $\mathbf{A}$ is a stable Metzler matrix from $\mathcal{M}^{n\times n}$. Metzler matrix is a matrix $\mathbf{A}$ for which

$$\mathbf{A} = (a_{ij}); \quad a_{ij} \geq 0, \quad i \neq j.$$

The following theorem holds

Theorem 3 (see [9]). *Non integer order positive system described by equation* (6) *is asymptotically stable if and only if eigenvalues of Metzler matrix* $\mathbf{A}$ *are all located in left complex half plane* ($\mathbf{A}$ *is a stable Metzler matrix).*

One of the properties of stable Metzler matrices is that it has at least one negative real eigenvalue witch has a positive eigenvector.

The ℓ_1 norm is considered:

$$\|\mathbf{x}\| = \sum_{i=1}^{n} |x_i|.$$

It will be shown that the equilibrium ($\mathbf{x} = 0$) of (6) is Mittag-Leffler stable. The exact solution of (6) has the form

$$\mathbf{x}(t) = \mathbf{E}_\alpha(\mathbf{A}t^\alpha)\mathbf{x}^0. \tag{7}$$

For the proof the following Lyapunov function is considered

$$V(\mathbf{x}) = \sum_{i=1}^{n} w_i |x_i|, \tag{8}$$

where w_i are certain positive constants and x_i are elements of the solution (7).

To verify the Mittag-Leffler stability of analysed system, it is necessary to check if the conditions (3) and (4) are fufilled. The first condition can be written as follows

$$\left(\min_{j\in\{1..n\}} w_j\right) \|\mathbf{x}\| \leq \sum_{i=1}^{n} w_i|x_i| \leq \left(\max_{j\in\{1..n\}} w_j\right) \|\mathbf{x}\| \tag{9}$$

so the constants α_1 and α_2 are respectively $\min\limits_{j\in\{1..n\}} w_j$ and $\max\limits_{j\in\{1..n\}} w_j$.

To verify condition (4), it is necessary to calculate non-integer derivative of order β of (8). Using (7) it can be written as

$$V(\mathbf{x}) = \sum_{i=1}^{n} w_i \left|\mathbf{e}_i^\mathsf{T}\mathbf{E}_\alpha(\mathbf{A}t^\alpha)\mathbf{x}^0\right|,$$

where $\mathbf{e}_i$ is the i-th unit vector.

Because the Mittag-Leffler function of Metzler matrix is a positive matrix, and initial conditions are in $\mathbb{R}_+^n$ then

$$V(\mathbf{x}) = \mathbf{w}\mathbf{E}_\alpha(\mathbf{A}t^\alpha)\mathbf{x}^0,$$

where $\mathbf{w} = [w_i]_{i=1..n}$. The order of derivative β is chosen as $\beta = \alpha$. After calculating the derivative it is obtained that

$${}_0^C\mathrm{D}_t^\alpha V(\mathbf{x}(t)) = \mathbf{w}\mathbf{A}\mathbf{E}_\alpha(\mathbf{A}t^\alpha)\cdot\mathbf{x}^0. \tag{10}$$

Choosing $\mathbf{w}$ as a left eigenvector of $\mathbf{A}$ (positive and associated with rightmost real eigenvalue) it can be observed that

$$\mathbf{w}\mathbf{A} = \lambda\mathbf{w}.$$

Remark 1. *Existence and properties of desired eigenvector are a consequence of Perron-Frobenius theorem.*

Then equation (10) takes form

$$ {}_0^C\mathrm{D}_t^\alpha V(\mathbf{x}) = \lambda \mathbf{w} \mathbf{E}_\alpha(\mathbf{A}t^\alpha) \cdot \mathbf{x}^0 $$

and because every element in (4.3) is positive

$$ {}_0^C\mathrm{D}_t^\alpha V(\mathbf{x}) = \lambda |\mathbf{w} \mathbf{E}_\alpha(\mathbf{A}t^\alpha) \cdot \mathbf{x}^0| = \lambda \sum_{i=1}^{n} w_i |\mathbf{e}_i^\top \mathbf{E}_\alpha(\mathbf{A}t^\alpha)\mathbf{x}^0| = \lambda V(\mathbf{x}). $$

From (9) and the fact, that λ is negative real number (from stability of system (6)) one can see that α_3 from (4) is

$$ \alpha_3 = |\lambda|\alpha_2. $$

Hence, the origin is Mittag-Leffler stable.

5 Conclusion

This paper presents the extensions of Lyapunov direct method for non-integer order systems. Some basics of non-integer calcusus are introduced along with the definition of Mittag-Leffler stability. Then the authors presented an extension of Lyapunov direct method and illustrated it with some examples.

From the three examples in that work, it can be easily seen that the method does not provide the same tool as in integer order case. In classic approach, there is no need to solve the differential equations in order to use Lyapunov function. In non-integer order systems lack of chain rule implies that the exact solution is indispensable for calculating the derivative. Therefore, that extension is not useful in case of some complicated nonlinear systems, where finding the solution is not straightforward.

Nonetheless, the results obtained in that paper show some usefulness of Lyapunov direct method for linear Caputo systems where the solution is known.

Acknowledgment. Work partially realised in the scope of project titled "Design and application of noniteger order subsystems in control systems". Project was financed by National Science Centre on the base of decision no. DEC-2013/09/D/ST7/03960.

References

1. Lyapunov function method: Non-integer order case. In: Materiały XXXVII Międzynarodowej konferencji z podstaw elektrotechniki i teorii obwodów IC-SPETO (2014)
2. Bania, P., Baranowski, J.: Laguerre polynomial approximation of fractional order linear systems. In: Mitkowski, W., Kacprzyk, J., Baranowski, J. (eds.) Theory & Appl. of Non-integer Order Syst. LNEE, vol. 257, pp. 171–182. Springer, Heidelberg (2013)

3. Bauer, W., Baranowski, J., Mitowski, W.: Non-integer order $PI^{\alpha}D^{\mu}$ control ICU-MM. In: Mitkowski, W., Kacprzyk, J., Baranowski, J. (eds.) Theory & Appl. of Non-integer Order Syst. LNEE, vol. 257, pp. 295–303. Springer, Heidelberg (2013)
4. Dziwiński, T., Bauer, W., Baranowski, J., Piąktek, P., Zagórowska, M.: Robust non-integer order controller for air heater. In: 2014 19th International Conference on Methods and Models in Automation and Robotics, MMAR (accepted 2014)
5. Kaczorek, T.: Selected Problems of Fractional Systems Theory. LNCIS, vol. 411. Springer, Berlin (2011)
6. Kawala-Janik, A., Podpora, M., Baranowski, J., Bauer, W., Pelc, M.: Innovative approach in analysis of eeg and emg signals? comparision of the two novel methods. In: 2014 19th International Conference on Methods and Models in Automation and Robotics, MMAR (accepted 2014)
7. Li, Y., Chen, Y., Podlubny, I.: Mittag–leffler stability of fractional order nonlinear dynamic systems. Automatica 45(8), 1965–1969 (2009)
8. Li, Y., Chen, Y., Podlubny, I.: Stability of fractional-order nonlinear dynamic systems: Lyapunov direct method and generalized mittag–leffler stability. Computers & Mathematics with Applications 59(5), 1810–1821 (2010), Fractional Differentiation and Its Applications
9. Mitkowski, W.: Remarks about fractional metzler systems. In: Busłowicz, M., Malinowski, K. (eds.) Advances in Control Theory and Automation: Mono- graph Dedicated to Professor Tadeusz Kaczorek in the Occasion of his Eightieth Birthday. Printing House of Białystok University of Technology, Białystok (2012)
10. Piątek, P., Baranowski, J.: P. Piątek and J. Baranowski. Investigation of Fixed-Point Computation Influence on Numerical Solutions of Fractional Differential Equations. Acta Mechanica et Automatica 5(2), 101–107 (2011)
11. Skruch, P., Mitkowski, W.: Fractional-order models of the ultracapacitors. In: Mitkowski, W., Kacprzyk, J., Baranowski, J. (eds.) Theory & Appl. of Non-integer Order Syst. LNEE, vol. 257, pp. 281–293. Springer, Heidelberg (2013)

Part VI
Applications

Non-integer Order Filtration of Electromyographic Signals

Jerzy Baranowski[1], Paweł Piątek[1], Aleksandra Kawala-Janik[2], Marta Zagórowska[1], Waldemar Bauer[1], and Tomasz Dziwiński[1]

[1] AGH University of Science and Technology, Department of Automatics and Biomedical Engineering, Al. Mickiewicza 30, 30-059 Kraków
{jb,ppi,zagor,bauer,tdz}@agh.edu.pl

[2] Faculty of Electrical Engineering, Automatic Control and Informatics, Opole University of Technology, ul. Proszkowska 76, 45-756 Opole, Poland
a.kawala-janik@po.opole.pl

Abstract. Electromyography (EMG) is recently of growing interest of doctors and scientists as it provides a tool for muscle performance verification. In this paper a new approach to EMG signal processing is considered. This approach is non-integer order filtering. Bi-fractional filter is designed and filtering occurs through exact computation.

1 Introduction

Electromyography (EMG) is recently of growing interest of doctors and scientists as it provides a tool for muscle performance verification. Medical sciences and physiotherapy are not the only possibilities to use EMG signals. It can be found also in sport, control science and computing. The main aim of the conducted research relied on application of the muscle generated power for the prospective use in robotics, where muscle work is exploited. First mentions of EMG were in 1950', when some very simple systems were build. Since then, the idea was constantly developed and led to commercial myoelectric prostheses [6, 17, 18].

Analysis of bio-signals causes various significant issues due to the deterministic nature of these signals and therefore implementation of these signals for the control purpose is a very challenging task. Analysis of both – EMG and EEG signals is the most difficult as the human brain has not been fully investigated and is not entirely known by the researchers [1, 10, 12].

For this study purpose surface EMG has been chosen, which has some limits, especially because of crosstalk and differentiations between joint motions and individual digits. As the chosen method is non-invasive, overcoming the existing limitation by using implantable myoelectric systems is impossible [18].

The second issue that must be taken into account is the dynamic asymmetry. It is widely known that robots have no differences between arms. However, human limbs function otherwise and this effect cannot be neglected when analysing prostheses. It is a complicated task, because very little research was conducted

© Springer International Publishing Switzerland 2015
K.J. Latawiec et al. (eds.), *Advances in Modeling and Control of Non-integer Order Systems*, Lecture Notes in Electrical Engineering 320, DOI: 10.1007/978-3-319-09900-2_21

regarding assessment of the dominant side. It is believed that the source has its origin in genetic or social development [1,3,7].

As mentioned above, EMG signals are a very popular data source applied for the purpose of external environments control. The have been implemented in numerous Human-Machine or Human-Computer Interaction interfaces. Their potential does not limit to HMI or HCI usage as they are also deployed in many clinical, industrial or interdisciplinary application [15].

In this paper a novel approach to analysing the EMG signal is presented. It is based on fractional calculus as tool for filtration. The concept of non-integer derivatives was first introduced by Liouville in XIX but as its usage were limited by technology. Therefore, its rapid development takes place since 1950' with increasing computational power [11].

2 Theoretical Background

EMG signals are bio-electrical signals generated in muscles during their activity – both voluntary and involuntary. This activity is usually controlled by the nervous system. Analysis of electromyography signals is a very challenging task due to the both internal and external artifacts present in signal. Internal artifacts are a result of physical and/or anatomical muscle properties, where external occur in noisy environment. The frequency range of the EMG signals is between 0 and 500 Hz [2,8,13,14].

EMG signals are controlled by the nervous system and depend on both physical and anatomic state of muscle, which generates these signals. It is able to provide significant source of information, which can be used for control. The assessment is also able to provide information regarding various neuromuscular disorders.

Discrete waveforms, known also as Motor Unit Action Potentials (MUPs), form EMG signals. The occurrence of MUPs results from emission of muscle fibres groups – Motor Units (MUs) [8,19]. It is possible to record EMG signals occurring during various muscle activity. It all depends on contraction force and as a result – the more force is being used, the more complex the signal becomes, what complicates the MUPs identification [13,14,19].

Any bio-signal, including EMG, is a set of electrical signals representing physical variable of interest. The signals are usually in form of time-function, what enables analysis of its amplitudes, phase or frequency [13,20]. In the Figure 1 a sample EMG data was presented.

It is also important to mention that traditional prosthesis systems for upper limb are usually based on EMG.

3 Non-integer Order Processing of EMG Signals

Because of presence of disturbances and noises (see figure 2) in the measured signals appropriate processing is required. Unfortunately standard integer order

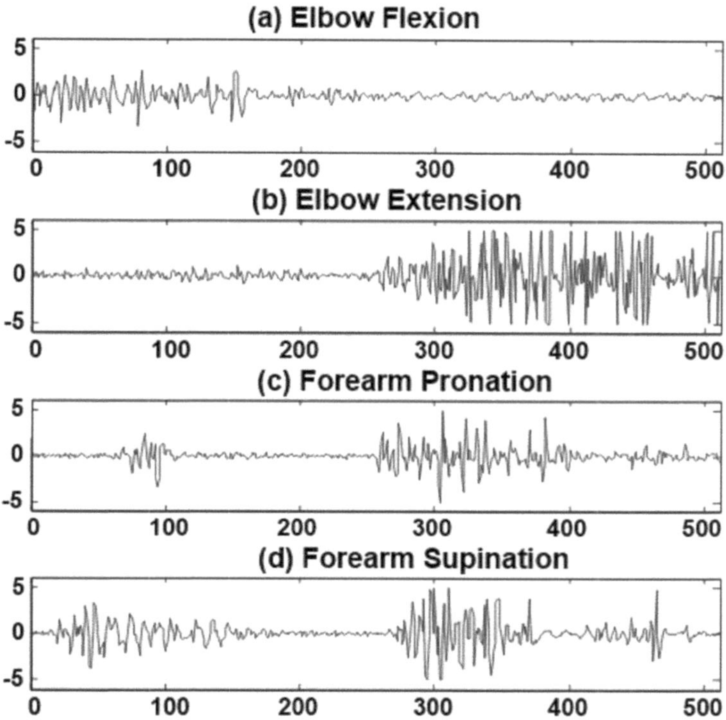

Fig. 1. Sample EMG data recorded during four different arm movements

filters have a strong damping effect with the increase of frequency. This is unfortunate, as EMG has very wide useful spectrum. Non integer order filters allow processing that introduces damming of less than 20 db/dec by the multiplier of its order α. One of important classes of non integer order filters are Bi-fractional filters.

3.1 Bi-Fractional Filters

Bi-fractional filters (BFF) are a class of non-integer filters fully characterised by three parameters:

- base order α
- damping coefficient b
- free coefficient c

and are given by the following transfer function

$$G(s) = \frac{c}{s^{2\alpha} + bs^{\alpha} + c} \tag{1}$$

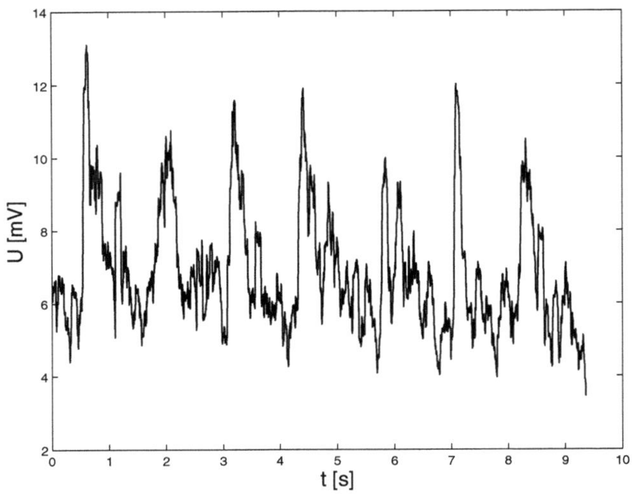

Fig. 2. Sample EMG data recorded during four different arm movements

In this paper filters with bounded impulse responses are considered, that is systems of base order $\alpha \in [1/2, 1)$. Another examples can be found in [5] or in [4].

Equivalent representation of (1) is the realisation in the form of a system of differential equations of order α. This system can take form (see [9])

$$\begin{aligned} {}_0^C\mathrm{D}_t^\alpha \mathbf{x}(t) &= \mathbf{A}\mathbf{x}(t) + \mathbf{B}u(t) \\ y(t) &= \mathbf{C}\mathbf{x}(t) \end{aligned} \tag{2}$$

with the following matrices

$$\mathbf{A} = \begin{bmatrix} 0 & 1 \\ -c & -b \end{bmatrix} \tag{3}$$

$$\mathbf{B} = \begin{bmatrix} 0 \\ 1 \end{bmatrix} \tag{4}$$

$$\mathbf{C} = \begin{bmatrix} c \; 0 \end{bmatrix} \tag{5}$$

In order to process the considered signal the BFF with following parameters was chosen:

$$\begin{aligned} \alpha &= 0.7 \\ b &= 11.1688 \\ c &= 124.7412 \end{aligned} \tag{6}$$

which locate the cutoff frequency at 5Hz. Frequency response of the filter is presented in the figure 3.

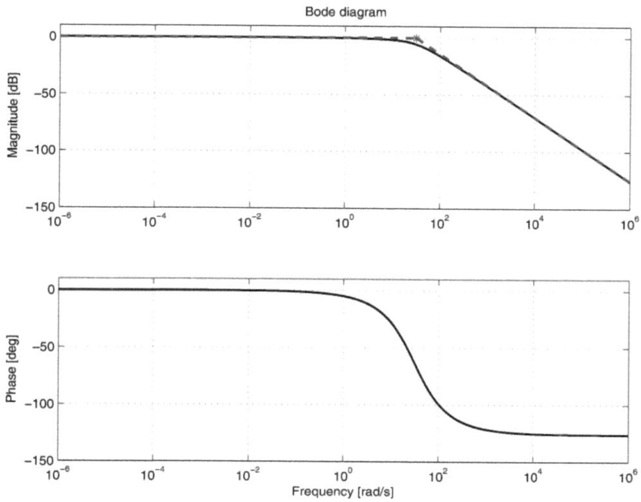

Fig. 3. Frequency response of bi-fractional filter with parameters given by (6)

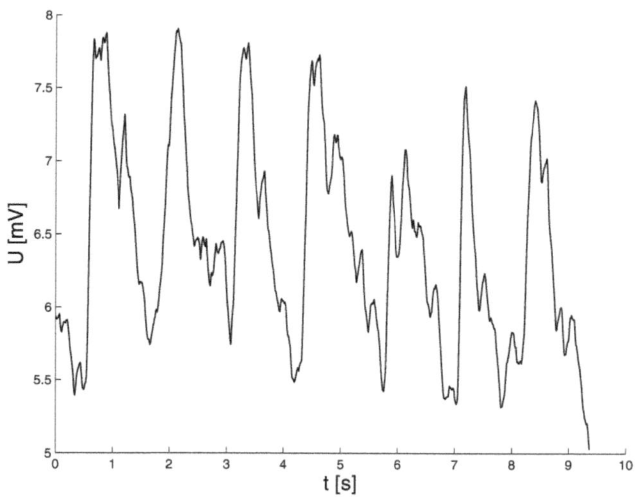

Fig. 4. Sample EMG data recorded during four different arm movements

Because considered processing does not have an online character one can benefit from finite length of signal and realize the filtration process exactly.

3.2 Realisation of Filtering

Filtering will be realized by a numerical method proposed in [16] based on Grünwald–Letnikov derivative

$$\mathbf{x}(t) = (I - h^\alpha \mathbf{A})^{-1} \left(h^\alpha \mathbf{B}u(t) - \sum_{k=1}^{p} c_k \mathbf{x}(t-kh) \right) \tag{7}$$

$$h = T/m,\ t = ph,\ p = 0,\ 1,\ \ldots,\ m \tag{8}$$

$$c_k = (-1)^k \binom{\alpha}{k}, \quad k = 1,\ 2,\ \ldots,\ m \tag{9}$$

$$u_f(t) = \mathbf{C}\mathbf{x}(t) \tag{10}$$

where $u(t)$ is the original signal and $u_f(t)$ is the filtered signal. Result of filtering is presentedd in the figure (4).

4 Conclusion

The paper presents a novel application of non-integer order filter - namely bi-fractional filter - to processing of EMG signals. Results show promise of the method and require further examination especially in the context of processed signal interpretation.

Acknowledgment. Work realised in the scope of project titled "Design and application of noniteger order subsystems in control systems". Project was financed by National Science Centre on the base of decision no. DEC-2013/09/D/ST7/03960.

References

1. Alkan, A., Gunay, M.: Identification of emg signals using discriminant analysis and svm classifier. Expert Systems with Applications 39(1), 44–47 (2012)
2. Amsuss, S., Gobel, P., Jiang, N., Graimann, B., Paredes, L., Self-correcting, D.F.D.: pattern recognition system of surface emg signals for upper limb prosthesis control. IEEE Transactions on Biomedical Engineering PP(99), 1–10 (2013)
3. Annett, M.: Left-handedness as a function of sex, maternal versus paternal inheritance and report bias. Behavior Genetics 29, 103–114 (1999)
4. Baranowski, J., Piątek, P., Bauer, W., Dziwiński, T., Zagórowska, M.: Bi-fractional filters, part 2: Right half-plane case. In: 2014 19th International Conference on Methods and Models in Automation and Robotics, MMAR (accepted, 2014)
5. Baranowski, J., Zagórowska, M., Bania, P., Bauer, W., Dziwiński, T., Piątek, P.: Impulse response approximation method for bi-fractional filter. In: 2014 19th International Conference on Methods and Models in Automation and Robotics, MMAR (accepted, 2014)
6. Borysiuk, Z., Zmarzly, D.: Surface electromyography (semg) as an innovative system of research work on information processes in sport. Polish Journal of Sports Medicine 22(1), 59–66 (2006)

7. Castiello, U., Stelmach, G.E.: Generalized representation of handwriting: Evidence of effector independence. Acta Psychologica 82(1-3), 53–68 (1993)
8. Devi, B.V.A., Priyadharsini, S.S.: Sdiagnosis of neuromuscular disorders using softcomputing techniques. International Journal of Soft Computing and Engineering 3(5), 105–111 (2013)
9. Kaczorek, T.: Selected Problems of Fractional Systems Theory. LNCIS. Springer (2011)
10. Kawala-Janik, A.: Efficiency Evaluation of External Environments Control Using Bio-Signals. PhD thesis, University of Greenwich (2013)
11. Kawala-Janik, A., Podpora, M., Baranowski, J., Bauer, W., Pelc, M.: Innovative approach in analysis of eeg and emg signals? comparision of the two novel methods. In: 2014 19th International Conference on Methods and Models in Automation and Robotics, MMAR (accepted, 2014)
12. Kawala-Janik, A., Podpora, M., Pelc, M., Baranowski, J., Pąitek, P.: Implementation of an inexpensive eeg headset for the pattern recognition purpose. In: The 7th IEEE International Conference on Intelligent Data Acquisition and Advanced Computing Systems: Technology and Applications (2013)
13. Michell, A.W.: Understanding EMG. Oxford University Press (2013)
14. Ozsert, M., Yavuz, O., Durak-Ata, L.: Analysis and classification of compressed emg signals by wavelet transform via alternative neural networks algorithms. Computer Methods in Biomechanics and Biomedical Engineering 14, 521–525 (2011)
15. Phinyomark, A., Phukpattaranont, P., Limsakul, C.: Feature reduction and selection for emg signal classification. Expert Systems with Applications 39(8), 7420–7431 (2012)
16. Piątek, P., Baranowski, J.: Investigation of Fixed-Point Computation Influence on Numerical Solutions of Fractional Differential Equations. Acta Mechanica et Automatica 5(2), 101–107 (2011)
17. Scheme, E., Lock, B., Hargrove, L., Hill, W., Uruganti, U., Englehart, K.: Motion normalized proportional control for improved pattern recognition-based myoelectric control. IEEE Transactions on Neural Systems and Rehabilitation Engineering 22(1), 149–157 (2014)
18. Sikdar, S., Eastlake, H.R.E.B., Hunt, I.A., Nelson, A.J., Devanathan, J., Shin, A., Pancrazio, J.J.: Novel method for predicting dexterous individual finger movements by imaging muscle activity using a wearable ultrasonic system. IEEE Transactions on Neural Systems and Rehabilitation Engineering 22(1), 69–76 (2014)
19. Subasi, A.: Classification of emg signals using combined features and soft computing techniques. Applied Soft Computing 12(8), 2188–2198 (2012)
20. Wojtczak, P., Amaral, T.G., Dias, O.P., Wolczowski, A., Kurzynski, M.: Hand movement recognition based on biosignal analysis. Engineering Applications of Artificial Intelligence 22, 608–615 (2009)

Reconstruction of the Thermal Conductivity Coefficient in the Time Fractional Diffusion Equation

Rafał Brociek, Damian Słota, and Roman Wituła

Institute of Mathematics, Silesian University of Technology
Kaszubska 23, 44-100 Gliwice, Poland
{rafal.brociek,damian.slota,roman.witula}@polsl.pl

Abstract. This paper describes reconstruction of the thermal conductivity coefficient in the time fractional diffusion equation. Additional information for the considered inverse problem was given by the temperature measurements at selected points of the domain. The direct problem was solved by using the finite difference method. To minimize functional defining the error of approximate solution the Fibonacci search algorithm was used.

Keywords: inverse problem, time fractional diffusion equation, identification, thermal conductivity.

1 Introduction

In recent years the applications of mathematical models using the fractional order derivatives become very popular in technical science. Different types of phenomena in physics, biology, viscoelasticity, heat transfer, electrical engineering, control theory, fluid and continuum mechanics can be modeled by using the fractional order derivatives [5,28,15,18]. The time fractional diffusion equation is related with a continuous time random walk and it models the anomalous diffusion in many fields, for example, the diffusion processes of contaminants in porous media [10].

One of the first paper in which the authors used the method of fractional calculus to solve classical inverse heat conduction problem is [2]. Authors of this paper used a non-integer identified model as the direct model for the estimation procedure in solving the inverse heat conduction problem. The first papers, in which the inverse problems for equations with fractional order derivatives were considered, are Murio's papers [20,22,24,23]. Murio in his works used the mollification method. In papers [20,23] there are considered the problems in semi-infinite domain. In the first one of these works the derivative with respect to space is of the first order, whereas the derivative with respect to time is the Riemann-Liouville fractional derivative. In the second paper the Gruünwald-Letnikov fractional derivative was used. In both papers the heat flux and the temperature on boundary of the domain were reconstructed in case of known

© Springer International Publishing Switzerland 2015
K.J. Latawiec et al. (eds.), *Advances in Modeling and Control of Non-integer Order Systems*, Lecture Notes in Electrical Engineering 320, DOI: 10.1007/978-3-319-09900-2_22

temperature measurements within the domain. In paper [22] the time fractional diffusion equation with the Caputo fractional derivatives was considered. In this paper as well, the heat flux and the temperature on boundary of the domain were reconstructed, however the additional information in the inverse problem was given by the double boundary condition on the other end of discussed interval (the temperature and the heat flux were known there). In paper [24], apart from the boundary conditions, the initial condition was additionally reconstructed.

Paper [19] describes identification of parameters in the Caputo type fractional initial problem, the analytical solution of which is known. In calculations for minimizing the quadratic criterion the Marquardt algorithm was used. On the other hand, in paper [26] several minimization algorithms (Levenberg-Marquardt algorithm, Gauss-Newton algorithm and Nelder-Mead method) were compared in application for identifying the parameters of the Riesz fractional advection-dispersion equation. In conclusion the authors stated that in case of the considered problem the best choice was the Nelder-Mead method. The identification problems are also considered in papers [29,32]. Moreover, identification of the diffusion systems by fractional models is described in papers [11,12].

The parameter identification problem for integer-order systems was analysed by many Authors. For solving problems of this kind several kinds of methods were applied, for example the genetic algorithm [7,9,13,27], the particle swarm optimization algorithm [1], the Taylor series approach [6], the functional identification approach [3], the sensitivity coefficients [17], and the finite-difference scheme combined with an iteration method [14].

In the current paper we deal with the inverse problem for the time fractional diffusion equation and we intend to reconstruct the thermal conductivity for this equation on the basis of the values of temperature measured in selected points of the domain. Basing on the given information about the temperature measurements the functional is created which defines the error of approximate solution. In order to minimize this functional the authors used the Fibonacci search algorithm [16], whereas to solve the direct problem the finite difference method was applied [21,4].

2 Formulation of the Problem

We consider a rod of length b (of circular cross section) with the insulated lateral surface. The rod is made from the material of known specific heat c and known density ϱ. Whereas the thermal conductivity coefficient λ of this material is unknown and needed to be determined. Temperature of the rod in the initial moment is known as well as the boundary conditions on its ends. Moreover, in order to solve the inverse problem we have given the temperature in internal point of this rod. We assume that the distribution of temperature within the rod can be modeled by using the time fractional diffusion equation. In paper [25] it is shown that in case of porous materials the fractional diffusion equation enables to reconstruct better the temperature distribution in comparison with the classical heat conduction equation.

Thus we discuss the following time fractional diffusion equation

$$c\,\varrho\,\frac{\partial^\alpha u(x,t)}{\partial t^\alpha} = \lambda\,\frac{\partial^2 u(x,t)}{\partial x^2}, \tag{1}$$

defined in area $D = \{(x,t) : x \in [0,b], t \in [0,t^*)\}$, where $\alpha \in (0,1)$ and c, ϱ and λ denote the specific heat, density and thermal conductivity coefficient, respectively. The initial condition is also posed

$$u(x,0) = \varphi(x), \qquad x \in [0,b], \tag{2}$$

as well as the Neumann and Robin boundary conditions

$$-\lambda\,\frac{\partial u}{\partial x}(0,t) = q(t), \qquad t \in [0,t^*), \tag{3}$$

$$-\lambda\,\frac{\partial u}{\partial x}(b,t) = h(t)\,\big(u(b,t) - u^\infty\big), \qquad t \in [0,t^*), \tag{4}$$

where h describes the heat transfer coefficient and u^∞ is the ambient temperature.

Fractional derivative with respect to time, which occurs in equation (1), will be the Caputo fractional derivative [28,8] determined in our case as follows

$$\frac{\partial^\alpha u(x,t)}{\partial t^\alpha} = \frac{1}{\Gamma(1-\alpha)} \int_0^t \frac{\partial u(x,s)}{\partial s}\,(t-s)^{-\alpha}\,ds, \tag{5}$$

where $\Gamma(\cdot)$ is the Gamma function [30].

Inverse problem, considered in this paper, consists in determination of the thermal conductivity λ in such a way that temperature in selected points of discussed area will adopt the preset values. So, the temperature measurements $((x_i,t_j) \in D)$:

$$u(x_i,t_j) = \widehat{U}_{ij}, \qquad i = 1,2,\ldots,N_1, \quad j = 1,2,\ldots,N_2, \tag{6}$$

are known, where N_1 denotes the number of sensors and N_2 means the number of measurements taken form each sensor.

For the fixed coefficient of thermal conductivity the investigated issue becomes a direct problem, solution of which is represented by the temperature values U_{ij} corresponding to the given value of the thermal conductivity coefficient. By using the computed temperatures U_{ij} and the measurement temperatures $\widehat{U}_{ij}$, the functional defining the error of approximate solution is created as follows:

$$J(\lambda) = \sum_{i=1}^{N_1}\sum_{j=1}^{N_2}\big(U_{ij} - \widehat{U}_{ij}\big)^2. \tag{7}$$

By minimizing this functional we intend to find the approximate value of the thermal conductivity coefficient.

3 Numerical Example

In the numerical example the following data are given: $\alpha = \frac{1}{2}$, $b = 0.2$, $t^* = 100$, $\varphi(x) = 900$, $q(t) = 0$, $c = 1000$, $\varrho = 2680$, $u^\infty = 300$ and

$$h(t) = 1200 \exp\left(\frac{45 - t}{705} \log \frac{24}{5}\right).$$

Exact value of the sought thermal conductivity coefficient λ is equal to 240.

Direct problem, defined by means of equations (1)–(4) for the fixed thermal conductivity coefficient, was solved with the use of implicit scheme of the finite difference method [21,4]. In result we obtained the distribution of temperature in considered region representing the benchmark for approximate results. From this distribution we select the values $\widehat{U}_{ij}$ of temperature simulating the measurements of temperature. In the following part of this paper we treat the values of temperature, obtained in this way, as the exact values.

In order to investigate the influence of measurement errors on the exactness of received results, and in consequence to examine the algorithm stability, we execute a commonly used numerical experiment [31], in which the solution of direct problem is perturbed by the pseudo-random error. Course of temperature, received in this way, is used as the input data for the inverse problem.

It is assumed that the temperature values are known at a single measurement point ($N_1 = 1$). It means that in the area a single thermocouple is located at a distance of 5 mm away from the boundary. From this thermocouple we obtained 100 measurements of temperature ($N_2 = 100$). The calculations were made for the exact values of temperature and the values perturbed by the pseudo-random error of sizes 0.5, 1, 2 and 5%.

The grid used in the finite difference method is defined as follows

$$S = \big\{(x_i, t_k),\ x_i = i\,\Delta x,\ t_k = k\,\Delta t, i = 0, 1, \ldots, N,\ k = 0, 1, 2, \ldots, M\big\}$$

where $\Delta x = b/N$, $\Delta t = t^*/M$. Fractional derivative is approximated by formula [21,4]:

$$D_t^{(\alpha)} u_i^k = \sigma(\alpha, \Delta t) \sum_{j=1}^{k} \omega(\alpha, j) \left(u_i^{k-j+1} - u_i^{k-j}\right), \tag{8}$$

where

$$\sigma(\alpha, \Delta t) = \frac{1}{\Gamma(1-\alpha)\,(1-\alpha)\,(\Delta t)^\alpha},$$
$$\omega(\alpha, j) = j^{1-\alpha} - (j-1)^{1-\alpha}.$$

Using approximations of the Neumann and Robin boundary conditions and the difference quotient for the second order derivative with respect to space, we get the following difference equations

$k \geq 1,\ i = 0$:

$$\left(\sigma(\alpha, \Delta t) + \frac{2\,a}{(\Delta x)^2}\right) u_0^k - \frac{2\,a}{(\Delta x)^2}\, u_1^k =$$
$$= \sigma(\alpha, \Delta t)\, u_0^{k-1} - \sigma(\alpha, \Delta t) \sum_{j=2}^{k} \omega(\alpha, j)\, \left(u_0^{k-j+1} - u_0^{k-j}\right) - \frac{2}{c\,\varrho\,\Delta x}\, q_k,$$

$k \geq 1,\ i = 1, 2, \ldots, N-1$:

$$-\frac{a}{(\Delta x)^2}\, u_{i-1}^k + \left(\sigma(\alpha, \Delta t) + \frac{2\,a}{(\Delta x)^2}\right) u_i^k - \frac{a}{(\Delta x)^2}\, u_{i+1}^k =$$
$$= \sigma(\alpha, \Delta t)\, u_i^{k-1} - \sigma(\alpha, \Delta t) \sum_{j=2}^{k} \omega(\alpha, j)\, \left(u_i^{k-j+1} - u_i^{k-j}\right),$$

$k \geq 1,\ i = N$:

$$-\frac{2\,a}{(\Delta x)^2}\, u_{N-1}^k + \left(\sigma(\alpha, \Delta t) + \frac{2\,a}{(\Delta x)^2} + \frac{2}{c\,\varrho\,\Delta x}\, h_k\right) u_N^k =$$
$$= \sigma(\alpha, \Delta t)\, u_N^{k-1} - \sigma(\alpha, \Delta t) \sum_{j=2}^{k} \omega(\alpha, j)\, \left(u_N^{k-j+1} - u_N^{k-j}\right) + \frac{2}{c\,\varrho\,\Delta x}\, h_k\, u^{\infty},$$

where $u_i^k = u(x_i, t_k)$, $h_k = h(t_k)$, whereas $a = \frac{\lambda}{c\,\varrho}$ is the thermal diffusivity coefficient.

The calculations are made on the grid of discretization intervals equal to $\Delta t = 1$, $\Delta x = b/100$. A reasonable change of the grid density did not have any significant influence on the obtained results. To avoid the inverse crime we executed the calculations in the inverse problem and in the direct problem, used for generating the pseudo-measurements, on the grids of different densities. For minimizing functional (7) the Fibonacci search algorithm [16] is used.

In Figures 1 and 2 the absolute and relative errors of the restored temperatures in measurement point are shown. In every case the reconstruction of temperature is very good. The maximal absolute error does not exceed the value of 0.124 K, while the relative errors are smaller than 0.014%. Reconstruction of the thermal conductivity coefficient is also very good (see Table 1). In case of the exact input data, the sought coefficient is restored with minimal errors (not exceeding 0.031%). For the perturbed input data, the errors of reconstructing the thermal conductivity coefficient are generally smaller than the values of error in the input. Only for the input data perturbed by error of size 1% the thermal conductivity coefficient is identified with the error greater than the input error. But it is still the acceptable error, particularly if we consider the error of temperature reconstruction in this case. Maximal relative error of the temperature restoration in this case is equal to 0.0113%, while the average value of this error is equal to 0.0092%. In Figures 3–5 distributions of the exact and reconstructed temperature as well as the error of this reconstruction in measurement point are displayed in case of the exact input data and the data perturbed by error of sizes 1 and 5%.

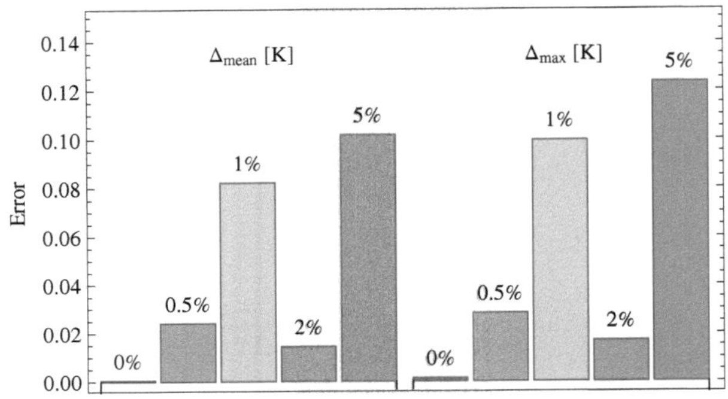

Fig. 1. Average and maximal absolute errors of the temperature reconstruction in measurement point for various perturbations of input data

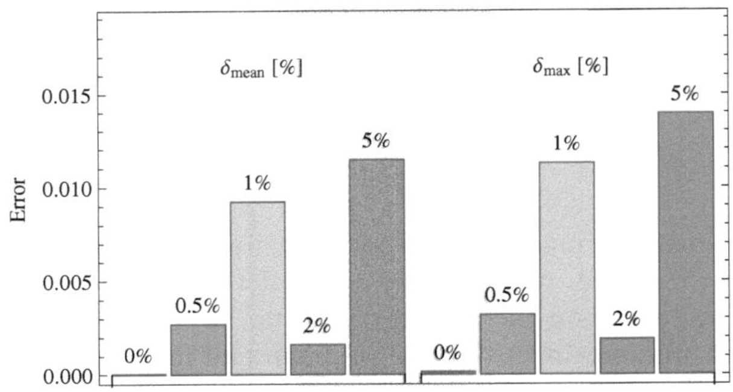

Fig. 2. Average and maximal relative errors of the temperature reconstruction in measurement point for various perturbations of input data

Table 1. Reconstructed values of thermal conductivity coefficient and errors of this reconstruction ($\overline{\lambda}$ – reconstructed value, δ_λ – percentage relative error)

Noise [%]	$\overline{\lambda}$	δ_λ [%]
0	240.074	0.031
0.5	241.095	0.456
1	236.593	1.420
2	240.684	0.285
5	235.775	1.760

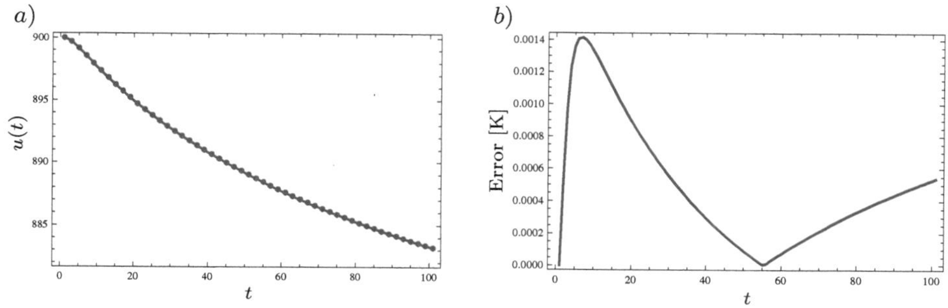

Fig. 3. Exact (solid line) and reconstructed (points) distribution of temperature in measurement point (*a*) and error of this reconstruction (*b*) for exact input data

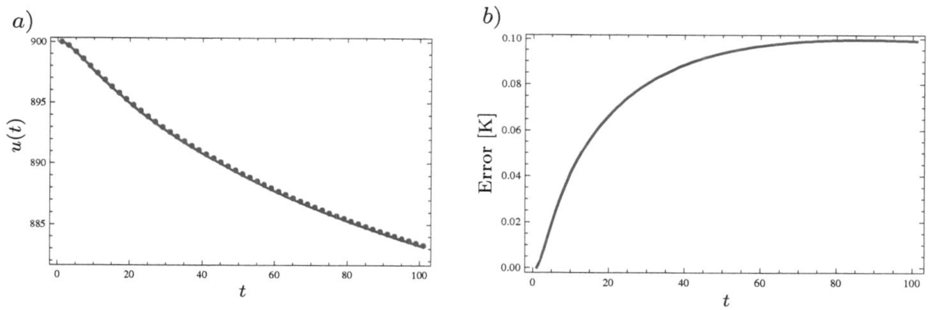

Fig. 4. Exact (solid line) and reconstructed (points) distribution of temperature in measurement point (*a*) and error of this reconstruction (*b*) for input data perturbed by 1% error

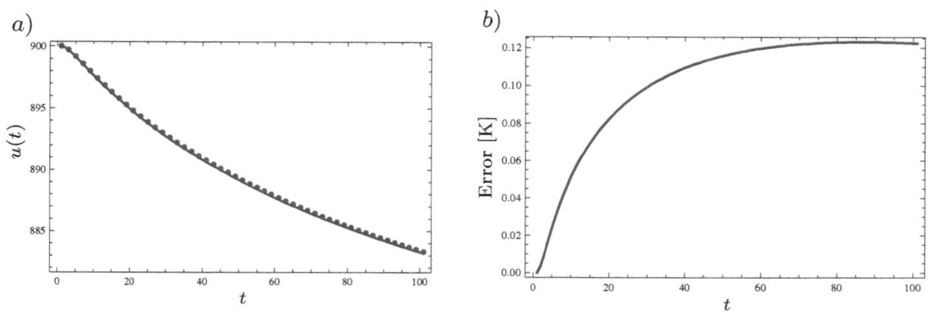

Fig. 5. Exact (solid line) and reconstructed (points) distribution of temperature in measurement point (*a*) and error of this reconstruction (*b*) for input data perturbed by 5% error

4 Conclusions

In the current paper we dealt with the inverse problem for the time fractional diffusion equation in which we intended to reconstruct the thermal conductivity coefficient on the ground of the values of temperature measured in selected points of considered domain. To solve the direct problem we used the finite difference method. Basing on the given information about the temperature measurements we created the functional defining the error of approximate solution. In order to minimize this functional we used the Fibonacci search algorithm.

Presented results show that in case of input data without perturbation the error of the thermal conductivity coefficient reconstruction is minimal. For perturbed input data the errors of reconstructing the thermal conductivity coefficient are generally smaller than the size of error in the input data. Only for 1% perturbation the thermal conductivity coefficient is reconstructed with error greater than error of input data. However it is still the acceptable error, particularly if we consider the error of temperature reconstruction. Reconstructions of temperature in every case are very good.

Acknowledgement. The Authors would like to thank the Referee for his constructive comments and suggestions.

References

1. Ardakani, M., Khodadad, M.: Identification of thermal conductivity and the shape of an inclusion using the boundary elements method and the particle swarm optimization algorithm. Inverse Probl. Sci. Eng. 17, 855–870 (2009)
2. Battaglia, J.L., Cois, O., Puigsegur, L., Oustaloup, A.: Solving an inverse heat conduction problem using a non-integer identified model. Int. J. Heat Mass Transfer 44, 2671–2680 (2001)
3. Borukhov, V.T., Tsurko, V.A., Zayats, G.M.: The functional identification approach for numerical reconstruction of the temperature-dependent thermal-conductivity coefficient. Int. J. Heat Mass Transfer 52, 232–238 (2009)
4. Brociek, R.: Implicite finite difference metod for time fractional diffusion equations with mixed boundary conditions. Zesz. Nauk. PŚ., Mat. Stosow. 4 (in press 2014)
5. Carpinteri, A., Mainardi, F.: Fractal and Fractional Calculus in Continuum Mechanics. Springer, New York (1997)
6. Chang, C.L., Chang, M.: Inverse estimation of the thermal conductivity in a one-dimensional domain by Taylor series approach. Heat Transfer Engineering 29, 830–838 (2008)
7. Czél, B., Gróf, G.: Inverse identification of temperature-dependent thermal conductivity via genetic algorithm with cost function-based rearrangement of genes. Int. J. Heat Mass Transfer 55, 4254–4263 (2012)
8. Diethelm, K.: The Analysis of Fractional Differential Equations. Springer, Berlin (2010)
9. Divo, E., Kassab, A., Rodriguez, F.: Characterization of space dependent thermal conductivity with a BEM-based genetic algorithm. Numer. Heat Transfer A(37), 845–875 (2000)

10. Dou, F., Hon, Y.: Numerical computation for backward time-fractional diffusion equation. Eng. Anal. Bound. Elem. 40, 138–146 (2014)
11. Gabano, J.D., Poinot, T.: Fractional modelling and identification of thermal systems. Signal Processing 91, 531–541 (2011)
12. Gabano, J.D., Poinot, T.: Estimation of thermal parameters using fractional modelling. Signal Processing 91, 938–948 (2011)
13. Imani, A., Ranjbar, A.A., Esmkhani, M.: Simultaneous estimation of temperature-dependent thermal conductivity and heat capacity based on modified genetic algorithm. Inverse Probl. Sci. Eng. 14, 767–783 (2006)
14. Kanca, F., Ismailov, M.: The inverse problem of finding the time-dependent diffusion coefficient of the heat equation from integral overdetermination data. Inverse Probl. Sci. Eng. 20, 463–476 (2012)
15. Klafter, J., Lim, S., Metzler, R.: Fractional dynamics. Resent advances. World Scientific, New Jersey (2012)
16. Kusiak, J., Danielewska-Tułecka, A., Oprocha, P.: Optimization. PWN, Warszawa (2009) (in Polish)
17. Majchrzak, E., Mendakiewicz, J., Piasecka-Belkhayat, A.: Algorithm of the mould thermal parameters identification in the system casting–mould–environment. J. Mater. Proc. Tech. 164-165, 1544–1549 (2008)
18. Mitkowski, W., Kacprzyk, J., Baranowski, J.: Advances in the Theory and Applications of Non-integer Order Systems. Springer Inter. Publ., Cham (2013)
19. Mitkowski, W., Obrączka, A.: Simple identification of fractional differential equation. Solid State Phenomena 180, 331–338 (2012)
20. Murio, D.A.: Stable numerical solution of a fractional-diffusion inverse heat conduction problem. Comput. Math. Appl. 53, 1492–1501 (2007)
21. Murio, D.: Implicit finite difference approximation for time fractional diffusion equations. Comput. Math. Appl. 56, 1138–1145 (2008)
22. Murio, D.: Time fractional IHCP with Caputo fractional derivatives. Comput. Math. Appl. 56, 2371–2381 (2008)
23. Murio, D.: Stable numerical evaluation of Grünwald-Letnikov fractional derivatives applied to a fractional IHCP. Inverse Probl. Sci. Eng. 17, 229–243 (2009)
24. Murio, D., Mejia, C.: Generalized time fractional IHCP with Caputo fractional derivatives. J. Phys. Conf. Ser. 135, 12074 (2008)
25. Obrączka, A., Kowalski, J.: Modeling of the heat distribution in ceramic materials by using the fractional differential equations. In: Szczygieł, M. (ed.) Proceedings of the XV Jubilee Symposium, Basic Problems in Power Electronics, Electromechanics and Mechatronics, PPEEm 2012, Gliwice, pp. 132–133 (2012) (in Polish)
26. Obrączka, A., Mitkowski, W.: The comparison of parameter identification methods for fractional partial differential equation. Solid State Phenomena 210, 265–270 (2014)
27. Orain, S., Scudeller, Y., Garcia, S., Brousse, T.: Use of genetic algorithms for the simultaneous estimation of thin film thermal conductivity and contact resistances. Int. J. Heat Mass Transfer 44, 3973–3984 (2001)
28. Podlubny, I.: Fractional Differential Equations. Academic Press, San Diego (1999)
29. Poinot, T., Trigeassou, J.C.: Identification of fractional systems using an output-error technique. Nonlinear Dynamics 38, 133–154 (2004)
30. Rabsztyn, S., Słota, D., Wituła, R.: Gamma and Beta Functions, vol. 1. Wyd. Pol. Śl, Gliwice (2012) (in Polish)
31. Szczygieł, I.: Analysis of Inverse Heat Convection Problems. Zesz. Nauk. PŚ., Ener. 140, 1–141 (2005) (in Polish)
32. Victor, S., Malti, R., Garnier, H., Oustaloup, A.: Parameter and differentiation order estimation in fractional models. Automatica 49, 926–935 (2013)

Robust Non-integer Order Controller for Air Heating Process Trainer

Tomasz Dziwiński, Waldemar Bauer, Jerzy Baranowski,
Paweł Piątek, and Marta Zagórowska

AGH University of Science and Technology,
Department of Automatics and Biomedical Engineering,
Al. Mickiewicza 30, 30-059 Kraków
{tdz,bauer,jb,ppi,zagor}@agh.edu.pl

Abstract. Robust non-integer control is an emerging field in control theory. Multiple works are focused on designing controllers with flat-phase and robust stability margins. In this paper design of non-integer order with robust properties is considered. This controller is designed and analysed for the model of air heating process trainer system belonging to the Department of Automatics and Biomedical Engineering of AGH University and Science and Technology.

1 Introduction

The designing of control system for a heat transmission line is associated with two major challenges. Firstly, the dynamical system of a plant is non-stationary, due to varying gain while the heater and transmission line is warming-up. Secondly, the study of stability becomes an infinite-dimensional problem, by the presence of continuous time-delay in a dynamical system. Therefore there is a strong need for a robust control system, resistant to changing parameters of the plant.

Bode in his famous work [4] discusses about an ideal shape of the loop transfer function, to make the closed-loop system robustly stable and insensitive for gain changes. The proposed transfer function has been later called Bode's ideal cut-off characteristic and it has a form (1) [1]

$$L(s) = \left(\frac{s}{\omega_{gc}}\right)^n \tag{1}$$

In the following paper, the robustness is achieved by increasing the controller complexity – as it was proposed i.a. by Oustaloup in [8]. The fractional-order PID (FOPID) controller provides more variables in tuning strategy and therefore can be a good way to achieve preset goals related to robust control. The method of tuning FOPID controller using multi-objective optimisation including robust stability conditions has been discussed by Meng and Xue in [7]. The tuning of $PI^{\alpha}D^{\mu}$ controller parameters has been considered among others by Bauer et al. in [5] or in [3], where integral absolute error is minimised. Another solution has been presented by Barbosa et al. in [2], where the constrained optimisation was

© Springer International Publishing Switzerland 2015
K.J. Latawiec et al. (eds.), *Advances in Modeling and Control of Non-integer Order Systems*,
Lecture Notes in Electrical Engineering 320, DOI: 10.1007/978-3-319-09900-2_23

used to achieve robust control system. In the following work three criteria of flatphase and robust stability margins have been selected and then used to form an objective function in optimisation procedure, by which the optimal controller setting have been obtained.

2 Problem Definition

2.1 Considered Linear Time Invariant Systems with Time Delay

There is given the Linear Time Invariant Systems (LTI) with Time Delay of air heating process, which simplified model can be expressed by a second order inertial system:

$$P(s) = \frac{K}{(T_1 s + 1)(T_2 s + 1)} e^{-s\tau}, \tag{2}$$

where k=18.8, T_1=7.783, T_2=0.0014 and τ=0.5842. The considered model has been precisely identified by the experiments on air heating process trainer system. The frequency response of the pointed system can be seen in Fig. 1

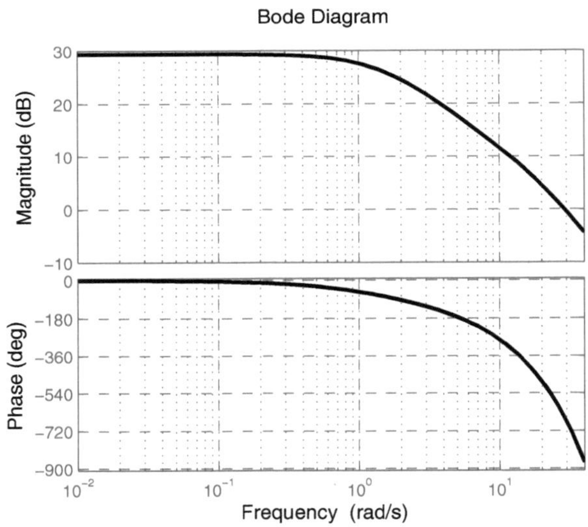

Fig. 1. The frequency response of the given two-inertia system

2.2 Fractional Order PID Controller

The fractional order controller $PI^{\alpha}D^{\mu}$ is generalised classical PID, firstly proposed by Podlubny in 1999 and represented by an equation (3)

$$C(s) = K_P + K_I \frac{1}{s^{\alpha}} + K_D s^{\mu} \quad \alpha, \mu > 0, \tag{3}$$

where K_P, K_I and K_D represent proportional gain, integral gain and derivative gain respectively.

A feedback control system block diagram is shown in Fig. 2.

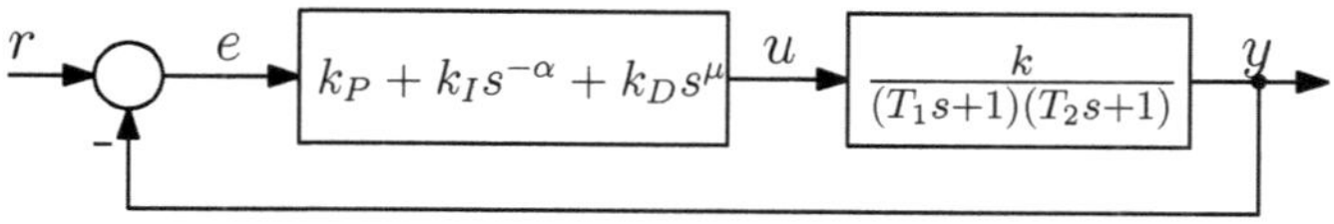

Fig. 2. Block diagram of the closed-loop system

There have been formulated the following controller design specification:

1. predefined phase margin ϕ_m and gain margin g_m,
2. robustness to variations in the gain of the plant,
3. high-frequency noise rejection,
4. output disturbance rejection,
5. steady-state error cancelation.

The stability of the closed-loop system is achieved by analysing the open-loop phase and gain margins. The higher stability margins, the more robust system is. Therefore the following criteria have been formulated, based on Bode's ideal transfer function [6]:

– Gain margin
$$|C(j\omega)P(j\omega)|\,|_{\omega=\omega_{pm}} = 12\text{dB} \tag{4}$$
– Phase margin
$$\arg\left(C(j\omega)P(j\omega)\right)|_{\omega=\omega_{gm}} = 45° \tag{5}$$
– Flat phase in gain crossover frequency (iso-damping property)
$$\left.\frac{d\arg\left(C(j\omega)P(j\omega)\right)}{d\omega}\right|_{\omega=\omega_{gm}} = 0 \tag{6}$$

Criteria (4) and (5) refer to robust stability margins and criterion (6) refers to an iso-dumping property, which makes a closed-loop system robust in the perspective of gain variations [2]. To achieve these goals, the Simulated Annealing optimisation method has been chosen for tuning $PI^{\alpha}D^{\mu}$ controller parameters. The optimisation process minimise an multi-objective function, that reflects how far the behaviour of FOPID controller is from the above assumptions. By optimising succeeded in obtaining an robust system, which Nyquist plot is shown in Fig. 4. An evaluation of the objective function is shown in Fig. 3.

To implement controller, the irrational transfer function have to be approximated with a rational function. There are several methods to approximate irrational transfer function arbitrarily close in the specified range $[\omega_{min},\omega_{max}]$ [1]. The Oustaloup continuous integer approximation is given by equation (7) [9].

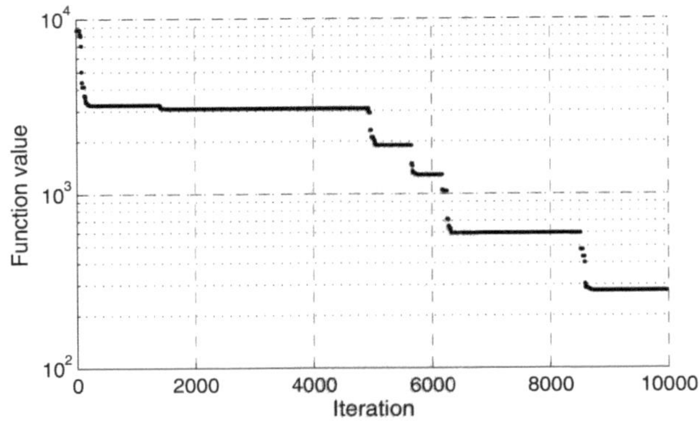

Fig. 3. The progression in simulated annealing optimisation

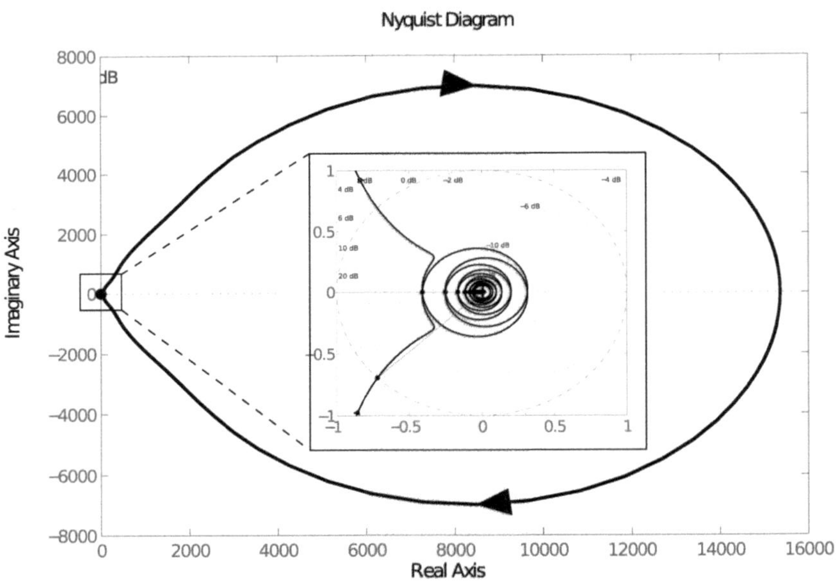

Fig. 4. The Nyquist plot with stability margins distinguished

Bode plots of the fractional system and its proposed rational Oustaloup approximation is shown in Fig. 5. In this approach, the $N = 8$ order of fractional-order integrator and derivative approximation has been selected in the range $\omega \in \left[10^{-6}, 10^{2}\right]$. It can be easily seen, that Oustaloup method gives satisfactory results on desired frequency range.

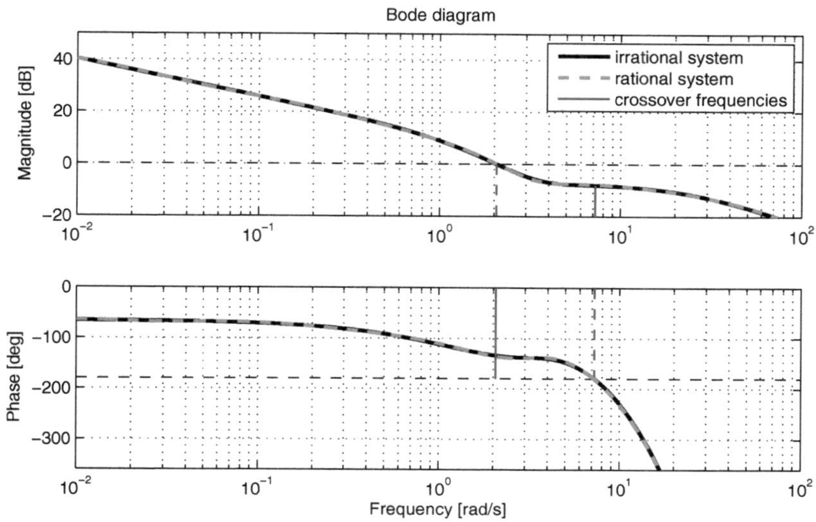

Fig. 5. Open-loop bode plot with phase crossover frequency and gain crossover frequency marked and the Oustaloup-Recursive approximation of designed controller

$$s^{\gamma} \approx K \prod_{k=1}^{N} \frac{s + \omega_k'}{s + \omega_k}, \quad \gamma > 0, \tag{7}$$

where poles, zeros and gain can be evaluated respectively as:

$$\begin{aligned} \omega_k' &= \omega_{min} \omega_u^{(2k-1-\gamma)/N} \\ \omega_k &= \omega_{min} \omega_u^{(2k-1+\gamma)/N} \\ K &= \omega_h^{\gamma} \\ \omega_u &= \sqrt{\frac{\omega_{max}}{\omega_{min}}} \end{aligned} \tag{8}$$

3 Sensitivity Functions

The sensitivity function S (9) determines how the disturbances are influenced by feedback. In other words, whether the distortions are amplified or suppressed by the closed loop. The complementary sensitivity function T (10) determines how the measurement noises are influenced by feedback.

$$T(s) = \frac{Y(s)}{R(s)} = \frac{C(s)P(s)}{1 + C(s)P(s)} \tag{9}$$

$$S(s) = \frac{E(s)}{R(s)} = \frac{1}{1 + C(s)P(s)} \tag{10}$$

The maximum sensitivities may be used as robustness measures (11).

$$M_S = \sup_{\omega} |S(j\omega)| \qquad M_T = \sup_{\omega} |T(j\omega)| \tag{11}$$

4 Results

The described optimisation method has been used for tuning both classical PID controller and FOPID controller. The optimal FOPID settings are collected in Tab. 2, while the indicators of robustness are presented in Tab. 1. The step response of closed-loop system is shown in Fig. 6. A Bode magnitude and phase plots of open-loop system is presented in Fig. 7. It can be seen, that the magnitude response of control system with FOPID has slope of 14.5 dB per decade in low-frequency side, while classical PID 20 dB per decade obviously. The detailed comparison between classical PID and FOPID is presented in Tab. 3, including maximum sensitivities (M_S) and (M_T), the peak overshoot (σ) and rise time (tr).

Table 1. Obtained stability margins

Phase margin	45.956°
Phase crossover frequency	1.15 Hz
Gain margin	-8.052 dB
Gain crossover frequency	0.32 Hz
Phase angle in ω_{cg}	-0.398°

Table 2. Obtained controller parameters

k_P	0.0035
k_I	0.1268
k_D	0.0206
α	0.7229
μ	0.7307

Table 3. Evaluation factor comparison for optimised classical PID and optimised FOPID

evaluation factor	PID	FOPID
M_S	1.64	1.66
M_T	1.50	1.30
σ [%]	28	11
tr [s]	0.72	0.68

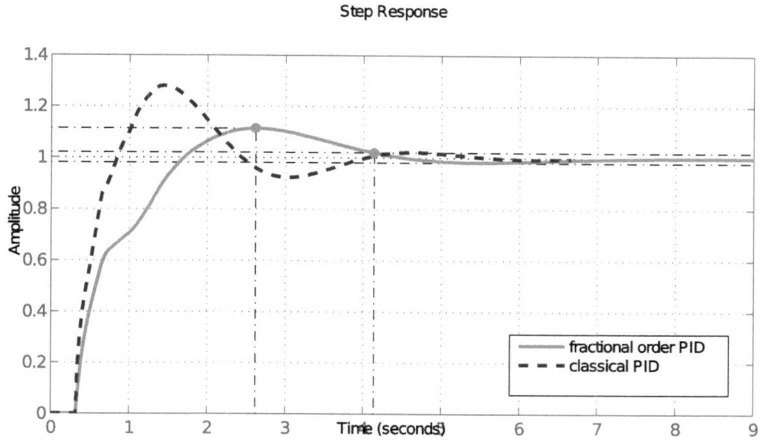

Fig. 6. The comparison of step responses

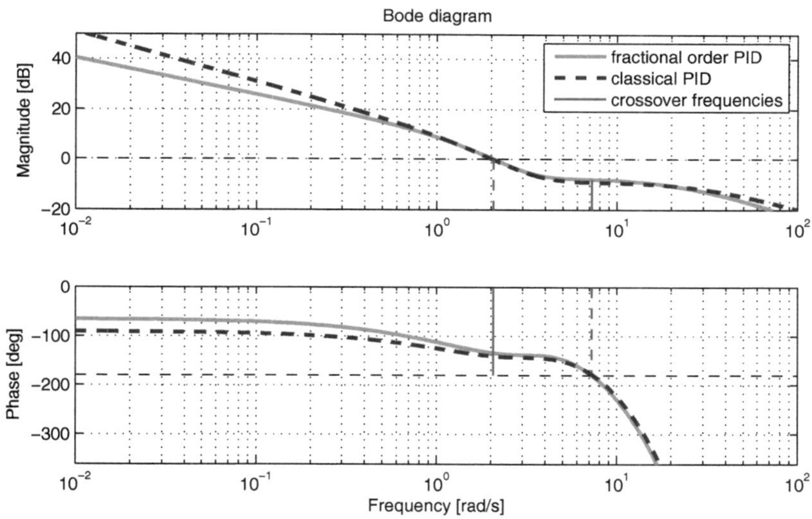

Fig. 7. The comparison of frequency responses

5 Conclusion and Further Research

Fractional-order controller allows for shaping the slope of both magnitude and phase characteristics at high and low frequencies. The designed controller has allowed to achieve stated assumptions: flat-phase and robust stability margins. It has been shown, that fractional-order $PI^{\alpha}D^{\mu}$ controller is suitable for control of the inertial system with time-delay and time-varying gain. The paper has shown that simulated annealing optimisation method could be helpful in the tuning process.

The further research is planned to implement $PI^{\alpha}D^{\mu}$ controller in digital real-time environment, based on RT-DAC board and MATLAB/RT-CON library, and to conduct experiments on physical plant.

Acknowledgement. Work realised in the scope of project titled "Design and application of non-integer order subsystems in control systems". Project was financed by National Science Centre on the base of decision no. DEC-2013/09/D/ST7/03960.

References

1. Åström, K.J.: Model uncertainty and robust control. Lecture Notes on Iterative Identification and Control Design, 63–100 (2000)
2. Barbosa, R.S., Machado, J.T., Ferreira, I.M.: Tuning of pid controllers based on bode?s ideal transfer function. Nonlinear Dynamics 38(1-4), 305–321 (2004)
3. Bauer, W., Baranowski, J., Mitowski, W.: Non-integer order $\mathrm{PI}^{\alpha}\mathrm{D}^{\mu}$ control ICU-MM. In: Mitkowski, W., Kacprzyk, J., Baranowski, J. (eds.) Theory & Appl. of Non-integer Order Syst. Lecture Notes in Electrical Engineering, vol. 257, pp. 295–303. Springer, Heidelberg (2013)
4. Bode, H.W.: Network analysis and feedback amplifier design (1945)
5. Dziwiński, T., Bauer, W., Baranowski, J., Piątek, P., Zagórowska, M.: Robust non-integer order controller for air heater. In: 2014 19th International Conference on Methods and Models in Automation and Robotics, MMAR (accepted 2014)
6. Luo, Y., Zhang, T., Lee, B., Kang, C., Chen, Y.: Fractional-order proportional derivative controller synthesis and implementation for hard-disk-drive servo system (2014)
7. Meng, L., Xue, D.: Design of an optimal fractional-order pid controller using multi-objective ga optimization. In: Control and Decision Conference, CCDC 2009, pp. 3849–3853. IEEE (2009)
8. Oustaloup, A.: La commande CRONE: commande robuste d'ordre non entier. Hermes (1991)
9. Oustaloup, A., Levron, F., Mathieu, B., Nanot, F.M.: Frequency-band complex non-integer differentiator: characterization and synthesis. IEEE Transactions on Circuits and Systems I: Fundamental Theory and Applications 47(1), 25–39 (2000)

Model–Based Fault Diagnosis with Fractional Models

Ryszard Kopka

Opole University of Technology,
ul. Sosnkowskiego 31, 45-272 Opole, Poland
r.kopka@po.opole.pl

Abstract. Paper presents the results of experimental research of fault detection based on difference signal of plant and model controllers output. The plant was simulated by electronic circuit with supercapacitor 'damaged' by discharging it by the current source. The model of the plant was described by fractional order differential equations. Such approaches allowed for some processes and the newest technologies, obtain better modelling accuracy. Model parameters was derived used identification procedures based on step response signal. The fault detection problem of this paper can be used as simple benchmark example to test new fault detection rules before applying it to real systems.

Keywords: fault detection, fractional calculus, fractional order model.

1 Introduction

One of the methods to meets reliability requirements is fault tolerant. To make these possible control systems need to have both solutions to detect and isolate failures, as well as possibility to continuation of the control process in their presence [1,9,10]. The use of new control methods is possible by the development of technology for sensors, transmission systems and actuators, but also the development of computational tools, in terms of computing speed and software solutions. The rapid development of technology, however, makes the existing mathematical rules not enough precisely to describe the process or phenomena taking places in the newest technological solutions [6]. Them therefore required new models that are much more accurate. The fractional order differential equations for example, could be used to describe more precisely some technological processes or used elements. These allow, in many cases, significantly more accurately describe the phenomena in some materials or elements [3,6]. It is particularly important in the model-based fault detection systems [4,10]. The concept of fractional calculus has drawn the attention of many famous mathematicians for hundreds years. But this kind of calculus was not popular until recent years when benefits stemming from using it became evident in various scientific fields, including system modeling and automatic control [5,15].

© Springer International Publishing Switzerland 2015
K.J. Latawiec et al. (eds.), *Advances in Modeling and Control of Non-integer Order Systems*,
Lecture Notes in Electrical Engineering 320, DOI: 10.1007/978-3-319-09900-2_24

2 Fractional Calculus

The fractional calculus have become very popular during last two decades. There are a lot papers and books presented their advantages and possibility of used [2,3,13,15]. Recent papers notion that fractional-order calculus should be employed where more accurate modeling and robust control are needed. Specifically, fractional order calculus found its possibility to resolve complex mathematical and physical problems and may be useful to modeling any system which has memory and/or hereditary properties. In the field of automatic control fractional calculus is used to obtain more accurate models, develop new control strategies and enhance the characteristics of control systems [15]. Below only the main definitions of fractional calculus used in this work are presented.

Among the several definition of fractional order differ-integral [3,5,16], the Grünwald-Letnikov was used.

Definition 1. *The continuous time Grünwald-Letnikov fractional differ-integral is defined as follow:*

$$ {}^{GL}{}_{a}D^{t}_{\alpha} f(t) = \lim_{h\to 0} \frac{1}{h^{\alpha}} \sum_{j=0}^{\left[\frac{t-a}{h}\right]} (-1)^{j} \binom{\alpha}{j} f(t-jh) \tag{1} $$

where $m-1<\alpha\leq m\in\mathbb{N}$ *and* $\alpha\in\mathbb{R}$ *(*$\mathbb{R}$ *is the set of real numbers) and while* $[\cdot]$ *means integer part.*

The binomial term in Definition 1 can be obtain by the following equation:

$$ \binom{\alpha}{j} = \begin{cases} 1 & \text{for } j=0 \\ \frac{\alpha(\alpha-1)\ldots(\alpha-1+j)}{j!} & \text{for } j>0 \end{cases} \tag{2} $$

In order to present the discrete fractional order model and controller the discrete form of Grünwald-Letnikov definition was used. The most commonly discretization schemes are the Euler expansion, Tustin rule, and the Al-Alaoui operator. The Euler expansion method is a simple generalization of the familiar Grünwald-Letnikov derivative

$$ \Delta^{\alpha}_{h} f(t)|_{t=kh} = \frac{1}{h^{\alpha}} \sum_{j=0}^{k} (-1)^{j} \binom{\alpha}{j} f(t-jh) \tag{3} $$

where $t=kh$ are the values of the discretized continuous time. Transforming samples of continuous time to the discrete time $k=0,1,\ldots$ the Eq. (3) can be rewrite as

$$ \Delta^{\alpha}_{h} f(t) = \frac{1}{h^{\alpha}} \sum_{j=0}^{t} (-1)^{j} \binom{\alpha}{j} f(t-j) \tag{4} $$

Using the simplest discretization scheme for the Grünwald-Letnikov derivative as in Eq. (3) and assuming for a sampling period $h=1$, the fractional-order difference in discrete time k can be defined as:

Definition 2. *Fractional order difference is given as follows:*

$$\Delta^{\alpha} x_k = \sum_{j=0}^{k} (-1)^j \binom{\alpha}{j} x_{k-j} \tag{5}$$

where $\alpha \in \mathbb{R}$, is a fractional degree, $\mathbb{R}$, is the set of real numbers and $k \in \mathbb{N}$ is the number of a sample for witch the approximation of the derivative is calculated.

For practical realization the number of samples taken into consideration has to be reduced to the predefined number L. In this case the Eq. (4) is rewritten as:

$$\Delta^{\alpha} x_k = \sum_{j=0}^{L} (-1)^j \binom{\alpha}{j} x_{k-j} \tag{6}$$

where L is a number of samples taken into account, called memory length and with assumption that $x_k = 0$ for $k < 0$ [3,16].

3 Model Based Fault Detection Mechanism

Different approaches for fault-detection using mathematical models have been developed in the last 20 years. The task consists of the detection of faults in the processes by using the dependencies between different measurable signals end mathematical process models outputs. But such approaches require the precisely knowledge of process model in both the form of a mathematical structure and the parameters [4,7,14].

Figure 1 shows the scheme of the measurement system. The Host and Target PCs were connected by Ethernet using xPC Matlab tools. All procedures prepared on Host PC, after compiled were send and run on Target PC. Fig. 2 shows the structure of model-based fault detection procedure. The reference signal $r(t)$ was applied to the input of both the model and the plant controllers. Without disturbing signal ($f(t) = 0$), the control signals of both controllers $u_P(t)$ and $u_M(t)$ are almost the same. In such case the difference signal $d(t)$ is close to zero. The fractional plant was simulated by supercapacitor. It store the electric

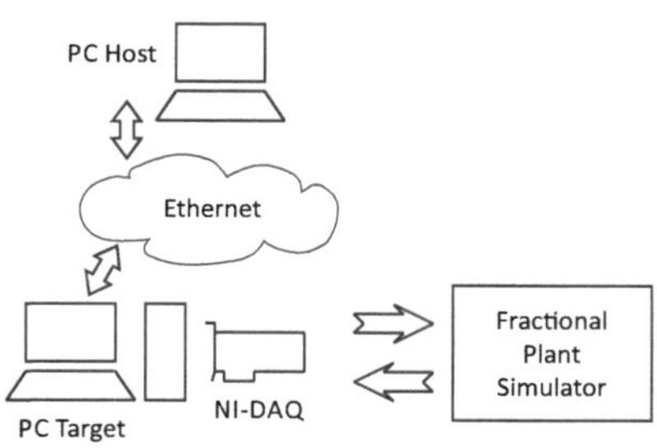

Fig. 1. Configuration of the measurement system

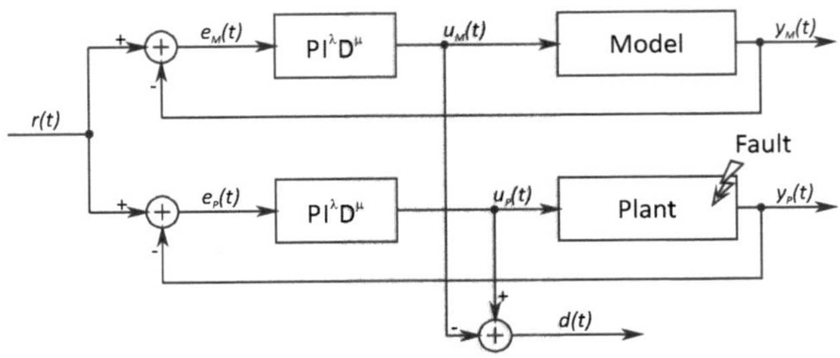

Fig. 2. Model-based fault detection mechanism

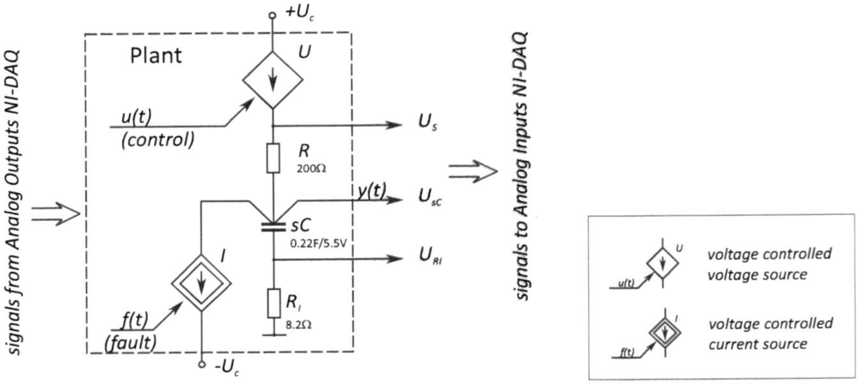

Fig. 3. Diagram of fractional plant with fault simulator

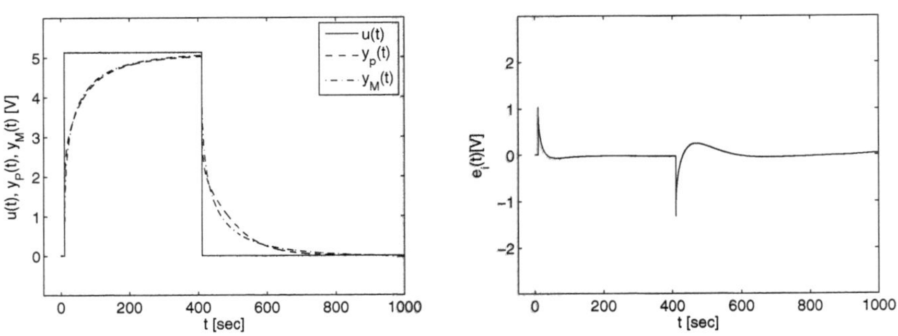

Fig. 4. Step response (a) of test setup and identified model and (b) identification error signal

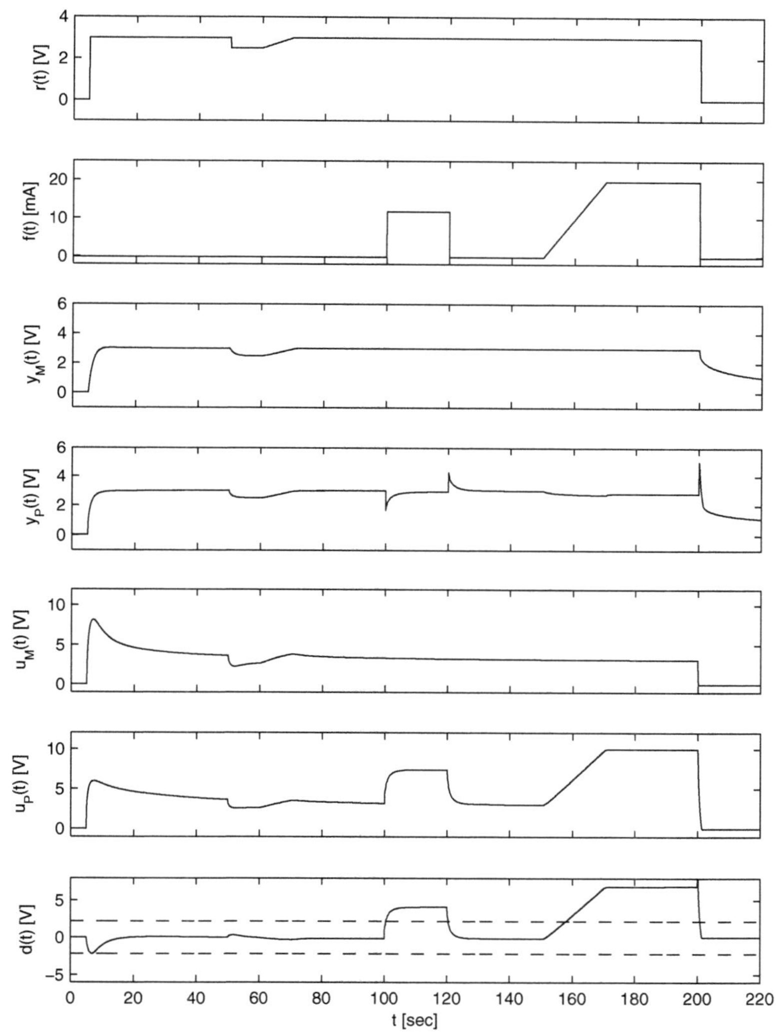

Fig. 5. Fault detection based on difference signal between plant and model controllers output

energy into an electrical double layer, which is formed at a porous solid electrode and electrolyte border. When a direct current voltage is applied, positive and negative ionic charges within the electrolyte are accumulated at the surface of the solid electrode[11]. Such special construction and specific materials generate a large capacity, which effects on its dynamics. The best approaches to mathematically modeling supercapacitor over a wide range of frequencies is using the fractional order calculus.

The electronic circuit of fractional plant with fault simulator was shown in Fig. 3. The voltage of supercapacitor was kept constant by voltage controlled voltage source. The fault was generated by voltage controlled current source discharged the supercapacitor. All voltages and currents were measured by analog inputs of NI-DAQ [8]. The experimental setup contained the supercapacitor of nominal capacity 0.22 F and nominal voltage 5.5 V.

4 Experimental Setup and Tests Results

Firstly the parametric identification based on discrete transfer function was performed. The method was precisely described in [3,12]. The comparison of the step response of continuous and identified discrete time model is presented in Fig. 4. As it can be seen the accuracy of the identification is very high. Next the identified model parameters were used to fault detection procedure according to diagram shown in Fig. 2. The test results are shown in Fig. 5.

Generated the fault signal discharging the supercapacitor, the plant controller had to generated the proportional controlling signal compensating the discharging effect, differ then the model controller. These resulted the difference signal $d(t)$ indicating fault occurrence. There were generated two types of fault signal $f(t)$: incipient and catastrophic. In both cases when the difference signal $d(t)$ crossing the threshold line, the fault was detected.

First curve shows the reference signal $r(t)$. It was applied to both model and plant controller. Next is the fault signal $f(t)$ steering the current source discharging the capacitor. Next two curves show the output signal of plant $y_P(t)$ end model $y_M(t)$, and next the output signal of plant $u_P(t)$ and model $u_M(t)$ controllers are presented respectively. Finally, the last curve shows the difference signal $d(t)$. Two dashed lines present the upper and lower limits. Crossed these lines by difference signal $d(t)$ was treated as fault.

5 Conclusion

The paper presents an example of the model based fault detection system with the model defined by fractional order equations. The plant was simulated by RC circuit with ultracapacitor charged by voltage controlled voltage source. The fault was generated by voltage controlled current source discharging supercapacitor. Using fractional calculus it is possible to obtain more accurate models. This is very crucial in model based method. Firstly the fractional order model of RC circuit with ultracapacitor was derived. The used transfer function method turned out to be good enough to plant identification. Next the model parameters were used to fault detection procedure. Presented fractional order model based method can be treat as a simple benchmark examples to test the new fault detection rules before applying it to real systems.

References

1. Blanke, M., Kinnaert, M., Lunze, J., Staroswiecki, M.: Diagnosis and Fault-Tolerant Control, 2nd edn. Springer, Berlin (2006)
2. Das, S.: Functional Fractional Calculus for System Identification and Controls. Springer, Berlin (2008)
3. Dzieliński, A., Sierociuk, D., Sarwas, G.: Some applications of fractional order calculus. Bulletin of the Polish Academy of Sciences 58(4), 583–592 (2010)
4. Isermann, R.: Model-based fault-detection and diagnosis status and applications. Annual Reviews in Control 29, 71–85 (2005)
5. Kaczorek, T.: Selected Problems of Fractional Systems Theory. Springer, Berlin (2011)
6. Kaczorek, T.: Singular fractional linear systems and electrical circuits. Int. J. Appl. Math. Comput. Sci. 21(2), 379–384 (2011)
7. Kopka, R.: Diagnostics of optical devices based on degradation processes. In: 7th Workshop on Advanced Control and Diagnosis, ACD 2009, Zielona Góra, November 19-20 (2009), http://www.issi.uz.zgora.pl/ACD_2009/program/Papers/87_ACD_2009.pdf
8. Kopka, R., Tarczynski, W.: Measurement system for supercapacitors equivalent parameters determination. Metrology and Measurement Systems XX(4), 581–590 (2013)
9. Korbicz, J., Kościelny, J.M., Kowalczuk, Z., Cholewa, W. (eds.): Fault Diagnosis. Springer, Berlin (2004)
10. Korbicz, J., Kościelny, J.M. (eds.): Modeling, Diagnostics and Process Control. Springer, Berlin (2010)
11. Li, Z., Chen, J.: An impedance-based approach to predict the state-of-charge for carbon-based supercapacitors. Microelectronic Engineering 85, 1549–1554 (2008)
12. Maiti D., Konar A.: Approximation of a Fractional Order System by an Integer Order Model Using Particle Swarm Optimization Technique. In: IEEE Sponsored Conference on Computational Intelligence, Control And Computer Visio In Robotics and Automation, IEEE CICCRA, pp. 149–152 (2008)
13. Monje, A.C., Chen, Y.Q., Vinagre, M.B., Xue, D., Feliu, V.: Fractional-order Systems and Controls. Springer, London (2010)
14. Noura, H., Theilliol, D., Ponsart, J.-C., Chamseddine, A.: Fault-tolerant Control Systems. Springer, London (2009)
15. Tepljakov, A.: Fractional-order Calculus based Identification and Control of Linear Dynamic Systems, Master thesis, Tallinn University of Technology, Tallinn (2011)
16. Stanisławski, R.: Advances in modeling of fractional difference systems - new accuracy, stability and computational results, Wydawnictwo Politechniki Opolskiej, Opole (2013)

Modeling of Elementary Fractional Order Plants at PLC SIEMENS Platform

Krzysztof Oprzedkiewicz, Marek Chochol, Waldemar Bauer, and Tomasz Meresinski

AGH University of Science and Technology,
Department of Automatics and Biomedical Engineering,
Al. Mickiewicza 30, 30-059 Krakow, Poland
{kop,mchochol,bauer}@agh.edu.pl,
tomasz.meresinski@gmail.com

Abstract. In the paper the problem of modeling elementary fractional order plants at PLC platform is presented. Models of considered plants were implemented at SIEMENS SIMATIC S7 300 platform with the use of STEP7 SCL language. Finally, comparison of simulation results and mathematical model is provided to demonstrate effectiveness of the proposed method.

Keywords: Fractional order plants, Oustaloup's approximation, PLC, Siemens.

1 Introduction

Past years brought many breakthroughs in many fields of science. Faster processors and hardware are able to compute advanced problems more efficiently. Thanks to artificial intelligence, self learning capabilities can be acquired. In control theory domain, all before mentioned elements are just tools, which improve algorithms in a sense of speed or additional capabilities. Most of current used solutions are based on rough but well optimized approximations of mathematical representations of real life problems. Although simpler models proved their value, next step for researchers in control theory, is to incorporate more accurate models to utilize subtle features that are characteristic during on-going processes. To tackle this idea, fractional calculus became an useful tool. Viscoelastic materials or anomalous diffusion processes can be described with the use of fractional calculus only. Although idea of non-integer order is not new, in past few years, many researches have been conducted [1] [2] [3]. Foundations of it are dated around end of seventieth century, but practical applications appeared around second half of twentieth century.

Main objective of this paper is to propose an implementation models elementary fractional order elements (derivative and integral) at PLC SIEMENS platform[4]. This paper is organized as follows. After introduction, a short description of basics behind the mathematical idea is described, together with

© Springer International Publishing Switzerland 2015
K.J. Latawiec et al. (eds.), *Advances in Modeling and Control of Non-integer Order Systems*,
Lecture Notes in Electrical Engineering 320, DOI: 10.1007/978-3-319-09900-2_25

fractional models of derivative and integral elements. Third part is a description of typical problems, which come with idea of hardware implementation fractional calculus, and Oustaloup's approximation. Next part is the Siemens PLC setup combined with functions and function blocks, which are proposed algorithms to solve predefined problem. Last two parts are simulation results and conclusion.

2 Fractional Calculus Basics

Starting point of fractional calculus idea is dated around 1695, when Leibniz and L'Hospital were discussing idea of derivatives of non-integer order and possible applications in various domains. First conclusions were indecisive and in 19th century, Liouville in his paper created strong foundations of fractional derivatives. Till second half of 20th century, idea of non-integer order was strictly theoretic, however in past few decades, fractionals proved to be adequate in solving more sophisticated problems.

In order to fully understand analysis of non-integer order, differentintegral operator ${}_a\mathbb{D}_t^\alpha$ needs to be explained.

2.1 Differintegral Operator

Differintegral [5] operator ${}_a\mathbb{D}_t^\alpha$ is an crucial element, while modeling dynamic objects.

Definition 1. *Differintegral is described as follows:*

$$ {}_a\mathbb{D}_t^\alpha f(t) = \begin{cases} \frac{d^\alpha}{dt^\alpha} & \alpha > 0 \\ 1 & \alpha = 0 \\ \int_a^t (d\tau)^\alpha & \alpha < 0 \end{cases} \tag{1} $$

where α is a real or complex order of derivative, a and t describe the range over which to compute the result.

There are three definitions of differintegral operator, which are: Caputo definition, Grünwald-Letnikov definition and Riemann-Liouville definition. In our paper, we would like to concentrate on the first one.

Definition 2. *Caputo differintegral*

$$ {}_a^C\mathbb{D}^\alpha f(t) = \frac{1}{\Gamma(\alpha - n)} \int_a^t \frac{f^{(n)}(\tau)\, d\tau}{(t-\tau)^{\alpha+1-n}} \tag{2} $$

where $n - 1 < \alpha < n$, and value of n is ceil of α and Γ is gamma function.

The above definition contains an element that must be discussed briefly, to understand whole idea of non-integer order calculus.

In addition to above definition, discrete form of fractional order derivative element should be presented.

Definition 3. *Fractional order, discrete derivative element*

Derivative element is described as follows:

$$y(t) = T_0 D^{\alpha} u(t) \tag{3}$$

where T_0 is constant parameter.

By applying approximation with backward difference:

$$y_n = h^{-\alpha} T_0 \sum_{j=0}^{n} \omega_j u_{n-j} \tag{4}$$

where h is sample time. By taking into account the short memory principle:

$$y_n = h^{-\alpha} T_0 \sum_{j=0}^{L} \omega_j u_{n-j} \tag{5}$$

2.2 Fractional Order Integrals

In agreement with Riemann-Liouville's conception, the notion of fractional order integral of order $\Re(\alpha) > 0$ is a natural consequence of Cauchy's formula for repeated integrals. This formula can be expressed as (see [3]):

Definition 4. *Fractional order integrals*

$$\mathbb{I}_t^{\alpha} f(t) = \mathbb{D}_c^{-\alpha} = \frac{1}{(\alpha - 1)!} \int_c^t (t-\tau)^{\alpha-1} f(\tau) \mathrm{d}\tau \tag{6}$$

where $t < c$ and $b \in \mathbb{Z}$.

As previously mentioned, discrete form of fractional order integral element has to be presented.

Definition 5. *Fractional order, discrete integral element*

Integral element is described as follows:

$$_0\mathbb{D}_t^{\alpha} y(t) = u(t) \tag{7}$$

By incorporating approximation with backward difference:

$$h^{-\alpha} \sum_{j=0}^{L} \omega_j y_{n-j} = u_n \tag{8}$$

After short transformations:

$$y_n = h^{\alpha} u_n - \sum_{j=1}^{L} \omega_j y_{n-j} \tag{9}$$

2.3 Fractional Order Transfer Function

In domain of control theory, transfer function is an element, which cannot be omitted. Transfer function, for fractional order analysis, just like differintegral operator can be described with three definitions. Once again, Caputo definition seems to be useful in most cases.

Definition 6. *Caputo transfer function*

$$\mathcal{L}\left\{{}_0\mathbb{D}_a^{\alpha} f(t)\right\} = s^{\alpha} F\left(s\right) - \sum_{k=0}^{n-1} s^{\alpha-k-1} f^{(k)}\left(0\right) \tag{10}$$

where n solves the $n - 1 \leq p < n$.

3 Oustaloup's Approximation

Before describing Oustaloup approximation method, possible issues of fractional calculus implementation in real-time applications need to be examined briefly [6]. Regular modeling with appropriate tools that depend on mathematical formalism developed by Riemann and Louville, can be applicable directly. Problem arises, when digital implementation on computer needs to be designed. Main concern lies in infinite memory, which from obvious reasons, cannot be implemented on recent machines. During calculation, integration from $-\infty$ to current point in time and non-integer differentiation of all past functions, takes place. Such instance is unrealistic and needs to be solved with the use of different approach.

As described previously, fractional calculus cannot be implemented directly at regular computer, because of its limitations. Non-integer integrator object requires infinite memory for calculation of historical data. However, band-limit implementation of fractional-order controller, can be represented by finite-dimensional approximation. For a s^{α}, where α is a real number, many approximations can be used. In this particular case, Oustaloup's approximation is described as follows [9]. It is often applied where a frequency band is considered within which the frequency domain responses should be fit by a bank of integer order filters to the fractional order derivative [9]. For $\alpha\colon 0 < \alpha < 1$ let us assume that frequency range to be fit is $[\omega_A, \omega_B]$, the term s/ω_u can be replaced by:

$$C_0 \frac{1 + s/\omega_b}{1 + s/\omega_h} \tag{11}$$

where

$$\sqrt{\omega_b \omega_h} = \omega_\mu \tag{12}$$

and

$$C_0 = \frac{\omega_b}{\omega_u} = \frac{\omega_u}{\omega_h} \tag{13}$$

Fitting quality around ω_b and ω_h may not be satisfactory. The Oustaloup's approximation model for s^α can be represented as follows [9]

$$\hat{H}(s) = \left(\frac{\omega_u}{\omega_h}\right)^\alpha \prod_{k=-N}^{N} \frac{1 + s/\omega_k'}{1 + s/\omega_k} \tag{14}$$

where

$$\omega_k' = \omega_b \left(\frac{\omega_h}{\omega_b}\right)^{\frac{k+N+1/2-\alpha/2}{2N+1}} \tag{15}$$

and

$$\omega_k = \omega_b \left(\frac{\omega_h}{\omega_b}\right)^{\frac{k+N+1/2+\alpha/2}{2N+1}} \tag{16}$$

are respectively the zeros and poles of rank. And $2N + 1$ is the total number of zeros or poles.

The ORA approximation was applied to check the correctness the build PLC models. Results will be presented in the example.

4 Implementation of Elementary Fractional Order Elements at Siemens PLC

At the Siemens PLC were implemented fractional modules:

- differentiator s^α
- integrator $\frac{1}{s^\alpha}$

Above elements were implemented on the Siemens PLC with CPU315. Sample time in order to calculate values by either integrator or differentiator module was not fixed and samples were calculated every 65ms with +/- 5ms margin. To collect results from the device a simple Scada system based on ProTool/Pro was used. Data was read by it with 500ms period and saved to .csv file. Data flow throughout algorithm can be described with the use of the block diagram show in figure 1. The source files of all elements of the program were written with the use of STEP7-SCL language, the whole program was assembled with the use of LD language in which Integrator and Differential block are incorporated.

The fractional-order differentiator described by discrete model (3)-(5) was implemented as function block FB4. The data of this FB were saved in Data Block DB4. Accordingly to that scheme, the fractional-order integrator, which was presented in equations (7)-(9), is computed in function block FB3, and data is stored in DB5. Actual history of signal is contained in data block DB2. Both

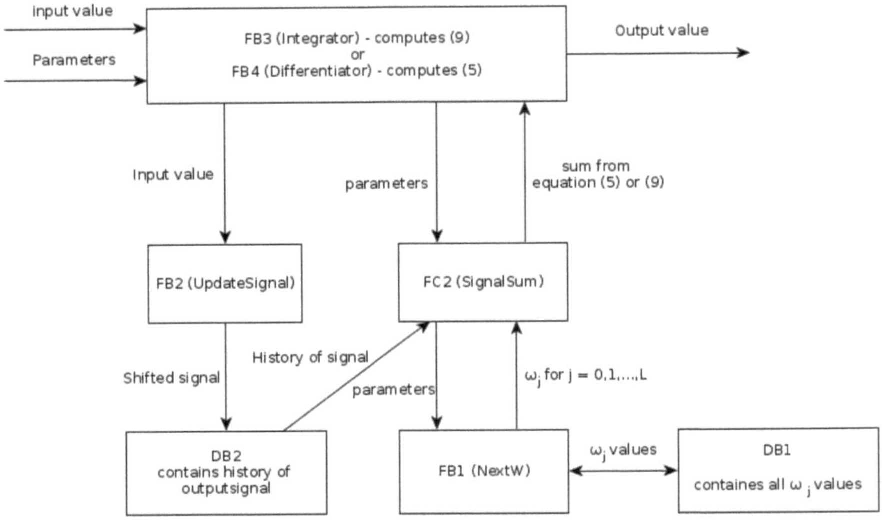

Fig. 1. PLC Program Block diagram

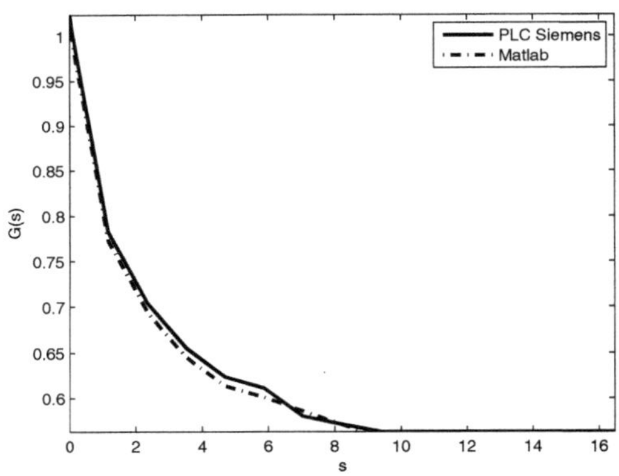

Fig. 2. Comparison of step response differentiation modules in Matlab and PLC implementation for $\alpha = 0.2$

function blocks use a group of common blocks. Firstly, function block FB1 is responsible for computing ω_j values from (5) and (9). It stores all computed values in DB1. Function FC2 computes $\sum_{j=a}^{L} \omega_j x_{n-j}$ where x is output signal with history of length L. Function block FB2 shifts signal from data block DB2 by one and adds new value to an end.

4.1 An Example

Tests of the implemented models were done via calculating the step response of each model and comparing it with the step response of the MATLAB model created with the use of ORA approximation.

Differentiator fractional module step response is shown in figures 2 and 3. In figures 5 and 4 step responses of fractional order integrator are shown. How can see step responses calculated with the use of MATLAB and PLC are similar.

SIEMENS SCL source code for differentiation module is presented below.

```
FUNCTION_BLOCK Differentiator

// Block Parameters
VAR_INPUT
  // Input Parameters
  in : REAL;
  alpha : REAL;
  h : INT;
  T : REAL;

END_VAR

VAR_OUTPUT
 // Output Parameters
    out : REAL;
END_VAR

VAR_TEMP
  // Temporary Variables
    END_VAR
VAR
  // Static Variables
  step : REAL := 0.0;
  value : REAL;

END_VAR
    step := EXP(alpha * LN(0.065));
    FB2.DB2(newValue := in); //updateSignal
    value := FC2(signal:= DB12.signal, ...
    length := 128, alpha := alpha, ...
    shift := 0); //SignalSum
    out := T * value / step; //equation (5)
  ;
END_FUNCTION_BLOCK
```

Source code in SCL language for integrator module is presented below.

```
FUNCTION_BLOCK Integrator

// Block Parameters
VAR_INPUT
  // Input Parameters
  in : REAL;
  alpha : REAL;
  h : INT;
  T : REAL;

END_VAR

VAR_OUTPUT
 // Output Parameters
    out : REAL;
END_VAR

VAR_TEMP
  // Temporary Variables
    value : REAL;
END_VAR
VAR
  // Static Variables
    step : REAL := 0.0;
END_VAR
    step := EXP(alpha * LN(0.07));
    value := FC2(signal:= DB2.signal, ...
    length := 128, alpha := alpha, ...
    shift := 1); //signalSum
    out := step * in - value; //equation (9)
    FB2.DB2(newValue := out); //updateSignal

END_FUNCTION_BLOCK
```

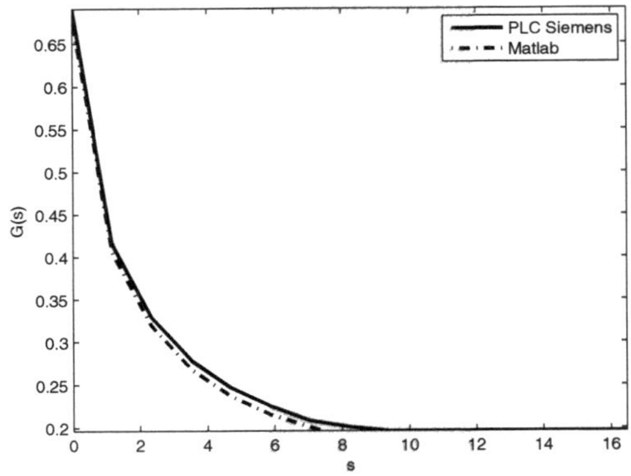

Fig. 3. Comparison of step response differentiator modules in Matlab and PLC implementation for $\alpha = 0.5$

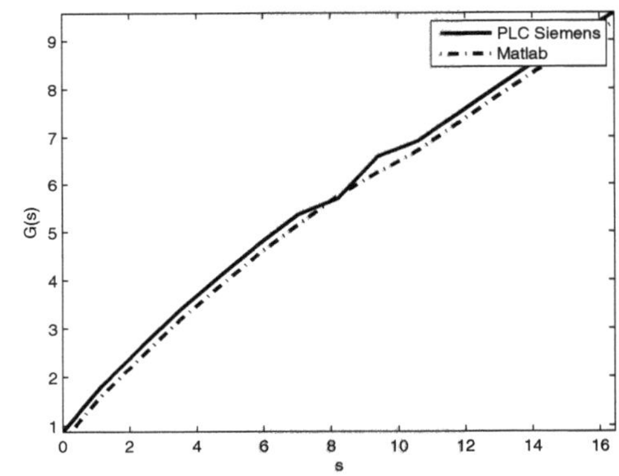

Fig. 4. Comparison of step response integral modules in Matlab and PLC implementation for $\alpha = 0.8$

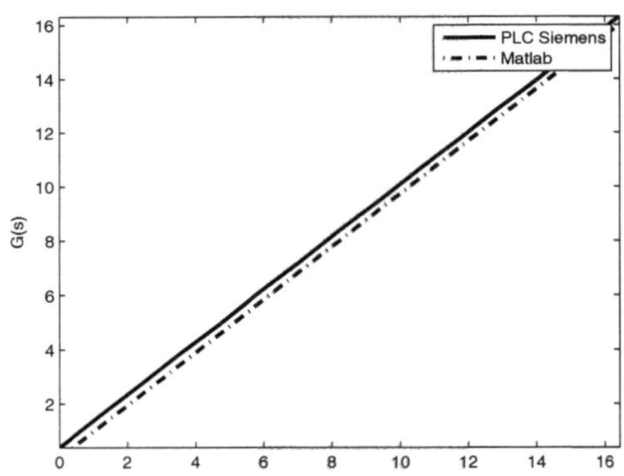

Fig. 5. Comparison of step response integral modules in Matlab and PLC implementation for $\alpha = 0.1$

5 Summary

Final conclusions to the paper can be formulated as follows:

- The elementary fractional order plants can be implemented at PLC platform with the use of normalized programming tools.
- The built elements can be applied to construct fractional order PID controllers.
- The another interesting problem is also the numerical complexity of built software. It can be tested with the use of methods proposed in [10]
- Differences between Matlab and SIEMENS PLC results are due to fact of different real number representations at both platforms. Additionally sampling time on Siemens PLC is not fixed, and it leads to small errors, which can be observed, when compared to Matlab fixed time response.

References

1. Manabe, S.: The non-integer integral and its application to control systems. JIEE (Japanese Institute of Electrical Engineers) Journal 80(860), 589–597 (1960)
2. Oustaloup, A.: Linear feedback control systems of fractional order between 1 and 2. In: Proc. of the IEEE Symposium on Circuit and Systems, Chicago, USA, vol. 4 (1981)
3. Chen, Y.Q., Moore, K.L.: Discretization schemes for fractionalorder differentiators and integrators. IEEE Trans. Circuits Syst. 49(3), 363–367 (2002)
4. Meresinski, T.: Modeling of elementary fractional order plants at PLC SIEMENS Platform, Bachelor of Science Thesis under supervision of K. Oprzedkiewicz, AGH EAIiIB 2013 (2013)
5. Podlubny, I.: Fractional Differential Equations. Academic Press, San Diego (1999)
6. Monje, C.A., Chen, Y., Vinagre, B.M., Xue, D., Feliu, V.: Fractional-order Systems and Controls, Fundamentals and Applications. Springer-Verlag London Limited (2010)
7. Siemens: SIMATIC Structured Control Language (SCL) for S7-300/S7-400 Programming
8. Siemens: SIMATIC Programming with STEP 7, 5 (2010)
9. Xue, D., Zhao, C., Chen, Y.: A Modified Approximation Method of Fractional Order System. In: Proceedings of the 2006 IEEE International Conference on Mechatronics and Automation, Luoyang, China, June 25 - 28 (2006)
10. Mitkowski, W., Oprzedkiewicz, K.: Application of fractional order time systems (in Polish), XLI KZM Zakopane (2012)

Space-Time-Fractional Advection Diffusion Equation in a Plane

Yuriy Povstenko

Institute of Mathematics and Computer Science,
Jan Dlugosz University in Częstochowa,
Armii Krajowej 13/15, 42-200 Częstochowa, Poland
j.povstenko@ajd.czest.pl

Abstract. The fundamental solution to the Cauchy problem for the space-time-fractional advection diffusion equation with the Caputo time-fractional derivative and Riesz fractional Laplace operator is considered in a case of two spatial variables. The solution is obtained using the Laplace integral transform with respect to time t and the double Fourier transform with respect to space variables x and y. Several particular cases of the solution are analyzed in details. Numerical results are illustrated graphically.

Keywords: Fractional calculus, advection diffusion equation, Caputo derivative, Riesz operator.

1 Introduction

The constitutive equation for the matter flux $\mathbf{j}$ (see, for example, [1])

$$\mathbf{j} = -a \operatorname{grad} c + \mathbf{v}c, \tag{1}$$

where $\mathbf{v}$ is the velocity vector, in combination with the balance equation for mass

$$\frac{\partial c}{\partial t} = -\operatorname{div} \mathbf{j} \tag{2}$$

gives

$$\frac{\partial c}{\partial t} = a\Delta c - \operatorname{div}(\mathbf{v}c) \tag{3}$$

or

$$\frac{\partial c}{\partial t} = a\Delta c - c \operatorname{div} \mathbf{v} - \mathbf{v} \cdot \operatorname{grad} c. \tag{4}$$

Supposing $\mathbf{v} = \text{const}$ (or $\operatorname{div} \mathbf{v} = 0$), we obtain the standard advection diffusion equation for the concentration

$$\frac{\partial c}{\partial t} = a\Delta c - \mathbf{v} \cdot \operatorname{grad} c. \tag{5}$$

Similarly to analysis of nonlocal generalizations of the classical Fick or Fourier law carried out in [2], [3], [4], [5] the time-nonlocal generalization of Eq. (1) can

© Springer International Publishing Switzerland 2015

K.J. Latawiec et al. (eds.), *Advances in Modeling and Control of Non-integer Order Systems*,
Lecture Notes in Electrical Engineering 320, DOI: 10.1007/978-3-319-09900-2_26

be considered. For example, the general time-nonlocal constitutive equation for the matter flux can be written as

$$\mathbf{j}\,(t) = \int_0^t K(t-\tau)\,[-a\,\mathrm{grad}\,c(\tau) + \mathbf{v}c(\tau)]\,\mathrm{d}\tau \tag{6}$$

and under the same assumptions as above leads to the advection diffusion equation with the general memory kernel $K(t-\tau)$:

$$\frac{\partial c}{\partial t} = \int_0^t K(t-\tau)\,[a\,\Delta\,c(\tau) - \mathbf{v}\cdot \mathrm{grad}\,c(\tau)]\,\mathrm{d}\tau. \tag{7}$$

The time-nonlocal constitutive equation for the matter flux with the "long-tail" power kernel

$$\mathbf{j}\,(t) = D^{1-\alpha}\,[-a\,\mathrm{grad}\,c(t) + \mathbf{v}c(t)]\,, \qquad 0<\alpha\leq 1, \tag{8}$$

results in the time-fractional advection diffusion equation

$$\frac{\partial^\alpha c}{\partial t^\alpha} = a\,\Delta c - \mathbf{v}\cdot \mathrm{grad}\,c. \tag{9}$$

Here $D^\alpha f(t)$ and $\frac{\mathrm{d}^\alpha f(t)}{\mathrm{d}t^\alpha}$ are the Riemann-Liouville and Caputo time-fractional derivatives, respectively [6], [7], [8]:

$$D^\alpha f(t) = \frac{\mathrm{d}^n}{\mathrm{d}t^n}\left[\frac{1}{\Gamma(n-\alpha)}\int_0^t (t-\tau)^{n-\alpha-1} f(\tau)\,\mathrm{d}\tau\right], \qquad n-1<\alpha<n, \tag{10}$$

$$\frac{\mathrm{d}^\alpha f(t)}{\mathrm{d}t^\alpha} = \frac{1}{\Gamma(n-\alpha)}\int_0^t (t-\tau)^{n-\alpha-1}\frac{\mathrm{d}^n f(\tau)}{\mathrm{d}\tau^n}\,\mathrm{d}\tau, \qquad n-1<\alpha<n\,, \tag{11}$$

where $\Gamma(\alpha)$ is the gamma function.

The space-time-fractional advection diffusion equation with the Caputo time-fractional derivative and the Riesz fractional Laplace operator has the following form:

$$\frac{\partial^\alpha c}{\partial t^\alpha} = -a(-\Delta)^{\beta/2}c - \mathbf{v}\cdot \mathrm{grad}\,c. \tag{12}$$

In this paper we will restrict ourselves to the orders of fractional derivatives $0<\alpha\leq 1$ and $1\leq\beta\leq 2$.

Space-time-fractional generalizations of advection diffusion equation were studied by many authors [9], [10], [11], [12], [13], [14]. In the papers mentioned above, one spatial coordinate was considered and several numerical schemes for solving the problems were proposed. In the present paper, we obtain analytical solution of the space-time-fractional advection diffusion equation in the case of two space variables.

2 Equation with Two Spatial Variables

Consider the space-time-fractional advection diffusion equation with two spatial variables

$$\frac{\partial^\alpha c}{\partial t^\alpha} = -a(-\Delta)^{\beta/2} c - v_x \frac{\partial c}{\partial x} - v_y \frac{\partial c}{\partial y},$$
$$-\infty < x < \infty, \quad -\infty < y < \infty, \quad 0 < t < \infty, \tag{13}$$

under the initial condition

$$t = 0: \quad c = p_0\, \delta(x)\, \delta(y), \tag{14}$$

where $\delta(x)$ is the Dirac delta function. The constant muptiplier p_0 in Eq. (14) has been introduced to obtain the nondimensional quantity $\bar{c}$ used in numerical calculations (see Eq. (35)).

It should be noted that the cumbersome aspects of space-fractional differential operators disappear when one computes their integral transforms. The two-dimensional fractional Laplace operator is defined as an operator having the following double Fourier transform with respect to the spatial coordinates x and y [8]:

$$\mathcal{F}\left\{-(-\Delta)^{\beta/2} f(x,y)\right\} = -\left(\xi^2 + \eta^2\right)^{\beta/2} \widetilde{f}(\xi,\eta), \tag{15}$$

where the tilde denotes the transform, ξ and η are the Fourier transform variables:

$$\mathcal{F}\left\{f(x,y)\right\} = \widetilde{f}(\xi,\eta) = \frac{1}{2\pi}\int_{-\infty}^{\infty}\int_{-\infty}^{\infty} f(x,y)\, \mathrm{e}^{ix\xi + iy\eta}\, \mathrm{d}x\, \mathrm{d}y, \tag{16}$$

$$\mathcal{F}^{-1}\left\{\widetilde{f}(\xi,\eta)\right\} = f(x,y) = \frac{1}{2\pi}\int_{-\infty}^{\infty}\int_{-\infty}^{\infty} \widetilde{f}(\xi,\eta)\, \mathrm{e}^{-ix\xi - iy\eta}\, \mathrm{d}\xi\, \mathrm{d}\eta. \tag{17}$$

It is obvious that Eq. (15) is a fractional generalization of the standard formula for the Fourier transform of the classical Laplace operator corresponding to $\beta = 2$:

$$\mathcal{F}\left\{\Delta\, f(x,y)\right\} = -\left(\xi^2 + \eta^2\right) \widetilde{f}(\xi,\eta). \tag{18}$$

Recall that for its Laplace transform rule the Caputo fractional derivative requires the knowledge of the initial values of the function $f(t)$ and its integer derivatives of order $k = 1, 2, \ldots, n-1$:

$$\mathcal{L}\left\{\frac{\mathrm{d}^\alpha f(t)}{\mathrm{d}t^\alpha}\right\} = s^\alpha f^*(s) - \sum_{k=0}^{n-1} f^{(k)}(0^+) s^{\alpha-1-k}, \quad n-1 < \alpha < n, \tag{19}$$

where the asterisk denotes the Laplace transform with respect to time and s is the transform variable.

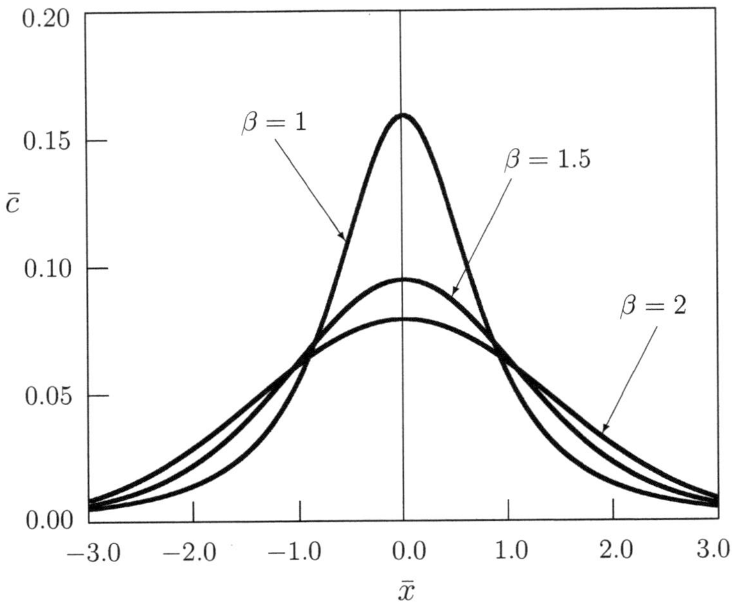

Fig. 1. Dependence of the solution on distance ($\alpha = 1$, $\bar{v}_x = \bar{v}_y = 0$)

The integral transform technique results in the solution of the Cauchy problem (13)–(14) in the transform domain

$$\widetilde{c}^*(\xi, \eta, s) = \frac{p_0}{2\pi} \frac{s^{\alpha-1}}{s^\alpha + a(\xi^2 + \eta^2)^{\beta/2} - i(v_x \xi + v_y \eta)}. \tag{20}$$

After inverting the integral transforms we get

$$c(x, y, t) = \frac{p_0}{4\pi^2} \int_{-\infty}^{\infty} \int_{-\infty}^{\infty} E_\alpha \left\{ -\left[a\left(\xi^2 + \eta^2\right)^{\beta/2} - i(v_x \xi + v_y \eta) \right] t^\alpha \right\} \times \mathrm{e}^{-ix\xi - iy\eta} \, \mathrm{d}\xi \, \mathrm{d}\eta, \tag{21}$$

where $E_\alpha(z)$ is the Mittag-Leffler function

$$E_\alpha(z) = \sum_{n=0}^{\infty} \frac{z^n}{\Gamma(\alpha n + 1)}, \quad \alpha > 0, \quad z \in C, \tag{22}$$

and the following formula [6], [7], [8]

$$\mathcal{L}^{-1} \left\{ \frac{s^{\alpha-1}}{s^\alpha + b} \right\} = E_\alpha \left(-b t^\alpha \right) \tag{23}$$

has been used.

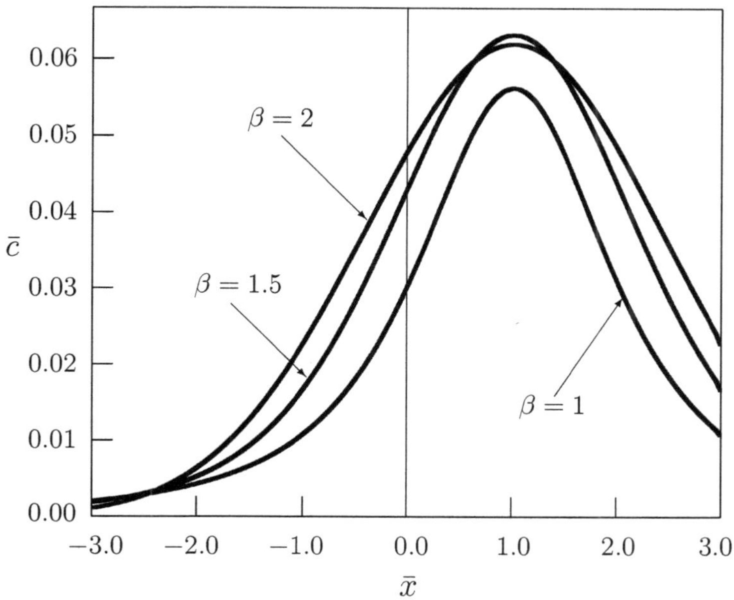

Fig. 2. Dependence of the solution on distance ($\alpha = 1$, $\bar{v}_x = \bar{v}_y = 1$)

3 Particular Cases of the Solution

If $\alpha = 1$, then

$$E_1(-z) = \mathrm{e}^{-z} \tag{24}$$

and

$$c(x,y,t) = \frac{p_0}{\pi^2} \int_0^\infty \int_0^\infty \exp\left[-a\left(\xi^2+\eta^2\right)^{\beta/2} t\right] \times \cos\left[(x-v_x t)\xi\right] \cos\left[(y-v_y t)\eta\right] \mathrm{d}\xi\, \mathrm{d}\eta. \tag{25}$$

Introducing the polar coordinate system in the (ξ, η)-plane

$$\xi = \rho\cos\theta, \quad \eta = \rho\sin\theta, \tag{26}$$

we obtain

$$c(x,y,t) = \frac{p_0}{\pi^2} \int_0^\infty \rho\, \mathrm{e}^{-at\rho^\beta}\, \mathrm{d}\rho \int_0^{\pi/2} \cos\left[(x-v_x t)\rho\cos\theta\right] \times \cos\left[(y-v_y t)\rho\sin\theta\right] \mathrm{d}\theta. \tag{27}$$

Substitution $w = \sin\theta$ with taking into account the integral (A.1) from Appendix allows us to arrive at the final result

$$c(x,y,t) = \frac{p_0}{2\pi} \int_0^\infty \mathrm{e}^{-at\rho^\beta}\, J_0\left[\rho\sqrt{(x-v_x t)^2 + (y-v_y t)^2}\right] \rho\, \mathrm{d}\rho. \tag{28}$$

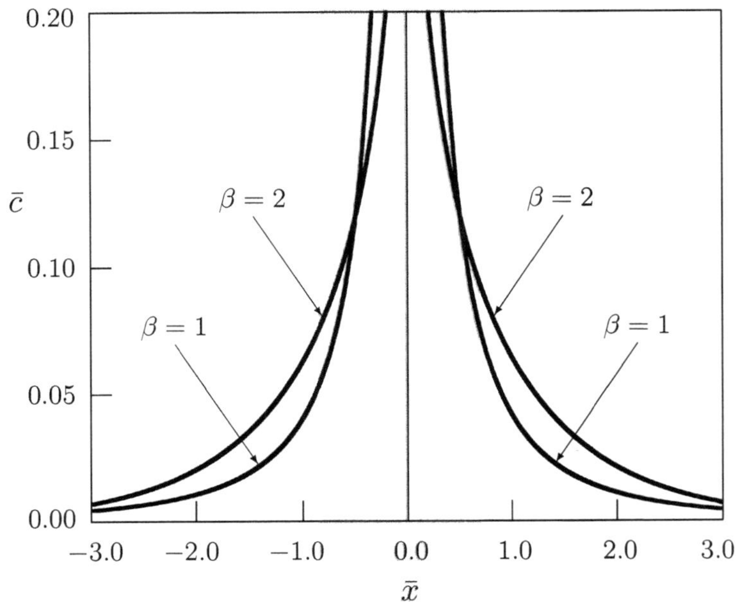

Fig. 3. Dependence of the solution on distance ($\alpha = 0.5$, $\bar{v}_x = \bar{v}_y = 0$)

The solution (28) simplifies in the following two particular cases:

i) $\alpha = 1$, $\beta = 2$

$$c(x, y, t) = \frac{p_0}{4\pi a t} \exp\left[-\frac{(x - v_x t)^2 + (y - v_y t)^2}{4at}\right]. \tag{29}$$

We have used Eq. (A.2) from Appendix to get the solution (29).

ii) $\alpha = 1$, $\beta = 1$

$$c(x, y, t) = \frac{p_0}{2\pi} \frac{at}{\left[a^2 t^2 + (x - v_x t)^2 + (y - v_y t)^2\right]^{3/2}}, \tag{30}$$

where Eq. (A.3) from Appendix has been used.

Another typical particular case is obtained for $\alpha = 1/2$. The corresponding Mittag-Leffler function can be expressed as

$$E_{1/2}(-z) = \frac{2}{\sqrt{\pi}} \int_0^\infty \mathrm{e}^{-u^2 - 2uz}\, \mathrm{d}u \tag{31}$$

and the solution reads

$$\begin{aligned} c(x, y, t) = \frac{p_0}{\pi^{3/2}} \int_0^\infty \int_0^\infty \exp\left(-u^2 - 2a\sqrt{t}u\rho^\beta\right) \\ \times J_0\left[\rho\sqrt{(x - 2uv_x\sqrt{t})^2 + (y - 2uv_y\sqrt{t})^2}\right] \rho\, \mathrm{d}\rho\, \mathrm{d}u \end{aligned} \tag{32}$$

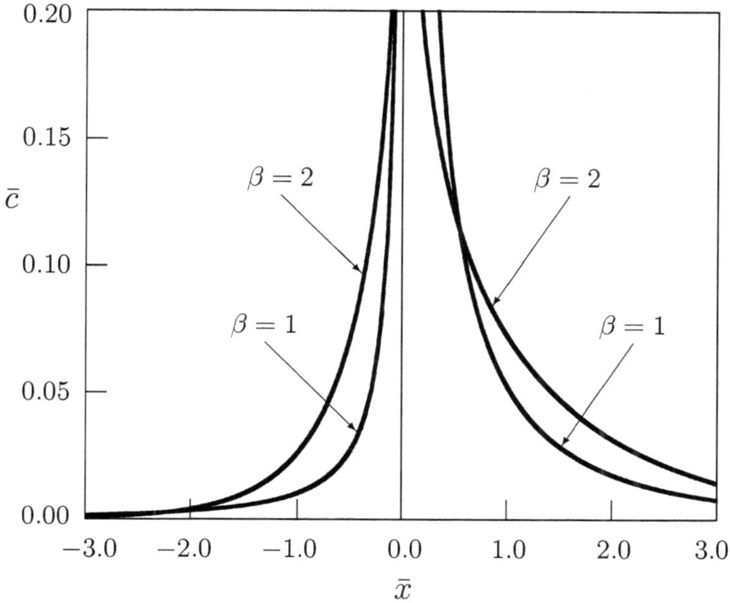

Fig. 4. Dependence of the solution on distance ($\alpha = 0.5$, $\bar{v}_x = \bar{v}_y = 1$)

with two particular cases:

i) $\alpha = 1/2$, $\beta = 2$

$$c(x,y,t) = \frac{p_0}{4a\sqrt{t}\pi^{3/2}} \int_0^\infty \frac{1}{u}\,\mathrm{e}^{-u^2} \times \exp\left[-\frac{(x-2uv_x\sqrt{t})^2 + (y-2uv_y\sqrt{t})^2}{8a\sqrt{t}u}\right] \mathrm{d}u. \tag{33}$$

i) $\alpha = 1/2$, $\beta = 1$

$$c(x,y,t) = \frac{2p_0 a\sqrt{t}}{\pi^{3/2}} \int_0^\infty u\,\mathrm{e}^{-u^2} \times \frac{1}{\left[(2a\sqrt{t}u)^2 + (x-2uv_x\sqrt{t})^2 + (y-2uv_y\sqrt{t})^2\right]^{3/2}}\,\mathrm{d}u. \tag{34}$$

The numerical results for the solution c are shown in Figures 1–6 for various values of the orders of fractional derivatives α and β and for various values of the drift parameter $\mathbf{v}$. In numerical calculations we have used the following nondimensional quantities:

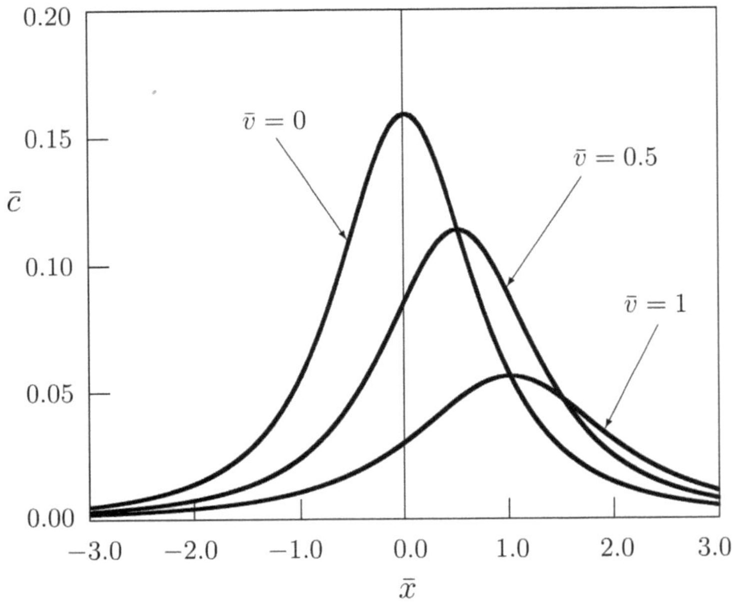

Fig. 5. Dependence of the solution on distance ($\alpha = 1$, $\beta = 1$; $\bar{v}_x = \bar{v}_y = \bar{v}$)

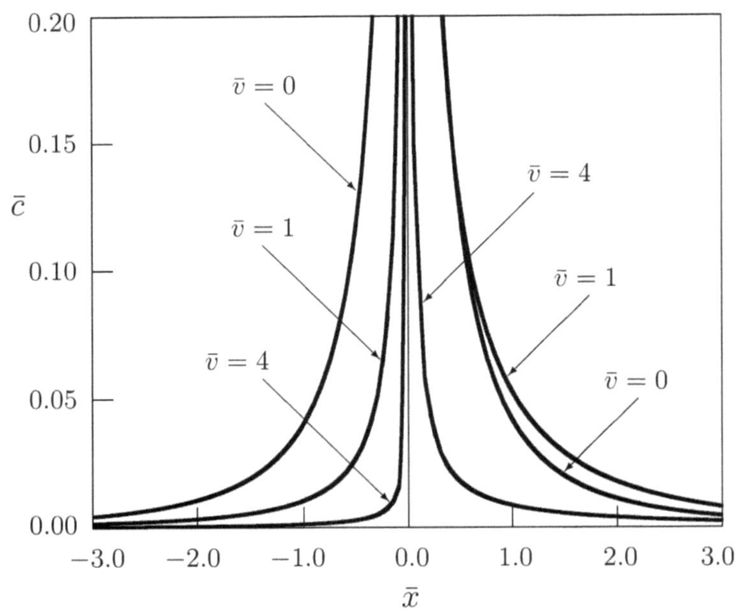

Fig. 6. Dependence of the solution on distance ($\alpha = 0.5$, $\beta = 1$; $\bar{v}_x = \bar{v}_y = \bar{v}$)

$$\bar{c} = \frac{a^{2/\beta} t^{2\alpha/\beta}}{p_0}\, c, \quad \bar{x} = \frac{x}{a^{1/\beta} t^{\alpha/\beta}}, \quad \bar{y} = \frac{y}{a^{1/\beta} t^{\alpha/\beta}},$$
$$\bar{v}_x = \frac{t^{\alpha(\beta-1)/\beta}}{a^{1/\beta}}\, v_x, \quad \bar{v}_y = \frac{t^{\alpha(\beta-1)/\beta}}{a^{1/\beta}}\, v_y. \tag{35}$$

In Figures we have taken $\bar{y} = 0$ and have shown the dependence of the solution on the coordinate $\bar{x}$.

4 Concluding Remarks

We have analyzed the solution of the space-time-fractional advection diffusion equation with the Caputo time-derivative of the order $0 < \alpha \le 1$ anf the Riesz fractional Laplace operator of the order $1 \le \beta \le 2$ in a plane. It should be emphasized that the fundamental solution to the Cauchy problem in the case $0 < \alpha < 1$ has the logarithmic singularity at the origin for all values of the order of the fractional Laplace operator $1 \le \beta \le 2$. Such a singularity disappears only for the advection diffusion equation with $\alpha = 1$. Comparison of the corresponding Figures shows that due to the logarithmic singularity of the solution at the origin for $0 < \alpha < 1$ the influence of the drift parametr $\mathbf{v}$ is quite different than in the case $\alpha = 1$. In the last case, the increase of $\mathbf{v}$ results in decrease of the maximum value of the solution and in its drift. For $0 < \alpha < 1$, the increase of $\mathbf{v}$ leads to approaching of the corresponding curves to the asymptote $\bar{x} = 0$.

Appendix

The following integrals [15], [16] are used in the paper:

$$\int_0^1 \frac{1}{\sqrt{1-x^2}} \cos\left(p\sqrt{1-x^2}\right) \cos(qx)\, \mathrm{d}x = \frac{\pi}{2}\, J_0\left(\sqrt{p^2+q^2}\right), \tag{A.1}$$

$$\int_0^\infty \mathrm{e}^{-px^2} J_0(qx)\, x\, \mathrm{d}x = \frac{1}{2p} \exp\left(-\frac{q^2}{4p}\right), \quad p > 0, \quad q > 0, \tag{A.2}$$

$$\int_0^\infty \mathrm{e}^{-px} J_0(qx)\, x\, \mathrm{d}x = \frac{p}{(p^2+q^2)^{3/2}}, \quad p > 0, \quad q > 0, \tag{A.3}$$

where $J_0(z)$ is the Bessel function of the first kind.

References

1. Kaviany, M.: Principles of Heat Transfer in Porous Media, 2nd edn. Springer, New York (1995)
2. Povstenko, Y.: Fractional Heat Conduction Equation and Associated Thermal Stress. J. Thermal Stresses 28, 83–102 (2005)

3. Povstenko, Y.: Thermoelasticity that Uses Fractional Heat Conduction Equation. J. Math. Sci. 162, 296–305 (2009)
4. Povstenko, Y.: Theory of Thermoelasticity Based on the Space-Time-Fractional Heat Conduction Equation. Phys. Scr. T(136), 014017 (2009)
5. Povstenko, Y.: Theories of Thermal Stresses Based on Space-Time-Fractional Telegraph Equations. Comp. Math. Appl. 64, 3321–3328 (2012)
6. Gorenflo, R., Mainardi, F.: Fractional Calculus: Integral and Differential Equations of Fractional Order. In: Carpinteri, A., Mainardi, F. (eds.) Fractals and Fractional Calculus in Continuum Mechanics, pp. 223–276. Springer, Wien (1997)
7. Podlubny, I.: Fractional Differential Equations. Academic Press, San Diego (1999)
8. Kilbas, A.A., Srivastava, H.M., Trujillo, J.J.: Theory and Applications of Fractional Differential Equations. Elsevier, Amsterdam (2006)
9. Zaslavsky, G.M., Edelman, M., Niyazov, B.A.: Self-similarity, Renormalization, and Phase Space Nonuniformity of Hamiltonian Chaotic Dynamics. Chaos 7, 159–181 (1997)
10. Saichev, A.I., Zaslavsky, G.M.: Fractional Kinetic Equations: Solutions and Applications. Chaos 7, 753–764 (1997)
11. Kolwankar, K.M., Gangal, A.D.: Local Fractional Fokker-Planck Equation. Phys. Rev. Lett. 80, 214–217 (1998)
12. Liu, F., Zhuang, P., Anh, V., Turner, I., Burrage, K.: Stability and Convergence of the Difference Methods for the Space-Time Fractional Advection-Diffusion Equation. Appl. Math. Comput. 191, 12–20 (2007)
13. Yıldırım, A., Koçak, H.: Homotopy Perturbation Method for Solving the Space Time Fractional Advection-Dispersion Equation. Adv. Water Res. 32, 1711–1716 (2009)
14. Abdel-Rehim, E.A.: Explicit Approximation Solutions and Proof of Convergence of Space-Time Fractional Advection Dispersion Equations. Appl. Math. 4, 1427–1440 (2013)
15. Prudnikov, A.P., Brychkov, Y.A., Marichev, O.I.: Integrals and Series. Elementary Functions, vol. 1. Gordon and Breach, Amsterdam (1986)
16. Prudnikov, A.P., Brychkov, Y.A., Marichev, O.I.: Integrals and Series. Special Functions, vol. 2. Gordon and Breach, Amsterdam (1986)

Variable Order Fractional Kalman Filters for Estimation over Lossy Network*

Dominik Sierociuk and Pawel Ziubinski

Institute of Control and Industrial Electronics
Warsaw University of Technology
Koszykowa 75, 00-662 Warsaw, Poland
{dsieroci,pawel.ziubinski}@ee.pw.edu.pl

Abstract. This paper presents generalization of Fractional Variable Order Kalman Filter (FvoKF) and Improved Fractional Variable Order Kalman Filter (ExFvoKF) algorithms for estimation of fractional variable order state-space systems over lossy network. This generalization is obtained for a state-space system based on one type of fractional variable order difference ($\mathcal{A}$-type) and assuming the knowledge about packets losing. The generalization of ExFKF algorithm based on the infinite dimensional form of a linear discrete fractional variable order state-space system and measurements equation augmentation. It required less restrictive assumptions than derivation of FKF. Finally, numerical simulations of proposed algorithms are presented and compared.

1 Introduction

The Kalman filter has played an important role in systems theory and has found wide applications in many fields such as signal processing, control, and communications. Recently there has been considerable interest in estimation and control over communication networks, so researches show how traditional Kalman filter or estimators can be applied to this kind of systems. In particular, there has been considerable effort in analyzing the effect of variable and random delays and packet loss (or dropout), such as in [1,2] or the effect of only packet loss such as in [3]. More detailed investigation of estimation and control issue over communication networks is presented in [4].

On the other hand, the fractional calculus, a natural generalization of traditional differential calculus for a case of derivatives and integrals in fractional (non-integer) orders, was found as a very efficient tool for modeling, control and signal processing. In [5], results of successful modeling for heat transfer process in heterogeneous materials were presented. In signal processing it is used, for example, to obtain new filter algorithms [6] and to describe complex noises dynamics [7]. In [8] generalization of traditional Kalman Filter algorithm for Discrete Fractional State Space System was proposed. Article [9] presents extension of this

* This work was supported by the Polish National Science Center with the decision number DEC-2011/03/D/ST7/00260.

© Springer International Publishing Switzerland 2015
K.J. Latawiec et al. (eds.), *Advances in Modeling and Control of Non-integer Order Systems*,
Lecture Notes in Electrical Engineering 320, DOI: 10.1007/978-3-319-09900-2_27

algorithm with least restrictive assumption. Moreover, it also presents generalization of this algorithm for estimation over lossy communication network. In [10] the use of fractional order Kalman Filter for state estimation in the case of existing fractional order noise was presented.

The case, when order of the derivative is changing in time, becomes recently intensively developed. In an article [11] the variable order equations are using to describe a history of drag expression. The fractional variable order algorithms can also be used to obtain variable order fractional noise [7], and to obtain new control algorithms [12]. In papers [13,14] the variable order interpretation of the analog realization of fractional orders integrators, realized as domino ladders was presented.

In this paper we present generalization of FKF and ExFKF algorithms, given in [9], for the Discrete Variable Order State Space Systems with measurements obtained by lossy communication network. Both algorithms based on one type of fractional variable order difference ($\mathcal{A}$-type). Generalization of ExFKF algorithm based on state vector augmentation, and require less restrictive assumptions to state and covariance matrix prediction than for FKF.

This article is organized as follows. In Section 2 the Discrete Fractional Variable Order State Space System is recall. In Section 3, the main contribution, generalization of FKF and ExFKF algorithms for Discrete Variable Order State Space System with measurements obtained by lossy communication network are proposed. In Section 4 numerical results for proposed algorithms are presented and compared.

2 Discrete Fractional Variable Order State Space System

The following definition constitutes a starting point for generalization of constant fractional order difference operators onto a variable order case. A constant fractional order difference operator is defined in the following way

$$\Delta^{\alpha} x_k \equiv \sum_{j=0}^{k} (-1)^j \binom{\alpha}{j} x_{k-j} \tag{1}$$

where

$$\binom{\alpha}{j} \equiv \begin{cases} 1 & \text{for} \quad j = 0\,, \\ \frac{\alpha(\alpha-1)\ldots(\alpha-j+1)}{j!} & \text{for} \quad j > 0\,. \end{cases}$$

At least 3 different definitions of variable order difference operator can be found in literature [15,16]. In this article we use only the first type, which is obtained by replacing a constant order α by variable order α_k.

Definition 1. *The $\mathcal{A}$-type of fractional variable-order difference is given by*

$$^{\mathcal{A}}\Delta^{\alpha_k} x_k \equiv \sum_{j=0}^{k} (-1)^j \binom{\alpha_k}{j} x_{k-j} \tag{2}$$

For such a definition, the discrete fractional $\mathcal{A}$-type variable order system will be defined as follows:

Definition 2. *[17]The linear Discrete Fractional Variable Order System in state-space representation, based on $\mathcal{A}$-type of fractional variable-order difference, is given as follows:*

$$ {}^{\mathcal{A}}\Delta^{\Upsilon_{k+1}} x_{k+1} = Ax_k + Bu_k + \omega_k \tag{3} $$

$$ y_k = Cx_k + \nu_k, \tag{4} $$

where

$$ x_{k+1} = {}^{\mathcal{A}}\Delta^{\Upsilon_{k+1}} x_{k+1} - \sum_{j=1}^{k+1} (-1)^j \Upsilon_{j,k+1} x_{k-j+1} \tag{5} $$

$$ \Upsilon_{j,k} = \text{diag}\left[\binom{\alpha_{1,k}}{j} \cdots \binom{\alpha_{N,k}}{j} \right] $$

$$ {}^{\mathcal{A}}\Delta^{\Upsilon_{k+1}} x_{k+1} = \begin{bmatrix} {}^{\mathcal{A}}\Delta^{\alpha_{1,k+1}} x_{1,k+1} \\ \vdots \\ {}^{\mathcal{A}}\Delta^{\alpha_{N,k+1}} x_{N,k+1} \end{bmatrix} $$

and $x_k \in \mathbb{R}^N$ is a state vector, $\alpha_{i,k} \in \mathbb{R}$ are time dependent (variable) orders of system equations (where i is a number of state variable and k is a time of the order) and N is the number of these equations.

□

More properties for the case of constant order (DFOSS) are presented in [18,19].

The system given by the Definition 2 can be rewritten as an m-finite dimensional form in the following way

Definition 3. *The linear discrete fractional variable order state-space system in m-finite form (the form with the new $m-th$ elements state vector) is defined as follows*

$$ \begin{aligned} \mathbb{X}_{k+1} &= \mathbb{A}_{m,k}\mathbb{X}_k + \mathbb{B}_m u_k + \mathbb{I}\omega_k \\ &\quad - \mathbb{I} \sum_{j=m+1}^{k+1} (-1)^j \Upsilon_{j,k} x_{k+1-j} \end{aligned} \tag{6} $$

$$ y_k = \mathbb{C}_m \mathbb{X}_k \tag{7} $$

where

$$ \mathbb{X}_k = \begin{bmatrix} x_k \\ x_{k-1} \\ \vdots \\ x_{k-m+1} \end{bmatrix}, \quad \mathbb{I} = \begin{bmatrix} I \\ 0 \\ \vdots \\ 0 \end{bmatrix} $$

$$\mathbb{A}_{m,k} = \begin{bmatrix} (A+\Upsilon_{1,k}) & -(-1)^2\Upsilon_{2,k} & \dots & -(-1)^m\Upsilon_{m,k} \\ I & \dots & 0 & 0 \\ \vdots & \vdots & \vdots & \vdots \\ 0 & \dots & I & 0 \end{bmatrix}$$

$$\mathbb{B}_m = \begin{bmatrix} B\, 0 \dots 0 \end{bmatrix}^T, \quad \mathbb{C}_m = \begin{bmatrix} C\, 0 \dots 0 \end{bmatrix}$$

being $I \in \mathbb{R}^{N\times N}$ the identity matrix.

■

2.1 Lossy and Delayed Network

For the case of measurements over lossy network some part of packets are losing during transmission, what has negative influence to efficiency of the estimation process. In order to improve estimation algorithms not only measurements values are needed but also information about packets losing γ_k is needed. The $\gamma_k \in \{1, 0\}$ has value 1 when packet y_k is obtained, and 0 when y_k is lost.

3 Fractional Kalman Filter over Lossy Networks (gFvoKF)

The Kalman filter for integer order systems with intermittent observations (packet loss) was presented in [20,21]. Now, extensions of the fractional Kalman filter algorithms for fractional order networked systems will be introduced.

Theorem 1. *For the discrete $\mathcal{A}$-type fractional variable order system, given by the Definition 2, with intermittent observations the simplified Kalman filter algorithm (called gFvoKF) is given by the following set of equations*

$$^{\mathcal{A}}\Delta^{\Upsilon_{k+1}}\tilde{x}_{k+1} = A\hat{x}_k + Bu_k \tag{8}$$

$$\tilde{x}_{k+1} = {}^{\mathcal{A}}\Delta^{\Upsilon_{k+1}}\tilde{x}_{k+1} - \sum_{j=1}^{k+1}(-1)^j \Upsilon_{j,k+1}\hat{x}_{k+1-j} \tag{9}$$

$$\tilde{P}_k = (A+\Upsilon_{1,k})\, P_{k-1}\, (A+\Upsilon_{1,k})^T + Q_{k-1} + \sum_{j=2}^{k} \Upsilon_{j,k} P_{k-j} \Upsilon_{j,k}^T \tag{10}$$

$$K_k = \tilde{P}_k C^T (C\tilde{P}_k C^T + R_k)^{-1} \tag{11}$$

$$\hat{x}_k = \tilde{x}_k + \gamma_k K_k (y_k - C\tilde{x}_k) \tag{12}$$

$$P_k = (I - \gamma_k K_k C)\tilde{P}_k, \tag{13}$$

where γ_k represents the knowledge of packet losses and initial conditions are

$$x_0 \in \mathbb{R}^N, \quad P_0 = \mathrm{E}[(\tilde{x}_0 - x_0)(\tilde{x}_0 - x_0)^T]$$

and ν_k and ω_k are assumed to be independent with zero expected value. ■

Proof:

Due to the limited length of this article, we are compelled to present only the most essential aspects of the proof. The main difference, in comparison to gFKF algorithm [9], is in prediction Equations (8) and (10). The prediction of state vector is obtained under the same simplifying assumption as in derivation of FKF [8] and gFKF [9] algorithms (updated is only the last state vector):

$$\begin{aligned}\tilde{x}_{k+1} &= \mathrm{E}[x_{k+1}|z_k^*] = \mathrm{E}[(Ax_k + Bu_k + \omega_k - \sum_{j=1}^{k+1}(-1)^j \Upsilon_{j,k+1} x_{k+1-j})|z_k^*] \\ &\approx A\hat{x}_k + Bu_k - \sum_{j=1}^{k+1}(-1)^j \Upsilon_{j,k+1}\hat{x}_{k+1-j}.\end{aligned}$$

The prediction of covariance error matrix is evaluated also with similar assumption that was used in FKF derivation [8] ($\mathrm{E}[(\hat{x}_l - x_l)(\hat{x}_m - x_m)^T] = 0$ for $l \neq m$):

$$\begin{aligned}\tilde{P}_k &= \mathrm{E}\left[(\tilde{x}_k - x_k)(\tilde{x}_k - x_k)^T\right] \\ &= (A - \Upsilon_1)\mathrm{E}[(\hat{x}_{k-1} - x_{k-1})(\hat{x}_{k-1} - x_{k-1})^T](A - \Upsilon_{1,k})^T \\ &+ \mathrm{E}[\omega_{k-1}\omega_{k-1}^T] + \sum_{j=2}^{k} \Upsilon_{j,k}\mathrm{E}[(\hat{x}_{k-j} - x_{k-j})(\hat{x}_{k-j} - x_{k-j})^T]\Upsilon_{j,k}^T \\ &= (A + \Upsilon_{1,k})\, P_{k-1}\, (A + \Upsilon_{1,k})^T + Q_{k-1} + \sum_{j=2}^{k} \Upsilon_{j,k} P_{k-j} \Upsilon_{j,k}^T\end{aligned}$$

The rest of the proof is obtained by minimizing the following objective function:

$$\hat{x}_k = \min_x[(\tilde{x}_k - x)\tilde{P}_k^{-1}(\tilde{x}_k - x)^T + \gamma_k(y_k - Cx)R_k^{-1}(y_k - Cx)^T] \tag{14}$$

The extremum is obtained from first order derivative of the objective function:

$$-2\tilde{P}_k^{-1}(\tilde{x}_k - \hat{x}_k) - 2C^T R_k^{-1}(y_k - C\hat{x}_k) = 0. \tag{15}$$

After rearranging the equation, applying Matrix Inversion Lemma and denoting that

$$K_k = \tilde{P}_k C^T (C\tilde{P}_k C^T + R_k)^{-1}, \tag{16}$$

estimation update relation (12) is obtained

$$\hat{x}_k = \tilde{x}_k + \gamma_k K_k (y_k - C\tilde{x}_k). \tag{17}$$

The covariance estimation error matrix is evaluated from its definition and has the following form

$$\begin{aligned}P_k &= \mathrm{E}\left[(\hat{x}_k - x_k)(\hat{x}_k - x_k)^T\right] = \mathrm{E}\left[(\tilde{x}_k + \gamma_k K_k(Cx_k + \nu_k - C\tilde{x}) - x_k)\right. &(18)\\ &\qquad \left.(\tilde{x}_k + \gamma_k K_k(Cx_k + \nu_k - C\tilde{x}) - x_k)^T\right] &(19)\\ &= (I - \gamma_k K_k H_k)\tilde{P}_k + (-\tilde{P}_k H_k^T + \gamma_k K_k H_k \tilde{P}_k H_k^T + \gamma_k K_k R_k)K_k^T\gamma_k &(20)\\ &= (I - \gamma_k K_k C)\tilde{P}_k. &(21)\end{aligned}$$

3.1 Improved Fractional Variable order Kalman Filter for Networked Systems (gExFvoKF)

In this section the generalization of Improved Fractional variable order Kalman Filter for variable order case will be given.

Theorem 2. *For the discrete fractional order system in a state-space representation introduced by the Definition 3 with intermittent observations the Kalman filter (referred to as gExFvoKF) is given by the following set of equations*

$$\tilde{\mathbb{X}}_{k+1} = \mathbb{A}_{m,k}\hat{\mathbb{X}}_k + \mathbb{B}_m u_k - \mathbb{I}\sum_{j=m+1}^{k+1}(-1)^j \Upsilon_{j,k}\hat{x}_{k+1-j} \tag{22}$$

$$\tilde{\mathbb{P}}_k = \mathbb{A}_{m,k}\mathbb{P}_{k-1}\mathbb{A}_{m,k}^T + \mathbb{Q}_{k-1} + \sum_{j=m+1}^{k}\mathbb{I}\Upsilon_{j,k}P_{k-j}\Upsilon_{j,k}^T\mathbb{I}^T \tag{23}$$

$$\hat{\mathbb{X}}_k = \tilde{\mathbb{X}}_k + \gamma_k\mathbb{K}_k(y_k - \mathbb{C}_m\tilde{\mathbb{X}}_k) \tag{24}$$

$$\mathbb{P}_k = (I - \gamma_k\mathbb{K}_k\mathbb{C}_m)\tilde{\mathbb{P}}_k \tag{25}$$

where

$$\mathbb{K}_k = \tilde{\mathbb{P}}_k\mathbb{C}_m^T(\mathbb{C}_m\tilde{\mathbb{P}}_k\mathbb{C}_m^T + R_k)^{-1}$$

and γ_k represents the knowledge of packet loss. Noises ν_k and ω_k are assumed to be independent and zero expected value, and matrices $\mathbb{P}_k$ and R_k are positive-defined. ■

Proof: The proof will based on the proof of gExFKF [9] algorithm, and also will present only the most essential aspects. The prediction of state vector is obtained under less restrictive assumption as in the derivation of gFvoKF and gFKF [9] algorithms, that $\mathrm{E}[x_{k+1-j}|z_k^*] = \mathrm{E}[x_{k+1-j}|z_{k+1-j}^*] = \hat{x}_{k+1-j}$ for $j > m$ (m newest state vectors are updated).

$$\begin{aligned}\tilde{\mathbb{X}}_{k+1} &= \mathrm{E}[\mathbb{A}_{m,k}\mathbb{X}_k + \mathbb{B}_m u_k + \mathbb{I}\omega_k - \mathbb{I}\sum_{j=m+1}^{k+1}(-1)^j\Upsilon_{j,k}x_{k+1-j}|z_k^*] \\ &= \mathbb{A}_{m,k}\hat{\mathbb{X}}_k + \mathbb{B}_m u_k - \mathbb{I}\sum_{j=m+1}^{k+1}(-1)^j\Upsilon_{j,k}\hat{x}_{k+1-j}\end{aligned}$$

The prediction of the covariance estimation error matrix is evaluated from its definition, under again less restrictive assumption that $\mathrm{E}[x_k x_j] = 0$ for $k \neq j$ and $k, j > m$, and has the form:

$$\begin{aligned}\tilde{\mathbb{P}}_k &= \mathrm{E}\left[(\tilde{\mathbb{X}}_k - \mathbb{X}_k)(\tilde{\mathbb{X}}_k - \mathbb{X}_k)^T\right] = \mathbb{A}_{m,k}\mathrm{E}[(\hat{x}_{k-1} - x_{k-1})(\hat{\mathbb{X}}_{k-1} - \mathbb{X}_{k-1})^T]\mathbb{A}_{m,k}^T \\ &+ \mathbb{I}\mathrm{E}[\omega_{k-1}\omega_{k-1}^T]\mathbb{I}^T + \sum_{j=m+1}^{k}\mathbb{I}\Upsilon_{j,k}\mathrm{E}[(\hat{x}_{k-j} - x_{k-j})(\hat{x}_{k-j} - x_{k-j})^T]\Upsilon_{j,k}^T\mathbb{I}^T \\ &= \mathbb{A}_{m,k}P_{k-1}\mathbb{A}_{m,k}^T + \mathbb{I}Q_{k-1}\mathbb{I}^T + \sum_{j=m+1}^{k}\mathbb{I}\Upsilon_{j,k}P_{k-j}\Upsilon_{j,k}^T\mathbb{I}^T\end{aligned}$$

The rest of the proof is fully analogical to the proof of the gFvoKF algorithm given by Theorem 1, and in this case the cost function has the following form:

$$\hat{\mathbb{X}}_k = \min_{\mathbb{X}}[(\tilde{\mathbb{X}}_k - \mathbb{X})\tilde{\mathbb{P}}_k^{-1}(\tilde{\mathbb{X}}_k - \mathbb{X})^T + \gamma_k(y_k - \mathbb{C}_m\mathbb{X})R_k^{-1}(y_k - \mathbb{C}_m\mathbb{X})^T] \quad (26)$$

and it is easy to check that this function implies the following relations

$$\hat{\mathbb{X}}_k = \tilde{\mathbb{X}}_k + \gamma_k\mathbb{K}_k(y_k - \mathbb{C}_m\tilde{\mathbb{X}}_k) \quad (27)$$

and

$$\mathbb{P}_k = (I - \gamma_k\mathbb{K}_k\mathbb{C}_m)\tilde{\mathbb{P}}_k \quad (28)$$

The rest of equations are the same.

4 Numerical Results

In this section some numerical simulations are given to show the effectiveness of the proposed algorithms.

4.1 Simulation Methodology

All simulation results included in this paper are obtained for the discrete fractional variable order state space system given by the following matrices:

$$A = \begin{bmatrix} 0 & -0.1 \\ 1 & 0.2 \end{bmatrix}, B = \begin{bmatrix} 0.2 \\ 0.3 \end{bmatrix}, C = \begin{bmatrix} 0.4 \; 0.3 \end{bmatrix}, \alpha_{1,k} = \alpha_{2,k} = 0.5 + 0.2\sin\left(\frac{2\pi k}{200}\right),$$

During experiments, we consider four sets of noises, as shown in Table 1, which includes the measured variance of output noise, ν_k, and covariance of internal noise, ω_k. As it could be seen, the correlations between noises $\omega_{1,k}$ and $\omega_{2,k}$ are much smaller than the value of noises variances, we can assume that noises are independent (as it was assumed in gFvoKF and gExFvoKF proofs). The transmission rate was equal to 85%, what means that 15% packets was lost. The estimation error is defined as follows:

$$e = \sum_{k=50}^{180} \left[(\hat{x}_k - x_k)(\hat{x}_k - x_k)^T\right] \quad (29)$$

The error e is defined for period $k = 50 \ldots 180$ in order to omit first part of estimation process which depends on initial conditions, because tested algorithms have a different form of initial conditions. For comparison purposes, the improvements of gExFvoKF algorithm with respect to gFvoKF are defined as:

$$e_{\%} = \frac{e_{\text{gFvoKF}} - e_{\text{gExFvoKF}}}{e_{\text{gFvoKF}}} 100\% \quad (30)$$

where e_{gExFvoKF} and e_{gFvoKF} are estimation errors of the gExFvoKF and gFvoKF filters, respectively.

Table 1. Description of noises used in experiments

Noise number	$E[\nu_k \nu_k^T]$	$E[\omega_k \omega_k^T]$
I	8.386210^{-5}	$\begin{bmatrix} 0.0436 & -0.0020 \\ -0.0020 & 0.0422 \end{bmatrix}$
II	0.0021	$\begin{bmatrix} 0.0980 & -0.0044 \\ -0.0044 & 0.0949 \end{bmatrix}$
III	0.0092	$\begin{bmatrix} 0.0903 & 0.0027 \\ 0.0027 & 0.0855 \end{bmatrix}$
IV	0.0399	$\begin{bmatrix} 0.2557 & 0.0230 \\ 0.0230 & 0.2309 \end{bmatrix}$

The parameters of the fractional Kalman filters are:

$$P_0 = 100I, \quad x_0 = [0,0]^T$$

The matrices Q_k and R_k have values of measured variances of noises presented in Table 1.

Example 1. Example of estimation and smoothing by using gFvoKF and gExF-voKF for variable order case.

In this example, we compare two estimation algorithms, the variable order gFvoKF and gExFvoKF, and we consider two different cases: in the first one, whose results are presented in Table 2, the influence of parameter m in estimation process is investigated for a fixed value of noise (noise I); the second case is a comparison for a fixed value of parameter m ($m = 20$) and different noises, which are included in Table 3.

Table 2. Estimation and smoothing results for gFvoKF and gExFvoKF filters for different values of m (noise I), for variable order case

Para-meter	gFvoKF Esti-mation	gExFvoKF algorithm					
		Estimation		Smoothing			
		$\hat{x}_k\|k$		$\hat{x}_{k-4}\|k$		$\hat{x}_{k-8}\|k$	
m	e	e	$e_{\%}$	e	$e_{\%}$	e	$e_{\%}$
5	7.71	7.48	2.97	5.86	23.97	–	–
10	7.71	7.32	5.04	5.70	26.08	5.67	26.42
20	7.71	7.29	5.43	5.67	26.48	5.64	26.83
50	7.71	7.44	3.50	5.81	24.66	5.78	25.04
100	7.71	7.44	3.47	5.80	24.75	5.78	25.10

As it can be seen on Table 2, the dependence between the parameter m and the estimation error is not proportional, and the best results, in presented case, are obtained for $m \in [10, 20]$.

Table 3. Estimation and smoothing results for gFvoKF and gExFvoKF filters ($m = 20$) for variable order case

Noise type	gFvoKF Estimation	gExFvoKF algorithm					
		Estimation		Smoothing			
		$\hat{x}_k\|k$		$\hat{x}_{k-4}\|k$		$\hat{x}_{k-8}\|k$	
	e	e	$e_{\%}$	e	$e_{\%}$	e	$e_{\%}$
I	7.71	7.29	5.43	5.67	26.48	5.64	26.83
II	18.81	18.10	3.75	14.58	22.48	14.54	22.72
III	18.38	18.14	1.32	16.62	9.60	16.55	9.95
IV	66.71	65.87	1.26	60.03	10.01	59.80	10.36

According to Table 3, it can be observed that the efficiency of the gExFvoKF algorithm decreases when variance of noise increases. For low value of variance, we obtain significant improvements in the smoothing action, whereas the improvements are not so significant for higher noise variance.

5 Conclusions

In this paper a generalization of Fractional Variable Order Klman Filter (gFvoKF) and Improved Fractional Variable Order Kalman Filter (gExFvoKF) algorithms for estimation of fractional variable order state-space systems over lossy network was presented. Both algorithms were obtained for $\mathcal{A}$-type of fractional variable order difference. Also both algorithm use assumed knowledge about packets losing. For both algorithms numerical results for different type of noises and different parameter m were obtained and compared. Results confirm more accurate estimation results of gExFvoKF algorithm, however, improvements were not so significant. More significant improvements were obtained in smoothing actions, for which only gExFvoKF only allows.

Acknowledgements. This work was supported by the Polish National Science Centre with the decision number DEC-2011/03/D/ST7/00260.

References

1. Shi, L., Xie, L., Murray, R.: Kalman filtering over a packet-delaying network: a probabilistic approach. Automatica 45, 2134–2140 (2009)
2. Moayedi, M., Soh, Y., Foo, Y.: Optimal kalman filtering with random sensor delays, packet dropouts and missing measurements. In: Proceedings of the American Control Conference (ACC 2009), pp. 3405–3410 (2009)
3. Epstein, M., Shi, L., Tiwari, A., Murray, R.: Probabilistic performance of state estimation across a lossy network. Automatica 44(12), 3046–3053 (2008)
4. Matveev, A.S., Savkin, A.V.: Estimation and Control over Communication Networks. Birkhäuser (2009)

5. Sierociuk, D., Dzielinski, A., Sarwas, G., Petras, I., Podlubny, I., Skovranek, T.: Modelling heat transfer in heterogeneous media using fractional calculus. Philosophical Transactions of the Royal Society A: Mathematical, Physical and Engineering Sciences 371 1990 (2013)
6. Sheng, H., Chen, Y., Qiu, T.: Signal Processing Fractional Processes and Fractional-Order Signal Processing. Springer, London (2012)
7. Sheng, H., Sun, H., Chen, Y., Qiu, T.: Synthesis of multifractional gaussian noises based on variable-order fractional operators. Signal Processing 91(7), 1645–1650 (2011)
8. Sierociuk, D., Dzieliński, A.: Fractional Kalman filter algorithm for states, parameters and order of fractional system estimation. Applied Mathematics and Computer Science 16(1), 129–140 (2006)
9. Sierociuk, D., Tejado, I., Vinagre, B.M.: Improved fractional Kalman filter and its application to estimation over lossy networks. Signal Processing 91(3, SI), 542–552 (2011)
10. Sierociuk, D., Ziubinski, P.: Fractional order estimation schemes for fractional and integer order systems with constant and variable fractional order colored noise. Circuits, Systems, and Signal Processing (accepted to 2014)
11. Ramirez, L., Coimbra, C.: On the variable order dynamics of the nonlinear wake caused by a sedimenting particle. Physica D-Nonlinear Phenomena 240(13), 1111–1118 (2011)
12. Ostalczyk, P., Rybicki, T.: Variable-fractional-order dead-beat control of an electromagnetic servo. Journal of Vibration and Control 14(9-10), 1457–1471 (2008)
13. Sierociuk, D., Malesza, W., Macias, M.: On the recursive fractional variable-order derivative – equivalent switching strategy, duality, and analog modeling. Circuits, Systems, and Signal Processing (accepted to 2014)
14. Sierociuk, D., Malesza, W., Macias, M.: Switching scheme, equivalence, and analog validation of the alternative fractional variable-order derivative definition. In: Proceedings of the 52nd IEEE Conference on Decision and Control, Florence, Italy, December 10-13 (2013)
15. Lorenzo, C., Hartley, T.: Variable order and distributed order fractional operators. Nonlinear Dynamics 29(1-4), 57–98 (2002)
16. Valerio, D., da Costa, J.S.: Variable-order fractional derivatives and their numerical approximations. Signal Processing 91(3, SI), 470–483 (2011)
17. Sierociuk, D.: System properties of fractional variable order discrete state-space system. In: Proceedings of the 13th International Carpathian Control Conference (ICCC), pp. 643–648 (2012)
18. Dzielinski, A., Sierociuk, D.: Stability of discrete fractional order state-space systems. JVC/Journal of Vibration and Control 14(9-10), 1543–1556 (2008)
19. Dzielinski, A., Sierociuk, D.: Reachability, controllability and observability of the fractional order discrete statespace system. In: Proceedings of the IEEE/IFAC International Conference on Methods and Models in Automation and Robotics, MMAR 2007, Szczecin, Poland, pp. 129–134 (2007)
20. Sinopoli, B., Schenato, L., Franceschetti, M., Poolla, K., Jordan, M., Sastry, S.: Kalman filtering with intermittent observations. IEEE Transactions on Automatic Control 49, 1453–1464 (2004)
21. Shi, L., Epstein, M., Murray, R.M.: Kalman filtering over a packet dropping network: A probabilistic approach. In: Proceedings of 10th Intl. Conf. on Control, Automation, Robotics and Vision Hanoi, Vietnam, ICARCV 2008, pp. 41–46 (2008)

Author Index

MIX
Papier aus verantwortungsvollen Quellen
Paper from responsible sources
FSC® C105338

If you have any concerns about our products,
you can contact us on
ProductSafety@springernature.com

In case Publisher is established outside the EU,
the EU authorized representative is:
Springer Nature Customer Service Center GmbH
Europaplatz 3, 69115 Heidelberg, Germany

Printed by Libri Plureos GmbH
in Hamburg, Germany